A Naturalist's Guide to the Arctic

Related Books by E. C. Pielou

After the Ice Age: The Return of Life to Glaciated North America (published by the University of Chicago Press)

The World of Northern Evergreens

A Naturalist's Guide to the Arctic

E. C. Pielou

The University of Chicago Press
Chicago and London

The University of Chicago Press, Chicago 60637
The University of Chicago Press, Ltd., London

Printed in the United States of America

14 13 12 11 10 09 08 07   7 8 9 10

ISBN: 0-226-66813-4 (cloth)
0-226-66814-2 (paper)

Library of Congress Cataloging-in-Publication Data

Pielou, E. C.
A naturalist's guide to the Arctic / E. C. Pielou.
p. cm.
Includes bibliographical references (p.) and index.
1. Natural history—Arctic regions. 2. Arctic regions.
I. Title.
QH84.1.P54 1994
508.98—dc20   94-2555

♾ The paper used in this publication meets the minimum requirements of the American National Standard for Information Sciences—Permanence of Paper for Printed Library Materials, ANSI Z39.48-1992.

For *Ruth and George,*
Richard and Joyce,
Frank

Contents

Preface

The vastness of Arctic North America is an inspiration to naturalists. It is the embodiment of remoteness and wildness. All the same, as travel becomes easier, the Arctic is becoming less inaccessible; and as naturalists become more numerous, more of them go there. An arctic trip is expensive, however, and travelers want to observe everything there is to observe. Birders don't want to concentrate exclusively on birds, or plant-watchers on plants; to do so would amount to wasting a precious opportunity to absorb the whole arctic experience. There is much to see and savor, and it is well worth making a determined effort to miss nothing.

This raises the problem of what guide books to take. It is impracticable to carry a book for every topic—one for plants, another for birds, a third for mammals, and others for insects, fishes, terrain structure, climate, and so on; adding a load of books to their other loads is out of the question for backpackers, kayakers, canoeists, and rafters. At the same time, no traveler wants the disappointment of realizing, after the chance for a closer look is gone, that opportunities have been missed: opportunities to observe, for example, the way arctic foxes find the loose, unfrozen sand of eskers especially suitable for their dens; or the way common eiders nest on tiny islands, beyond the reach of land-based predators (provided break-up of the sea ice doesn't come too late); or the way butterflies settle so as to absorb the sun's warmth most efficiently.

This book is designed to simplify the problem. It covers *all* fields of natural history, as a glance at the table of contents shows. In a book designed to be easily portable, no subject can be covered exhaustively, of course. I chose the material to include on each topic by posing, and answering, the following question: what information is needed at the time, and on the spot, to ensure that you see, understand, and appreciate what is there before you? Conversely, material was excluded if it seemed likely that you would be content to look it up after returning home: for instance, the gestation period of caribou (about 230 days) or the date when the Great Auk became extinct (about 1844).

Some sections of the book are identification guides for different groups of organisms (plants, birds, mammals, fish). Others are brief accounts of arctic natural history, using that term in its broadest sense. Natural history is more than ecology. It deals with the nonliving as well as the living, and naturalists who ignore inanimate things—land, sea, and sky—are missing much that the world has to offer. This is especially true of the Arctic world, where the low vegetation leaves the very shape of the ground exposed to view and to interpretation; where the sea is more diverse than in warmer latitudes, being solid ice as often as it is liquid water; and where the midnight sun makes the summer scene unique.

I hope therefore that, besides using the book as a guide to the identification of particular plants and animals, naturalists will also find it an interesting read when one is needed: when savage weather confines them in camp, or when the midnight sun makes sleep impossible.

Denman Island, 1994

Introduction

The ancient Greeks regarded the known world as composed of four elements: earth, air, fire, and water. If they had known of the polar regions, they would undoubtedly have added a fifth element, ice. For ice is more than simply frozen water; it is a solid that shapes land and sea and affects the existence of all living things in the polar regions. The Arctic can be defined as the region centered on the North Pole, where ice dominates the landscape, more precisely as the region where the soil is permanently frozen and where trees cannot grow. The boundaries of the North American Arctic, the subject of this book, are shown on the map.

It is an area where naturalists can expand their interests as they can nowhere else. Earth, air, water, and ice—even fire, in the shape of the midnight sun—can be seen as more than simply a backdrop to the living world. At the same time, the world of living things is less complicated than it is in warmer climates, making it feasible for naturalists to appreciate all of it. To cover so much material systematically, this book is divided into nine chapters each dealing with a separate topic.

Chapters 5, 6, and 7 contain field guides to arctic plants, birds, and mammals, respectively. The species in each group are described by families (e.g., the Saxifrage Family, the Plover Family, the Bear Family, and so on). The format differs slightly in each chapter because of the very different numbers of species in the respective groups and their different degrees of uniqueness.

Chapter 5 (on plants) begins with several sections on plant natural history in general, followed by field guide sections on arctic flowers (sec. 5.13) and ferns (sec. 5.14).

Chapter 6 (on birds) begins with comparatively few sections on birds in general because bird families differ from one another in activities as well as in structure so that, compared with plants, fewer generalizations are possible. The family descriptions in the field guide (sec. 6.6), however, are longer than those for plants in chapter 5. For each bird family, details on species identifications are followed by further notes on the family's natural history and behavior.

Chapter 7 (on mammals) begins with a single section (sec. 7.1) of general comments, after which each family is given a numbered section of its own (sec. 7.2: the Bear Family; sec. 7.3: the Cattle Family; and so on).

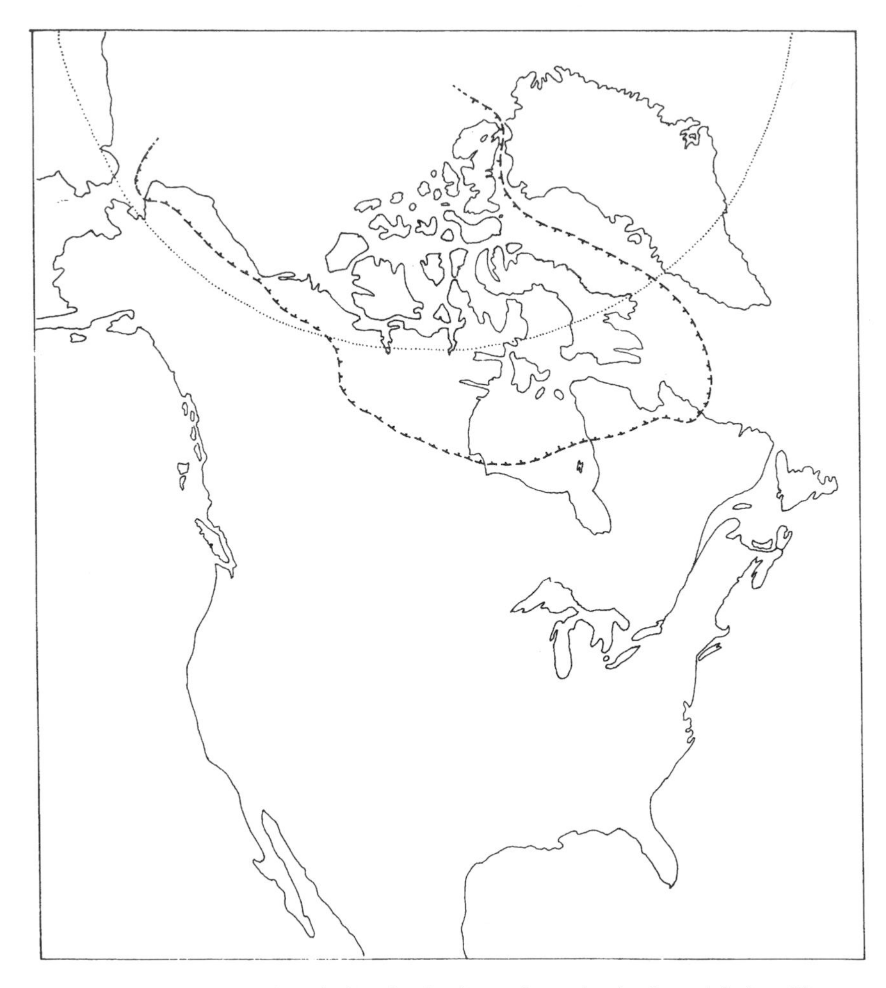

The limits of the region described in this book are shown by the flagged dashes. The dotted line is the Arctic Circle.

The field guide sections should enable a naturalist to identify nearly all the distinctive plants, birds, and mammals of the area. An illustration is given wherever one would be helpful (the majority of cases). But there are cases where words are better than pictures. An example: chickweed plants all look much alike, but one species can be recognized by the clammy, moist feel of its leaves and stems.

To keep this book portable, references have been kept to a minimum. Books and magazine articles with colored photos describing particular topics in more detail than is possible in a book to be carried in a backpack are listed at the end of each relevant section.

English names for plants and animals are used throughout the book except in the field guide sections, where the scientific (Latin) name of each species is given, following its English name, when it is first mentioned. Latin names are always, by convention, printed in italics. Every species has a two-word name; the first word always has a capital initial, the second a lower-case initial. (When species are divided into subspecies, which is not done in this book, a third word is added, identifying the subspecies.) For more detail on Latin names, see the preliminary notes in section 5.13 (the field guide to arctic flowers).

Serious naturalists use Latin, the less serious don't. The latter should not be put off by Latin names, however, nor should they think that using them is merely pedantic. The merit of the Latin names is that they are universal, the same in all languages, and the same in all geographic variants of a single language. For instance, the bird known scientifically as *Stercoriarius parasiticus* is called a parasitic jaeger in North America and an arctic skua in Britain, but either way it's the same species. Worse, a number of English names are sometimes used for the same species even in one region. For instance, the plant *Epilobium latifolium* is known by three English names in North America: river beauty, dwarf fireweed, and broad-leaved willowherb. Use of Latin names therefore prevents confusion. They also reveal something about relationships among similar species. For instance, the English names of most gull species consist of the word "gull" preceded by a descriptive adjective. Nothing in these names shows that among them is a subgroup of closely related, visibly similar gulls, distinct from all the others. The Latin names reveal the relationship at once; the gulls in the subgroup all have *Larus* as the first word of their names.

To conclude: all naturalists find the Arctic a very special place, whether they are visiting it for the first or the fiftieth time. It is at risk, however, like all wild places on earth, because of the human population explosion. To protect it requires knowledge and determination. Naturalists hoping to help defend and conserve the arctic wilderness can do so best by first learning all they can about its natural history.

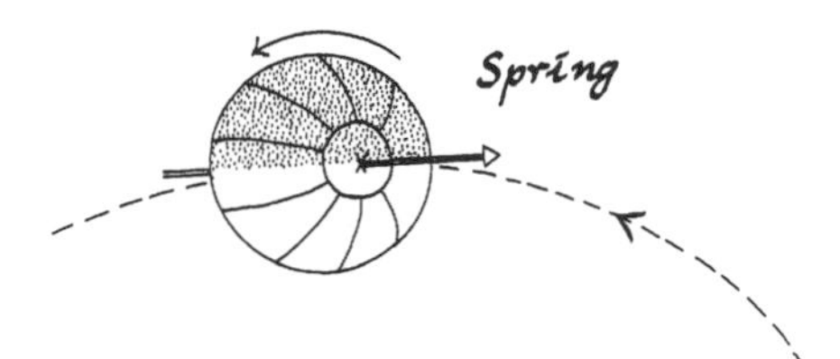

1 Sky

1.1 The Land of the Midnight Sun

"The Arctic" means different things to different people. To those who pay more attention to the sky than to the ground (which includes astronomers, of course), it means all that part of the earth, centered on the North Pole, where, at least once in the year, the sun remains above the horizon for a full 24 hours without setting so that (if it isn't cloudy) the sun shines at midnight on at least one day of the year.

An understanding of why there are days in summer when the sun doesn't set (and days in winter when it doesn't rise) cannot be gained without concentrating closely on astronomical matters, specifically on the way the earth rotates around a tilted axis as it makes its yearly journey round the sun. Readers not in the mood for heavy thinking on these topics should skip to section 1.3.

The lowest latitude at which the midnight sun is ever seen is the Arctic Circle, at 66½° N (or at 23½° from the Pole; the two angles add up to a right angle). The reason for this can be understood from the diagram (fig. 1.1) showing how the axis on which the earth spins (joining the north and south poles) is tilted at an angle of 66½° to the earth's orbit round the sun. The day of the year on which the tilted axis points most nearly toward the sun is known as the summer solstice; it comes on June 21. On that day, as the diagram shows, every point within the Arctic Circle is in sunlight for the whole of the earth's rotation, that is, the whole day long. At the other end of the year is the winter solstice (December 22) when, at every point within the Arctic Circle, the sun is below the horizon the whole day long.

Contemplation of figure 1.1 shows that the times of sunrise and sunset must change with the seasons. The exact times, taken from the Nautical Almanac, were used to draw figure 1.2. The four diagrams show how the number of days during which the sun never sets varies from zero at 60° N (below the Arctic Circle) to over 2 months at 70° N, nearly 4 months at 80° N, and finally 6 months (half the year) at the pole itself. These periods of continuous sunlight begin and end at the dates shown, respectively, on the left and right sides of each diagram, where it is

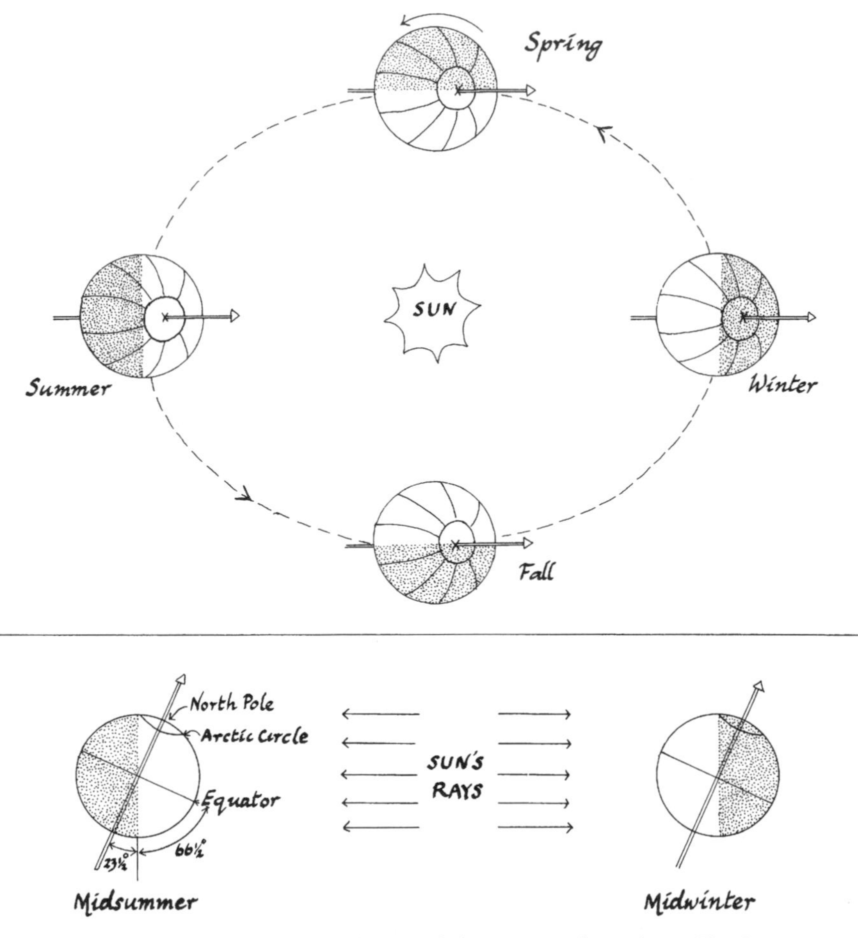

Figure 1.1. *Above:* The earth in its orbit round the sun, seen from above. The direction the earth spins is shown at the top. The arrow through each globe is the earth's axis, with the arrowhead pointing north. The axis is tilted, with the arrowhead raised up from the page through an angle of 66½°. The circle around the pole (marked X) is the arctic circle. Shaded: dark side of the earth, not in sunlight. *Below:* Side views of the earth at midsummer (left) and midwinter (right).

crossed by the "sunrise line" (the line separating the white part from the rest of the diagram).

The diagrams also show the duration of two kinds of twilight: *Civil twilight,* which begins when the sun sets and continues until it is 6° below the horizon; during this period "operations requiring daylight" can be carried on, as the legal description puts it. And *nautical twilight,* which begins when civil twilight ends and lasts until the sun is 12° below the horizon, by which time all the stars used for celestial navigation have "come out." In the morning, of course, the order is reversed: first comes nautical twilight and then civil twilight, which ends at sunrise.

There is a third level of twilight, so dim it is treated as darkness in the diagrams. It prevails while the sun is between 12° and 18° below the horizon, and is called *astronomical twilight.* During astronomical twilight, fainter stars appear, and the outlines of large objects and the horizon remain just visible. When the sun is lower than 18°, darkness is total (except for moonlight, starlight, the aurora, and nightglow, the faint radiance of the night sky itself!).

The diagrams can be used to give (roughly) the times of sunrise and sunset, and of the beginnings and ends of the two kinds of twilight, on a chosen date at each of the latitudes shown. For example, a horizontal line through April 1 on the 70° N diagram, shows that on that day, at that latitude, nautical twilight begins at about 2 A.M. and ends at about 10 P.M. Civil twilight begins at about 4 A.M. and ends at about 8 P.M. Sunrise and sunset are at 5 A.M. and 7 P.M., respectively. On the same day at 80° N, civil twilight lasts all night, even though the sun is below the northern horizon for the 3½ hours centered on midnight. And at the pole, the sun doesn't set at all on that day.

The dates at which (at a particular latitude) the sun sets for the long winter night and rises at the end of it, can also be read, approximately, from the relevant diagram. They are the dates shown on the right and left sides of the diagram (respectively) where it is cut by a horizontal line across it, drawn so as to touch the bottom of the dip in the "sunrise line." At 70° N, for example, the sun disappears at the end of November and reappears in the middle of January. The corresponding dates at 80° N are late October and late February. The special conditions at 90° N—the Pole itself—are described later.

Note that, when the long winter night comes to an end, the sun first shows itself on the *southern* horizon. The first "day" lasts from a few seconds before noon to a few seconds after, when the sun is in the south. Astronomical considerations clearly show that the first sunrise of the year is never in the east even though writers accustomed to temperate latitudes sometimes forget this.

To return to the midnight sun: Even when the sun doesn't set, there is a noticeable difference in temperature between midday (when the sun is due south) and midnight (when it is due north) in all but the highest latitudes. The words "day" and "night" are often used, despite the continuous daylight, to label the warmer and cooler halves of the 24-hour day. The temperature difference is because at midday the sun is at the highest point in its path through the sky, and at midnight at its lowest point. The contrast is most marked on the Arctic Circle, where, on its

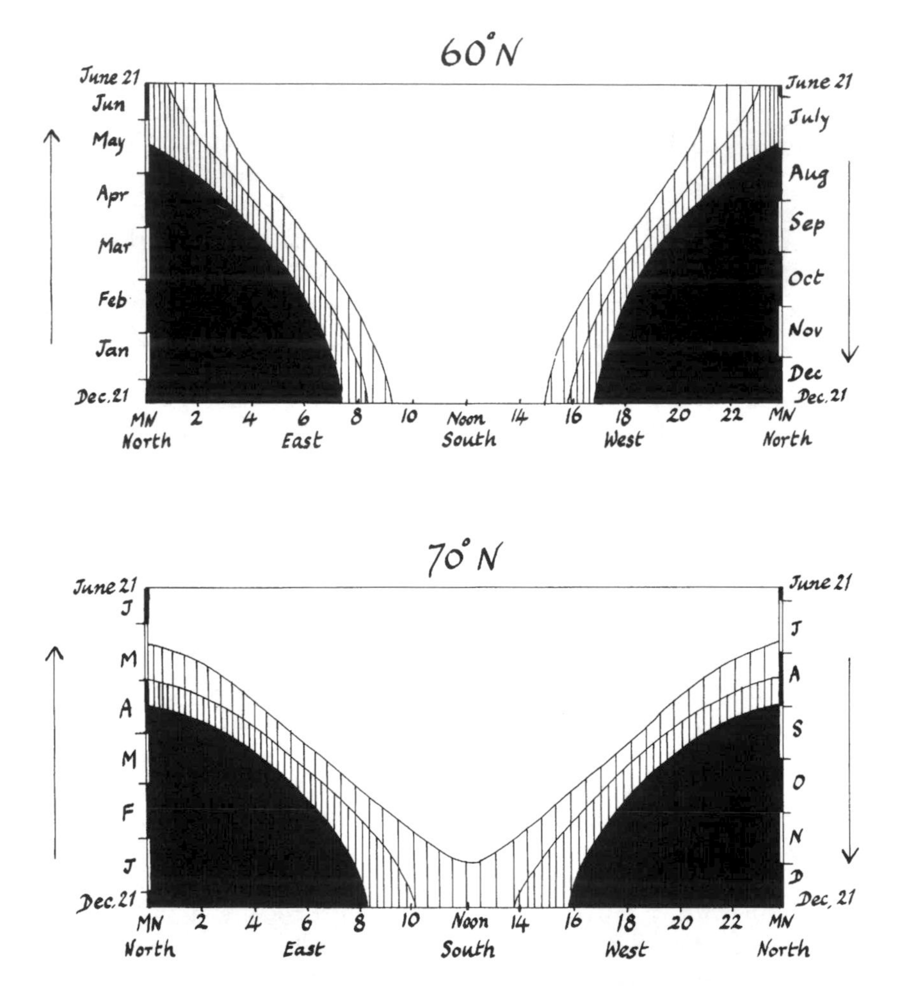

Figure 1.2. The times of twilight, sunrise and sunset at four latitudes, 60° N, 70° N, 80° N, and 90° N (the North Pole). The 24 hours from midnight (MN) to midnight are shown along the bottom in each diagram; the labels North, East, South and West show the direction of the sun at different times through the day (from the North Pole, all directions are South). Dates are shown on the left and right sides, following the arrows:

single annual appearance (June 21), the midnight sun is right on the horizon. The contrast between day and night becomes less and less as the midnight sun period becomes longer and longer, that is, the farther north you go: as the latitude increases, the height of the sun above the horizon at midday becomes progressively less, and its height at midnight becomes progressively greater, until at the North Pole itself they become equal.

The course of events right at the pole deserves special attention (see

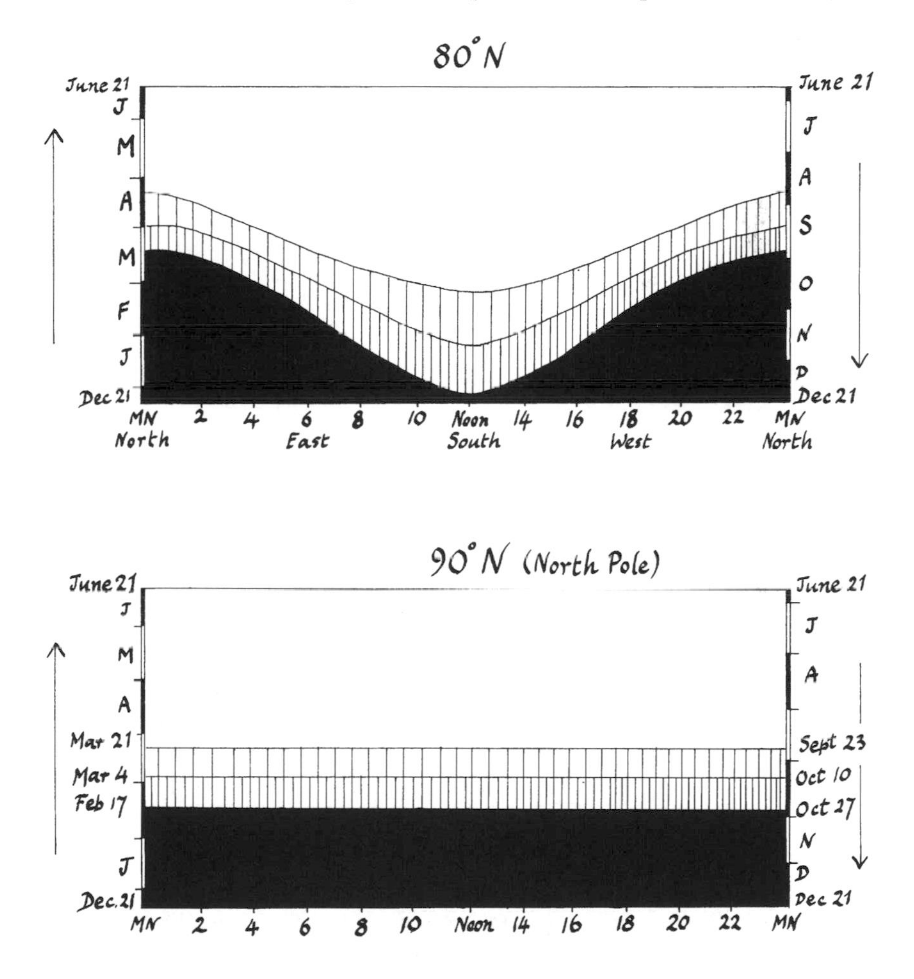

from midwinter (December 21) upward through spring to midsummer (June 21) on the left; from midsummer downward through fall to midwinter on the right. For any date, *read all the way across the diagram* for the course of events on that day. Full darkness is shown black; nautical twilight by close rulings, civil twilight by wide rulings. (More details in the text.)

fig. 1.2). After months of winter darkness (relieved only by the moon; see sec. 1.2), a day comes when a faint glow becomes just noticeable at some point on the horizon. The glow does a complete rotation, along the horizon, every 24 hours as day succeeds day (measuring days on a clock). By February 17, many of the fainter stars have become invisible in the brightening sky and the horizon can just be seen; the sun is 12° below the horizon and nautical twilight has begun. The light continues to increase gradually, as the sun slowly rises; it does not show itself, but its position below the horizon is obvious from the glow in the sky. By March 4 the sun is 6° below the horizon and civil twilight has begun. At last, on March 21 (the spring equinox), the sun, or at least the upper edge of it, appears. It takes more than 30 hours, circling all the time, to rise sufficiently to clear the horizon, and once up it continues to circle for the next six months. As it circles it slowly ascends, but too slowly for the change in its elevation in a single day to be noticeable to an observer without astronomical instruments; it seems, rather, as though the sun's daily path is a horizontal circle in the sky, parallel with the ground. The ascent goes on until the summer solstice (June 21) when the sun reaches its highest point, 23½° above the horizon. Then it begins its descent and the events just described all happen in reverse order. Figure 1.2d shows the dates of sunset, and of the end of civil and nautical twilight, on the right side of the diagram. In sum, a 6-month day is succeeded by a 6-month night but because the twilight periods of spring and fall each last for more than a month, the 6-month night is not all dark. And besides twilight, there is the moon to consider, as we see in the following section.

1.2 What about the Midday Moon?

Summer visitors to the Arctic, experiencing continuous daylight, seldom wonder what the moon is doing. The motions of the moon deserve a few moments' thought. As with the sun, the height of the moon in the sky depends on the observer's latitude. And the phase of the moon at any moment—the shape of its sunlit side as seen by an observer on earth—depends on the relative positions of sun, moon, and earth. At full moon, we see the whole of the moon's sunlit side; near new moon (just before or just after), we see only a thin, crescent-shaped sliver of the sunlit side.

The moon revolves around the earth, and goes through its cycle of phases from new moon through first quarter (a half-moon shaped like a D) to full moon, and on through third quarter (a backward D) to the next new moon, in the course of one lunar month (about 29 days). Its orbit around the earth is almost (within about 5°) in the same plane as that of the earth's orbit around the sun. As a result, there are what could be called "moon seasons" of the same pattern as the familiar "sun seasons"; but the full cycle of moon seasons takes only a month instead of a whole year.

At latitudes greater than 72° N (i.e., 5° or more north of the Arctic Circle) there is a period in each month during which the moon never sets, remaining above the horizon for more than 24 hours; this is the moon's equivalent of the sun's midnight-sun period. During this period the moon goes round and round in the sky for days on end, progressing through several phases without disappearing. Likewise, there is a period in each month during which the moon never rises, remaining below the horizon for more than 24 hours; this is the moon's equivalent of the "endless night" of arctic winter. These continuous "moon-stays-up" and "moon-stays-down" periods alternate with periods during which the moon rises and sets daily, just as it does in temperate latitudes. The moon-stays-up and moon-stays-down periods last longer the farther north you go until, at the pole itself, each lasts for half a month.

The next question is: how do the moon's phases change in the moon-stays-up and moon-stays-down periods? The answer can be figured out by recalling that the moon is full when it is on the opposite side of the sky from the sun. And it is "new," colloquially speaking (a thin crescent shaped like a backward C), or "old" (a thin C-shaped crescent) when it is close to the sun, to the left (east) or the right (west) of it, respectively. Strictly speaking, new moon comes while the moon is invisible, halfway between its disappearance just to the west of the sun and its reappearance just to the east of the sun.

Consequently, in midwinter, the full-moon phase comes in the middle of the moon-stays-up period (when moonlight is continuous); and the new moon phase in the middle of the moon-stays-down period (when the moon is below the horizon). The order of events is reversed in midsummer; then, the new moon phase comes in the middle of the moon-stays-up period, and the full-moon phase in the middle of the moon-stays-down period. The result is that the moon is seldom visible to midsummer travelers in the Arctic. For much of the time that the midnight sun is shining, the moon is either below the horizon, or, if above it, too close to the sun to be visible in the sun's glare. When it can be seen, near first or third quarter, it is usually inconspicuous—pale against a pale sky.

In figures 1.3 and 1.4 are sky maps showing the moon's paths through the sky, in a midwinter month and a midsummer month, as seen from latitude 73° N. Viewing each map from the left edge across to the right

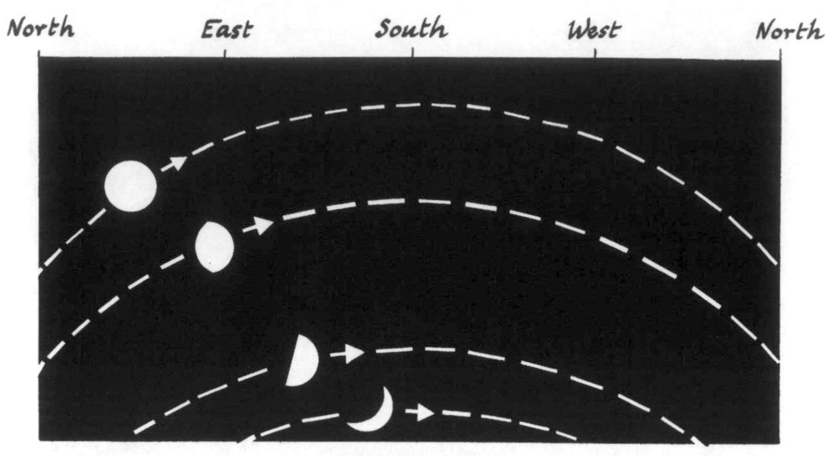

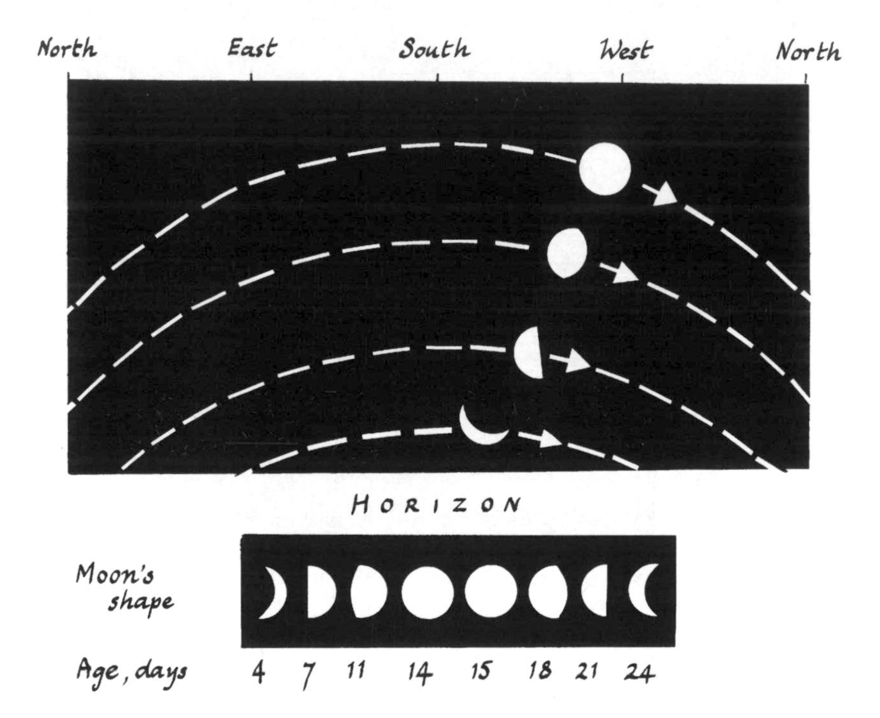

Figure 1.3. The two maps together show the path of the moon in its apparent daily circuits round the sky, during a midwinter month (when the sun never rises), as seen from 73° N latitude. For clarity, only 8 of these circuits are shown. The lower edge of each map is the horizon. In periods when the moon never sets its path cuts the left and right edges of the map instead of the horizon. The dashed lines in the upper map show (from right to left ascending) the moon's path and shape when it is 4, 7, 11 and 14 days old, respectively. The dashed lines in the lower map show (from right to left descending) the moon's path and shape when it is 15, 18, 21 and 24 days old, respectively. The outlines at the bottom show the moon's shape as it ages from a 4-day old moon to a 24-day old moon, i.e., through all that part of the month when it is above the horizon.

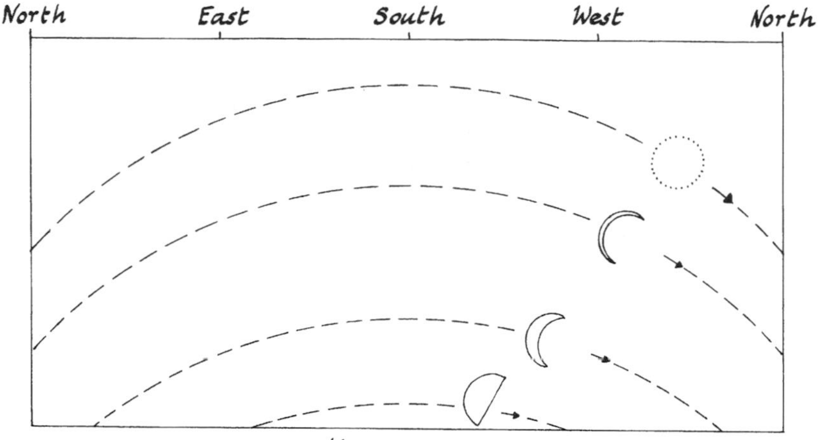

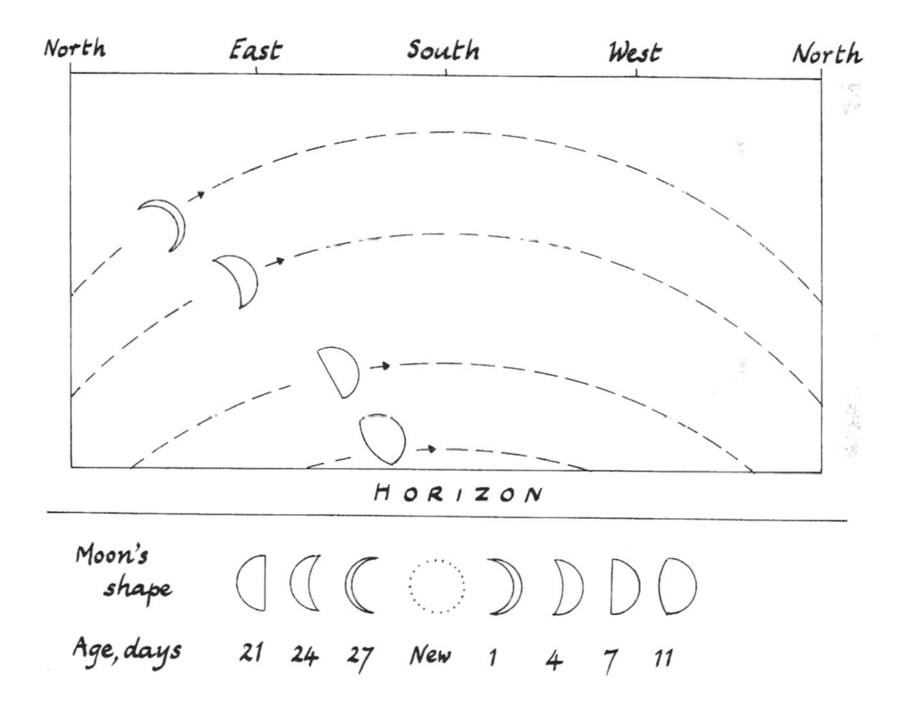

Figure 1.4. As fig. 1.3, but for a midsummer month, when the sun never sets. The dashed lines in the upper map show (from left to right ascending) the moon's path and shape when it is 21, 24, and 27 days old and at New Moon, respectively. (At New Moon the moon is too near the sun to be visible anywhere on earth; in the diagram it is shown as a dotted outline.) The dashed lines in the lower map show (from left to right descending) the moon's path and shape when it is 1, 4, 7 and 11 days old, respectively, in the lunar month starting with New Moon in the upper map. The outlines at the bottom show the moon's shape as it ages from a 21-day old moon to an 11-day old moon in the following lunar month, i.e. through all that part of the month when it is above the horizon. Note that when it is a thin crescent, the moon is so close to the sun (which doesn't set) as to be invisible in its glare.

edge is equivalent to looking at the sky from north round through east, south, west, and back to north again. Notice that in moon-stays-up periods the paths of the moon do not reach the horizon but remain above it, cutting through the vertical edges of the map; as the moon crosses the northern meridian in real life, it disappears off the right edge and immediately reappears at the left edge on the map. Another point to notice is that on maps showing the sky as it would appear from the North Pole exactly, the paths of the moon would be barely distinguishable from straight lines parallel with the horizon.

As remarked, the figures relate to midwinter and midsummer. To complete the story: the middle of the moon-stays-up period coincides with a first-quarter moon in spring, and with a third-quarter moon in fall.

1.3 A Compass near the Magnetic Pole

As arctic travelers well know, a compass is an unreliable direction indicator at high latitudes. In the vicinity of the earth's North Magnetic Pole among the Arctic Islands (fig. 1.5), it is so unreliable as to be completely useless. But this is not because the compass needle fails to point true north, toward the true North Pole; it rarely does. The magnetic pole, to which the compass does point, is more than 1400 km distant from the true (or geographic) pole, which is the endpoint of the earth's axis of rotation. Consequently, the compass indicates true north only when you happen to be standing at a point from which the two poles are in line with each other. From any other point, the magnetic pole is to one side of the true pole—east of it or west of it—and to find true north with a compass, the user needs to know what correction to apply.

Then why, you may ask, cannot the appropriate correction be applied wherever you happen to be? It looks straightforward enough. The reason it doesn't work is that near the magnetic pole the point of the compass needle—itself a magnet—is pulled downward as well as in a horizontal direction. As you approach the magnetic pole, the vertical component of the magnetic force becomes progressively stronger and the horizontal component progressively weaker until the latter is too weak to overcome the friction in the bearings of even the most expensive compass. The needle moves sluggishly and where it stops is merely a matter of chance. At the magnetic pole itself, the magnetic pull is wholly downward and there is no horizontal component. This is how its location is recognized: it is the site at which the *magnetic dip* (the angle that the magnetic force makes to the horizontal) is 90°: in other words, where a suitably suspended compass needle, able to turn freely in any direction, points straight down.

The whole earth is itself a magnet, with a magnetic pole at each end; the surprising thing is that the north and south magnetic poles do not coincide with the north and south geographic poles (poles of rotation). The earth's magnetism is caused (chiefly) by electric currents flowing in its hot liquid core, turning the whole planet into a huge electromagnet. This magnet is not tidily aligned with the axis of rotation because its position is affected by (among other things) the arrangement of magnetic rocks in the earth's solid crust. Moreover, the North Magnetic Pole doesn't stay put: since its position was first discovered in 1831, by the explorer James Clark Ross, it has drifted more than 750 km to its present position (as determined in 1994) at 78.3° N, 104° W (see fig. 1.5). This drift is much too fast to be accounted for by plate tectonics. The reason for the movement is believed to be that the big electromagnet modifies

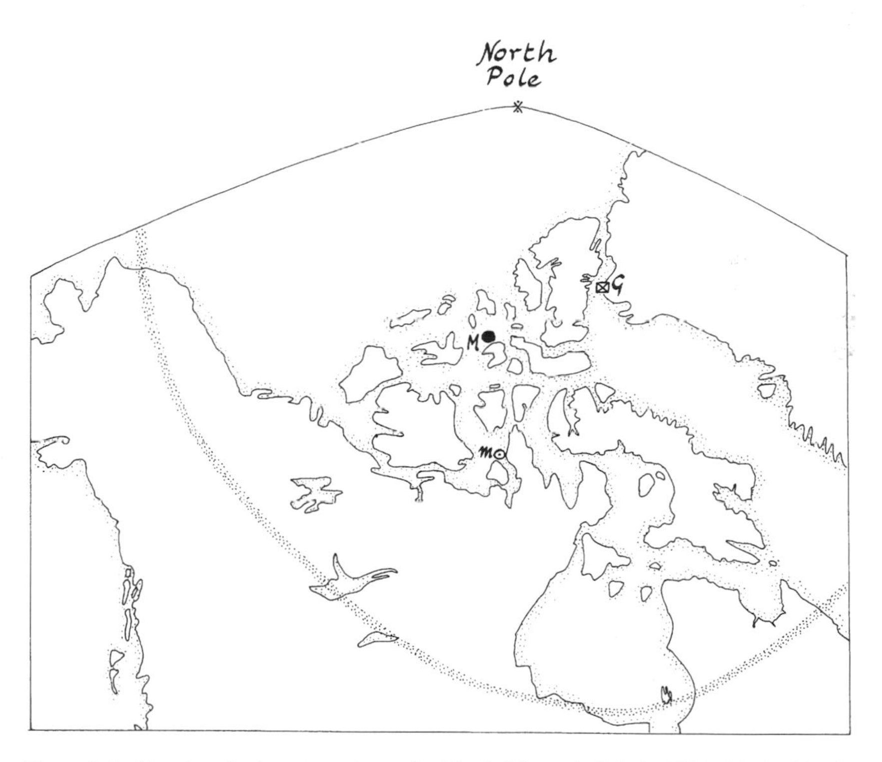

Figure 1.5. Showing the location of: *m*, the North Magnetic Pole in 1831. *M*, the North Magnetic Pole at present (1990s). *G*, the North Geomagnetic Pole (see sec. 1.4). The stippled band is the Auroral Oval (see sec. 1.5).

the electric currents in the core, which then modify the electromagnet itself, which then modifies the currents again, which then . . . , and so on. The explanatory theories devised by physicists involve electromagnetism, hydrodynamics, viscosity, electrical and thermal conductivity, and much else.

Besides "drifting," the magnetic pole also "wobbles": every day it travels round a closed path, roughly elliptical in shape and often more than 100 km across. This it does in response to the electomagnetic effect of electrical currents in the upper atmosphere, caused by electrified particles pouring in from the sun.

1.4 The Geomagnetic Pole

Some maps of the north polar regions show a North Geomagnetic Pole as well as a North Magnetic Pole. This needs explanation. First note that the pole described in the preceding section is the North Magnetic Pole, sometimes called the North "Dip Pole" because it is the spot at which the magnetic dip is 90°. There is also, in Antarctica, a South Magnetic Pole or South Dip Pole, where the opposite end of the suitably suspended compass needle described in the preceding section would point straight down. But the North and South Magnetic Poles (the latter at 65° S, 139° E) are not exactly opposite each other: a line joining them would miss the earth's center by several hundred kilometers instead of passing right through it.

The earth's magnet is, in fact, slightly distorted. The pattern of magnetic forces it creates at ground level is not precisely like that of an ordinary bar magnet—a magnetized iron rod with magnetic poles at the ends and the magnetic forces symmetrically arranged all around it; rather, it has bends and twists.

Now suppose you wished to construct a model of the earth with an ideal (undistorted) bar magnet going through its center and having magnetic properties as similar as possible to those that actually exist a long way above ground level, beyond the influence of "earthly" irregularities—asymmetrically arranged continents and the like. (A geophysicist studying the earth's magnetism *would* so wish.) Where would the poles of this imaginary magnet be? The answer, arrived at by esoteric mathematical calculations, is that they would be several thousand kilometers above the surface, directly above 79° N, 71° W in the northern hemisphere (see fig. 1.5), and 79° S, 109° E in the southern. These are the points shown on a map as the North and South Geomagnetic Poles, respectively; unlike the "dip" poles, they are at opposite ends of a line going through the earth's center.

There is nothing to observe when you arrive at, or rather under, the North Geomagnetic Pole; it is about as theoretical as anything can be. It drifts, though slowly, moving approximately northwest at about 2 km per year. And its position is related to the aurora.

1.5 The Aurora Borealis, or "Northern Lights"

When the arctic sky is totally dark—which it never is, of course, in high summer—the chances are favorable for seeing a display of the aurora borealis, or northern lights. In good displays, the glowing, shimmering, flickering lights sometimes take the form of arcs or bands, sometimes of rippling draperies, and sometimes of rapidly pulsating patches of light streaming upward through the sky. The color is usually pale green, very occasionally red or violet.

Contrary to legend, the aurora makes no sound: it is silent. Reports of shushing sounds probably mean that the observer was hearing the swish of dry snow blown over hard snow crust, and of crackling sounds that the observer was hearing the static crackle of dry woolen or synthetic clothes. (It is noteworthy that reports of audible auroras seem always to refer to displays on intensely cold nights, and never to those on warm nights, in populous, mid-latitude regions.)

The aurora is caused by electrically charged particles streaming earthward from the sun and striking atoms and molecules (chiefly of oxygen and nitrogen) in the rarefied upper atmosphere. Subatomic particles are dislodged from the atoms and molecules hit, and energy is liberated in the form of light. The same thing happens in a glowing neon sign, in which molecules of the rare gas neon are the targets. The colors of the aurora depend on the varying energy of the incoming particles (the missiles) and on the chemical nature of the atmospheric atoms and molecules struck (the targets). The common pale green aurora is emitted by oxygen; the rarer reds and violet come from both oxygen and nitrogen, struck by particles with different energies. This electrical activity commonly takes place at between 100 and 300 km above the ground.

The paths taken by the incoming particles from the sun (the missiles) are controlled, when they get near the earth, by the form of the earth's magnetic field far above the ground, which is there almost the same as the "ideal" field described in the preceding section. (As an added complication the particles also affect the magnetic field as well as being affected by it.) To make a long story short—and this is where the geomagnetic pole comes into it—the missile particles are deflected from their original straight line paths in a way that makes most of them converge toward a more or less oval ring, centered on the geomagnetic (not the magnetic) pole. This oval is called the *auroral oval* (see fig. 1.5). It marks the zone where auroras are most often seen and are at their most splendid; they become progressively less frequent and less splendid as you go away from the oval, either inward or outward. Near the center of the oval, directly below the geomagnetic pole, auroras are no more common than they are over, for example, Montreal or Duluth. Outside the oval, auroras become less and less frequent until they almost peter out altogether. In a nutshell, look for the most spectacular auroras in North America along the southern border of the Low Arctic tundra, a suitably inspiring setting.

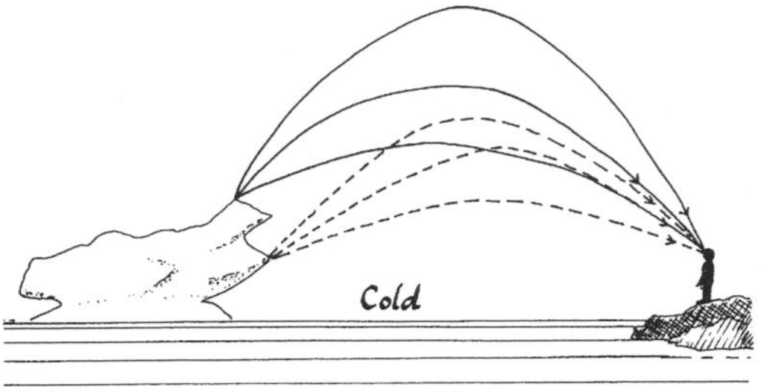

2 Climate and Atmosphere

2.1 The Arctic Climate

The arctic climate's most obvious characteristic is that it is cold. Less well known is the fact that it is also dry. Over most of the Arctic Islands, indeed everywhere except on the high mountains of the eastern islands, the annual precipitation (rain and snow combined) is less than 200 mm, less, that is, than anywhere else in North America except for the southwestern desert. Figure 2.1 gives a summary of conditions: precipitation on the central map and temperatures through the year for a selection of recording stations on the accompanying "clock faces." The number of months in which the average daily temperature never rises above freezing increases the farther north you go.

The sun's failure to rise high in the sky at any time of year (see fig. 1.1) is the cause of arctic cold. At high latitudes, the sun's rays always strike the ground obliquely; therefore, a given amount of solar energy is spread more "thinly" (over a larger patch of ground) than it would be in lower latitudes where the sun rises higher in the sky. Moreover, the rays from a low sun have passed through a greater thickness of atmosphere than have those from a high sun and lose much of their energy by absorption before ever reaching the ground. It is the height of the sun in the sky, not the fraction of the year that it is above the horizon—which is the same (exactly one-half) everywhere on earth—that controls the arctic climate.

Height of the sun is not the only control, however: the cold is tempered by warmth reaching the Arctic from the south, warmth in the form of warm air masses and warm ocean currents. These movements of air and sea have the effect of evening out temperature extremes; with-

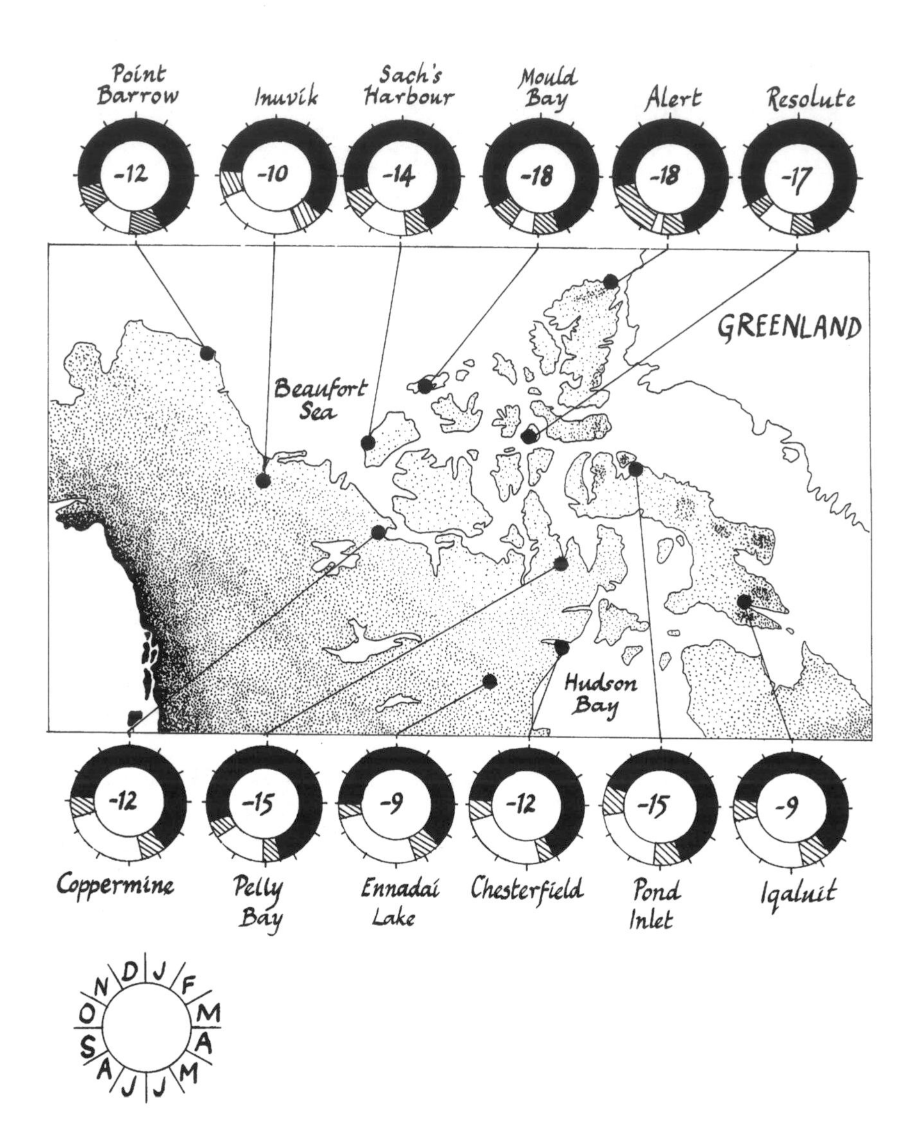

Figure 2.1. The Arctic climate. Stippling on the map shows how precipitation varies, from a low of less than 100 mm per year at Mould Bay, to a high of over 600 mm east of Iqaluit. The "clock faces" above and below the map show how mean daily temperatures vary through the year at selected stations (circle below shows how the months are arranged). Black: mean daily maximum temperature below freezing (winter). White: mean daily minimum above freezing (summer). Shaded: maxima above freezing and minima below freezing ("warm-up" and "freeze-up").

out them, the Arctic would be colder, and the tropics warmer, than in fact they are. Unless the climate is undergoing long-term change (which, of course, it probably is), the amount of solar energy the earth gains in any one year is balanced by the amount of heat it loses by radiation into space. This balance relates to the earth considered as an indivisible whole. It has been calculated that at low latitudes (between 37° N and 37° S) the earth's surface receives more energy from the sun than it loses by radiation, while at higher latitudes (north and south) it loses more than it gains. Consequently, if it were not for cross-latitude movements of air masses and ocean currents, the polar regions would cool down, and the tropics warm up until a new equilibrium was reached.

During the long night of arctic winter when the sun never rises (a night which varies in length from a single day on the Arctic Circle to 6 months at the Pole), no solar energy reaches the ground at all; if it were not for the arrival of air masses from the south, the temperature would drop to unimaginably low levels. Southerly air masses do come, however, bringing amelioration. Instead of dropping uninterruptedly for weeks or months, temperatures wander up and down, though not, of course, on a 24-hourly basis, for there are no nights and days. The immediate warming effect comes less from the warmth of the arriving air masses, which are often at a considerable height above the ground, than from the moisture they contain: the moisture condenses into clouds and the clouds exert their familiar "blanketing" effect. Just as in temperate latitudes a cloudy night is milder than a clear one, likewise in the Arctic: the long arctic "night" has cloudy periods, when it is comparatively mild, and clear periods, when the cold is bitter, but the lengths of these periods are irregular and unpredictable.

In summer, of course, the temperature does vary on a 24-hourly basis. Even in the midnight-sun period when it never sets, the sun is higher when it is to the south (at "midday") than when it is to the north (at "midnight"), and temperatures rise and fall correspondingly. But the difference between the sun's altitude above the horizon at "midday" and "midnight," and likewise the difference in temperature, become less and less the nearer the pole you go. At Alert on the northern shore of Ellesmere Island at 82° N, the difference between midday and midnight temperatures on a typical July day averages a mere 1° C: the difference between "day" and "night" is imperceptible.

2.2 The Seasons

Four climatic seasons make up the arctic year, as in temperate latitudes. But in the Arctic, winter is long and summer short, and the seasons between them could more aptly be described as "warm-up" and "freeze-up" than as spring and fall. These seasons correspond with the four periods marked on the clock face diagrams in figure 2.1. Winter is black; warm-up is the cross-hatched sector in spring; summer is white; freeze-up is the cross-hatched sector in fall.

Winter consists of the polar night when the sun never rises plus those periods in which, though the sun is above the horizon for some hours each day, it isn't up long enough and doesn't rise high enough to have an appreciable warming effect. The sun's rays cannot warm the earth as long as the surface is covered with a brilliant white snow layer that reflects all but a small fraction of solar energy back to the sky. And incoming warmth cannot raise temperatures until it has first melted the ice in the soil and the layer of ice covering sea and freshwater; in the process of melting, ice "consumes" a large amount of heat without any rise in temperature. Not until melting is complete does the water produced start to warm up. In the far north, winter doesn't end before June.

When the warm-up finally comes, it doesn't last long. The midday sun is at, or past, its highest for the year; temperatures are above freezing by day and below freezing by night, causing crusts on the snow. Even so, the sun's warmth penetrates, the ground thaws and the snow cover melts; shallow pools melt too, and rivers begin to flow. The wetness of the landscape increases fast. Water evaporates, clouds form, falling rain washes away more snow, and summer comes quickly.

Summer is the period when the temperature remains above freezing all through the 24-hour day (this is an average: frost is possible at any time). The thawed soil above the permafrost (the active layer) is as deep as it's going to get. The land is snow-free except where permanent snowbanks persist. Meltwater pours off ice caps and glaciers. Monthly precipitation, most of it as rain, reaches its maximum; the amount of rain is small, however, and increasing evaporation allows the land to become a little less wet. In spite of the aridity of the climate, skies are often cloudy. But this does not mean that there are not plenty of memorable days, even weeks, with clear skies and 24 hours of continuous sunshine. In the warmest month, July, temperatures can become truly summery. At the three northernmost, coldest sites shown in figure 2.1, the record highs in 30 years of temperature records were 20° C for Alert, 16° for Mould Bay, and 18° for Resolute. (It would give a wrong impression not to list the lowest July lows in the same 30 years; they were, respectively, −6° C, −4°, and −3°.)

Summer ends when freeze-up begins—in August or September in most places, but as early as July in Alert. With the sun getting lower in the sky every day, the ground starts to lose heat: more heat is radiated away from the surface than comes in. Everything starts to freeze: soil, marshes, bogs, ponds, lakes, rivers, and finally the sea.

Then winter, and with it the snow, returns. The amount of snow is not great, and the heaviest falls usually come in October, after which the air becomes so cold it can carry only negligible amounts of water vapor and precipitation stops. Thereafter the snow is constantly re-arranged—swept off the ridges and drifted in hollows—whenever the wind rises.

2.3 The Climate near the Ground

Up to this point we have considered climate as meteorologists measure it, with air temperatures recorded in standardized instrument boxes (Stevenson screens) at a standard height of 1 m above the ground. The climate closer to the ground, as experienced by low-growing plants, low-flying insects, ground-nesting birds, and small mammals is quite different. For them, arctic summers are considerably warmer.

The reason is that when dry ground is warmed by the sun, the temperature of the air in contact with it rises well above that at the standard 1-m level. Going down from top to bottom of the lowest 1-meter-thick layer of air, the temperature increases; the difference in temperature between air at 1 m, and air at 1 cm, above a dry, sunlit surface can be as much as 7° or 8° C.

In temperate latitudes, this warming of the lowermost "skin" of air happens in just the same way, during daylight; but after sunset, the ground cools off rapidly, especially under a clear sky. The temperature of the lowermost air layer then falls well below that of the air at the standard 1 m above the surface, and a ground frost is perfectly possible even when the "official" temperature remains above freezing. This reversal of the normal temperature gradient, with the temperature becoming higher at increasing heights above the ground (rather than lower as it normally does), is a temperature *inversion.*

When the midnight sun shines in the Arctic, temperature inversions seldom happen. Instead, "daytime" conditions, with warm ground, persist through the 24 hours. This can cause plants to grow faster at high latitudes than at low: because inversions become rarer the higher the latitude, a plant growing at a high latitude may accumulate "growing hours" (hours above a threshold temperature) faster than one at a lower latitude. Similarly, the development of insects, from eggs to adults, can proceed without any interruptions due to low temperatures. The lack of night frosts is a boon to life at ground level.

The lack of dew may also have some subtle ecological effects. It is rare in summer, because the midnight sun keeps the ground warm. Dew forms when moisture in the air condenses on cold surfaces, and in the summer Arctic the air is dry, and cold surfaces uncommon.

The midnight sun produces other strange effects, because at night the sun shines from the north. In hilly or mountainous country, south-facing slopes and north-facing slopes have their warmest part of the day 12 hours apart, the former at midday, the latter at midnight. The rest periods of animals whose activities slow down during the cool hours are likewise 12 hours out of phase.

On a north-facing hillside which happens to have just the right slope (its angle to the *vertical* equal to the latitude), the sun's rays strike the ground at exactly the same angle at midday and midnight. This doesn't ensure a constant temperature, however, because the rays have a longer path through the atmosphere when the sun is in the north (and not far above the horizon) than when it is in the south (and higher above the horizon).

In winter, the climate near the ground is crucially important to animals too small to maintain their body warmth without the protection of snow, for example, lemmings, voles, ermines, ptarmigans, and redpolls. The deep, fluffy snow of the northern forests—the best kind, for insulation—is seldom available in the Arctic. Not much snow falls: the annual snowfall at the stations shown in figure 2.1 averages less than 130 cm, equivalent to 13 cm of rain; and snow-laden blizzards, rearranging what snow there is, sweep exposed ground bare and deposit their loads in sheltered spots as firmly packed snowdrifts. But although this snow is not the best kind for insulation, it is evidently adequate to keep the habitat endurable for the warm-blooded animals living under it. Given light, fluffy snow, a layer only 20 cm deep will prevent loss of heat from the ground and keep the temperature at ground level nearly constant, but packed snow has to be twice as deep or more to insulate as effectively.

2.4 Arctic Mirages

Mirages are to be seen in any latitude. They appear when the temperature of the air near the ground changes greatly over a short vertical distance. If the ground surface is very hot—as it is when a hot sun shines on a hardtop road, for instance—the air temperature will be found to drop rapidly if you move a thermometer upward, from the hot layer of air in contact with the ground into progressively cooler layers above. If the surface is very cold—as it is over arctic sea ice—air temperature rises as you go upward, from the cold layer of air in contact with the ice into progressively warmer layers above; in other words, there is an inversion

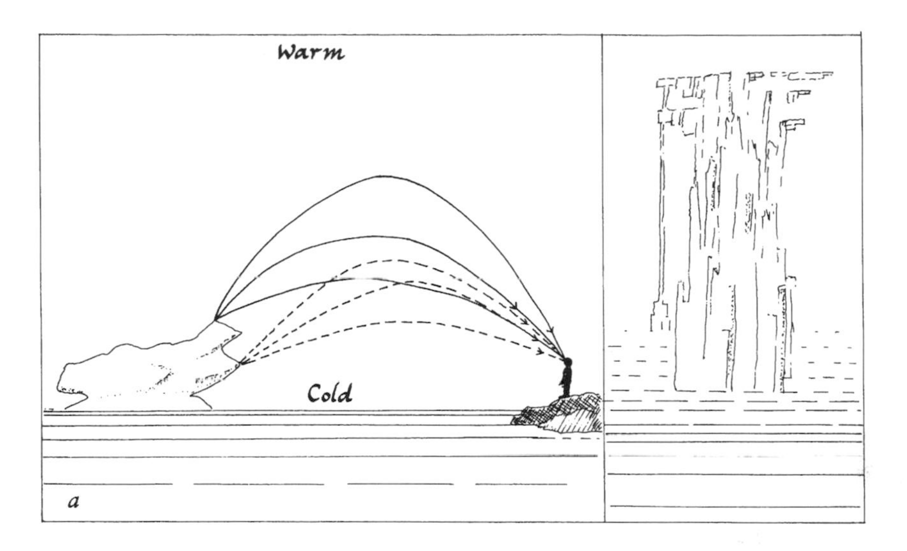

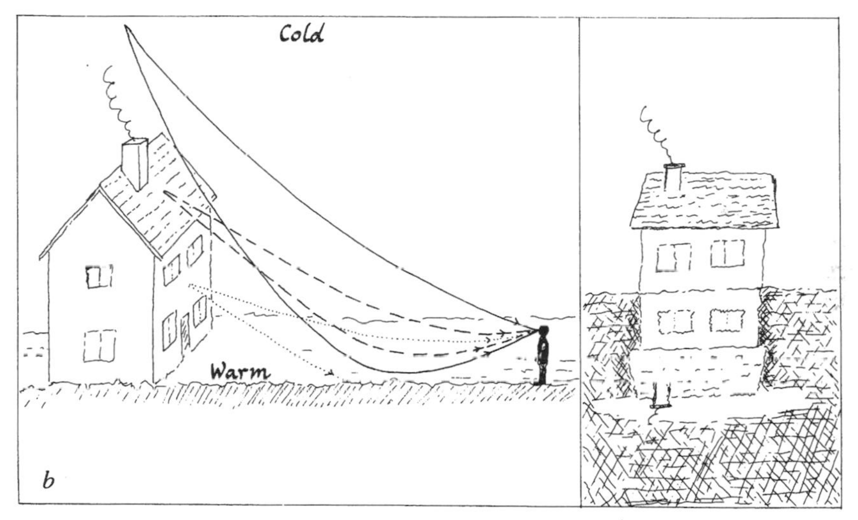

Figure 2.2. (a) A superior mirage (the fata morgana). (b) An inferior mirage. In both cases the diagram on the left shows the curved paths of light rays travelling from an object to an observer; two or more rays starting from the same point on the object are shown by lines of the same style (solid, dashed or dotted). An observer instinctively assumes the rays are coming in a straight line, and sees each point in the view as though it lay straight behind the arrowhead showing the ray's direction when it reaches his/her eye. The picture to the right of each ray diagram shows the view the observer sees. (Note the imaginary "pool" below the inverted roof.)

(see sec. 2.3). In either case, it is the changing air temperatures that produce mirages. Light rays traveling through such "layered" air are traveling through a nonuniform medium and are curved as a result; only in a uniform medium does light travel in straight lines. Of course, even in mild weather, the temperature always changes from ground level upward (it usually falls slightly) so that, in theory, every scene is a mirage. But for a mirage to become noticeable, the temperature gradient has to be fairly steep.

First, note that to someone looking at an object in the distance, the object seems to be in the direction from which light rays are arriving from it, which is not necessarily in the "true," or straight line, direction (see fig. 2.2). Therefore, an object seen through layered air appears displaced and distorted. Sometimes light rays from every point on the object reach the eye from several directions at once. Then the object appears duplicated, perhaps more than once, with some of the images upside down.

Light travels slightly faster through warm air than through cold, causing a light ray, as it passes from cooler to warmer air, to curve back toward the cooler air. This means that to somebody observing a distant object, say an ice floe, through air that is warmer above than below, the floe will seem higher than it really is; it will also be elongated vertically, making it seem taller than it really is, because single points are drawn out into vertical lines (fig. 2.2). This is a *superior* mirage, seen across a cold surface, and the kind likely to be seen in the Arctic.

When the air is hotter below than above, an *inferior* mirage appears. (An inferior mirage is less likely in the Arctic, but is described here to make the contrast clear.) It looks entirely different: a distant object, say a house, will appear shorter than it really is, because the horizon seems to have come closer, cutting off the bottom of the house; under that, the top of the house is often seen repeated, upside down, and under that again, a patch of sky (fig. 2.2). What looks like a pool of water on the highway ahead on a hot summer's day is really the displaced image of a patch of blue sky.

In the Arctic, a superior mirage is often to be seen on still, clear summer days, sometimes across sea ice, sometimes across a calm sea in which floes are floating, or across a sea dotted with low rocky islands. The effect is magical. Distant ice floes—even quite small chunks of ice—appear to have grown upward into tall columns like the towers, turrets, and spires of a fairy castle seen from afar, in shimmering blue and white. (The apparition is romantically known as the *fata morgana,* the Italian name of Fairy Morgan, King Arthur's sister, who undoubtedly lived in such a castle.) If the towers are faint, their apparent distance, and

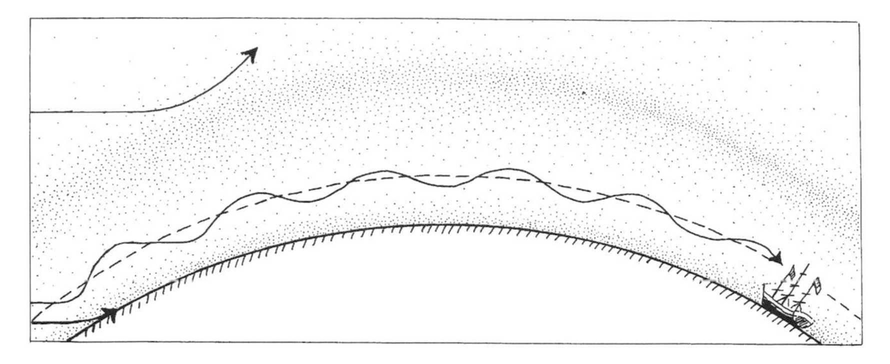

Figure 2.3. The Novaya Zemlya effect. Some of the suns's rays (arrows) reach the ship around the curvature of the earth while the sun is still geometrically below the horizon. Denser stippling shows warmer air. Rays reaching the ship are "trapped" in the layer of cool air enclosing the level of minimum temperature (dashed line).

consequently their height, are even more exaggerated. The scene changes as time passes; the towers lengthen and shorten as temperatures in the stacked layers of air rise and fall. Likewise with rocky islands: low islands seem to have grown up into smooth, precipitous cliffs, or into several cliffs with the upper ones overhanging the lower.

The fata morgana not only elongates distant objects: it raises them skyward. The curvature of the light rays compensates for the curvature of the earth, causing distant scenes that "should" be below the horizon to rise into view. The "new" horizon is abnormally far away. From a point where a distant landmass, seen over water, is the farthest thing in view, the displaced water horizon seems to have "climbed up" the landmass.

A rare and remarkable form of mirage, known as the Novaya Zemlya effect, has enabled observers to see as much as 400 km over the horizon. The name refers to the Siberian Arctic island where the effect was first recorded, in 1596; near the end of winter, sailors saw the sun rise above the horizon about 2 weeks earlier than it "should" have done if it had conformed to astronomical calculations—or 2 weeks earlier than it would have done if light traveled in straight lines. Figure 2.3 shows what makes the horizon appear to recede to such enormous distances.

For the effect to happen, there has to be a temperature inversion (a layer of air in which temperature rises with increasing height) starting at some distance above the ground. Below the inversion temperature rises with *de*creasing height above the ground in the usual way. This means that at the level where (going upward) the temperature stops falling and starts rising, it must reach a minimum: the air at this level—

the "floor" of the inversion—is colder than the air above and below. On the rare occasions when everything is perfectly positioned, light rays become trapped close to the inversion's floor: rays directed away from the floor, either upward or downward, are bent back toward it as they start to enter warmer air. In this way the rays are held close to a level from which they cannot escape. The mechanism is exactly the same as that by which an optical fiber works. (Note: Any inversion must also have a "ceiling," a level at which the temperature reaches a maximum before beginning to fall again as you go still higher. This ceiling plays no part in the Novaya Zemlya effect, however.)

For more details on mirages and other atmospheric optical effects, see *Rainbows, Halos, and Glories,* by Robert Greenler (Cambridge: Cambridge University Press, 1980).

2.5 The Air Made Visible

The arctic tundra, like the prairies, is "big sky" country. The sky is part of the scenery, and there are some typically arctic features to look for in that scenery.

Two of them are *ice blink* and *water sky.* Ice blink is a streak of dazzling brightness on the underside of distant clouds, where they are lit from below by sunlight reflected up from a surface of snow or ice. On cloudy days at every latitude, the sunlight that penetrates the clouds is continually reflected back and forth between the ground surface and the undersides of the clouds. Most people know that the sun's heat rays do this: it explains why a winter day is warmer under a cover of cloud than under a clear sky. The fact that the sun's light rays behave in exactly the same way usually goes unnoticed, as there is normally nothing much to see. But if a pure white patch of surface—say, a field of floating pack ice—forms a bright island in dark surroundings—the dark, unfrozen sea—then the clouds directly over the pack ice will be brilliantly lit, and will contrast strongly with the dimly lit clouds all around. The effect is especially striking when the reflecting ice is over the horizon, and the shining cloud is seen above dark sea. This is ice blink at its best.

Water sky is the converse of ice blink. A dark patch appearing on the otherwise bright ceiling of cloud over a snow-covered world shows that there must be something dark below; that something is nearly always a patch of open water in the pack ice. The darkness is especially striking if the open water causing it lies over the horizon, so that the dark patch is seen above a shining field of white ice. When conditions are right, water sky and ice blink combine to make a precise map of the white and dark pattern of the ground on the underside of the overlying cloud cover.

Over the centuries, many ships that might have become inadvertently trapped in pack ice have been steered to open water by navigators relying on water sky to guide them.

During the arctic spring, *ice fog* often forms, a mist of tiny ice crystals floating in the air. Typically the fog reduces visibility at ground level without obscuring the sun. In these conditions, an observer standing in the fog will often see spectacular halos round the sun and other optical displays such as sun dogs and sun pillars. The patterns are often quite elaborate. The form they take depends on the shape of the ice crystals, which may be flat, hexagonal plates or pencil-shaped hexagonal prisms, and on the arrangement of the crystals, which may be aligned with one another or may be at random. The colors depend on whether the sunlight is reflected by the ice crystals, giving pure white columns and halos, or refracted through them, giving rainbow-colored halos and arcs. The physics is complicated, but it is well worth making an accurate sketch of unusual displays, so that you can investigate the details on your return from the Arctic.

The air itself becomes visible in late winter and early spring, in the phenomenon known as *arctic haze*. The blue-white haze greatly reduces visibility on calm, clear, sunny spring days. It results from the fine particles always floating in the air, particles which, in the Arctic, are up to 40 times as abundant in winter as in summer. The particles of summer are of natural origin: they consist mostly of wind-blown dust and tiny crystals of sea salt. Those of winter are man-made pollutants—dusts and gases—wafted across the Pole from the industrial regions of northern and eastern Europe including Russia; not to mince words, the haze is smog. It is beyond doubt that the pollutants originate mainly in the Old World; the prevailing wind over arctic North America is from the northwest, from northern Eurasia, all year long. This does not mean, however, that North Americans are free of blame for arctic air pollution. The pollutants from New World industries are carried on westerly winds to the Eurasian Arctic. The industries of the Old World are much "dirtier" than those of the New, however, and cause more than twice as much air pollution.

The pollutants don't remain suspended in the atmosphere, of course. The rain and drizzle of summer wash them out of the dry winter air. This has been demonstrated by scientists who found that the amount of dilute sulphuric acid in the atmosphere goes up and down in step with decreases and increases in the amount of low-level cloud. Numerous other pollutants are washed from the smog to the ground: heavy metals, agricultural pesticides including DDT (which is still used in Asia), and a nasty blend of organochlorines. These toxins flow into the food chain, becoming more and more concentrated as they ascend the chain to the top predators, which include polar bears, grizzlies, humans, and wolves.

Natural air pollution is also to be found in the Arctic, as a local curiosity. It happens in the Smoking Hills, on the Arctic Ocean coast about 350 km east of the Mackenzie Delta, where the very ground itself is on fire: pyrite (which is rich in sulphur) and low-grade coal in the shaley rock are undergoing spontaneous combustion. The hills have probably been smoldering for hundreds of years; they were first discovered by the explorer John Richardson, in 1828, and he explained the process thus: "the shale takes fire in consequence of its containing a considerable quantity of sulphur in a state of such minute division, that it very readily attracts oxygen from the atmosphere, and inflames." Present-day scientists agree. Columns of pungent smoke, visible from afar, rise from the cliffs where the hills meet the sea. Sulphur dioxide in the smoke has altered the chemistry of the soil in the surrounding area, and only a few pollution-resistant plants are able to grow there.

For details on the magnificent optical effects caused by ice fog, see *Rainbows, Halos, and Glories,* by Robert Greenler (Cambridge: Cambridge University Press, 1980). For details on arctic air pollution, see "The Not-So-Pristine Arctic," by Karen Twitchell, *Canadian Geographic,* Feb.–March 1991.

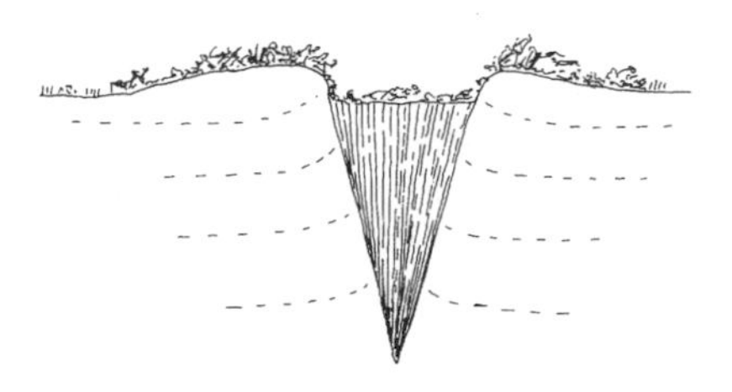

3 Terrain

3.1 Ice Caps and Glaciers

A smooth, swelling skyline of pure white, high up against a clear blue sky, is often a land traveler's first view of an arctic ice cap; it is a quintessentially arctic sight. The permanent ice that covers 5 percent of the total arctic land area is the true heart of the terrestrial Arctic, just as the permanent polar ice pack is the heart of the marine Arctic. A map of the arctic ice caps and glaciers is given in figure 3.4 (next section). They are found only on the islands, not on the mainland.

Ice caps cover highlands whose summits are above the permanent snowline and where winter snowfall is comparatively heavy. Snow accumulates at these sites, gradually changing from light, new snow to dense, granular, old snow known as *firn;* as the firn becomes deeper, its weight compresses the underlying layers into glacier ice, and the ice cap grows. Ice is not rigid, and as snow is added to the top, the cap does not simply grow upward; distorted by its own weight, it also expands outward until, where the margin intersects the borders of the highland, the ice begins its downhill movement toward the lowlands. The shape of the land surface molds the slowly moving ice into tongues that fill the valleys and creep down them under the pull of gravity. These outward flowing tongues are known as *outflow glaciers,* and they form all around the periphery of an ice cap.

The outward and downward expansion is finally brought to a halt by *ablation;* the word means loss of ice from any cause. The three common causes are: melting; sublimation, which is the conversion of ice directly into vapor without the intermediate formation of liquid water; and, if the margin reaches the sea, calving, which is the breaking off of marginal ice as icebergs. Ice is occasionally lost because the ice margin creeps to the brink of a precipice where chunks break away.

Somewhere along its length, each outflow glacier is crossed by an invisible *equilibrium line;* upslope of the line is the accumulation zone of the glacier, where snow and ice accumulate faster than they disappear; downslope of the line is the ablation zone, where ablation exceeds accumulation. The line itself is where accumulation and ablation exactly balance (see fig. 3.1).

A glacier in mountains of the temperate zone usually has its equilibrium line close to the permanent snowline. But in an outflow glacier descending from an arctic ice cap the equilibrium line is a long way downslope from the permanent snowline. This is because an arctic glacier accumulates additional ice by a process called refreezing. Below the permanent snowline, snow on the glacier's surface inevitably melts in summer, but the meltwater doesn't simply flow away as it would on a temperate glacier; instead, it soaks down into the surface slush and refreezes where it comes into contact with the cold, arctic ice beneath. The ice so formed is known as *superimposed* ice. The equilibrium line is the line where the accumulation of superimposed ice, formed by refreezing, is balanced by ablation from all causes. Because of refreezing, arctic glaciers are long; they often extend far down into the lowlands.

Refreezing happens only in arctic and antarctic glaciers, which are "cold" glaciers. Disregarding the slush on top in summer, they consist of ice at a temperature far below the freezing point of water (equivalently, the melting point of ice). The ice of "warm" glaciers—the mountain glaciers of temperate and tropical lands—is at a temperature close to freezing point from the surface right down to the bottom, where ice rests on bedrock. Indeed, warm glaciers, but not cold ones, have internal streams of meltwater ("melt-streams") at every level throughout their thickness, and torrents of water flow out from under the ice at a warm glacier's terminal or "toe." In contrast, the water flowing from a cold glacier consists entirely of run-off from the top and through shallow tunnels not far below the surface.

Some glaciologists classify cold glaciers further; they recognize two kinds of cold glacier, subpolar and polar. Refreezing occurs only in subpolar glaciers; polar ones are too cold to melt at all, so there is nothing to refreeze.

There is another contrast between cold glaciers (both kinds) and warm ones. At the "sole" of a warm glacier (the bottom of the ice), the temperature is so near melting point that the ice doesn't freeze to the rock. As a result it can move by sliding over the rock, and movement may even be lubricated by a layer of meltwater between ice and rock. The sole of a cold glacier, on the other hand, is frozen hard to the underlying rock, making sliding impossible. The ice oozes downhill because of internal shearing movements within it.

The meltwater formed on a glacier's surface in summer below the equilibrium line does not refreeze; it streams down over the surface slush eroding winding channels as it goes which, combined with a multitude

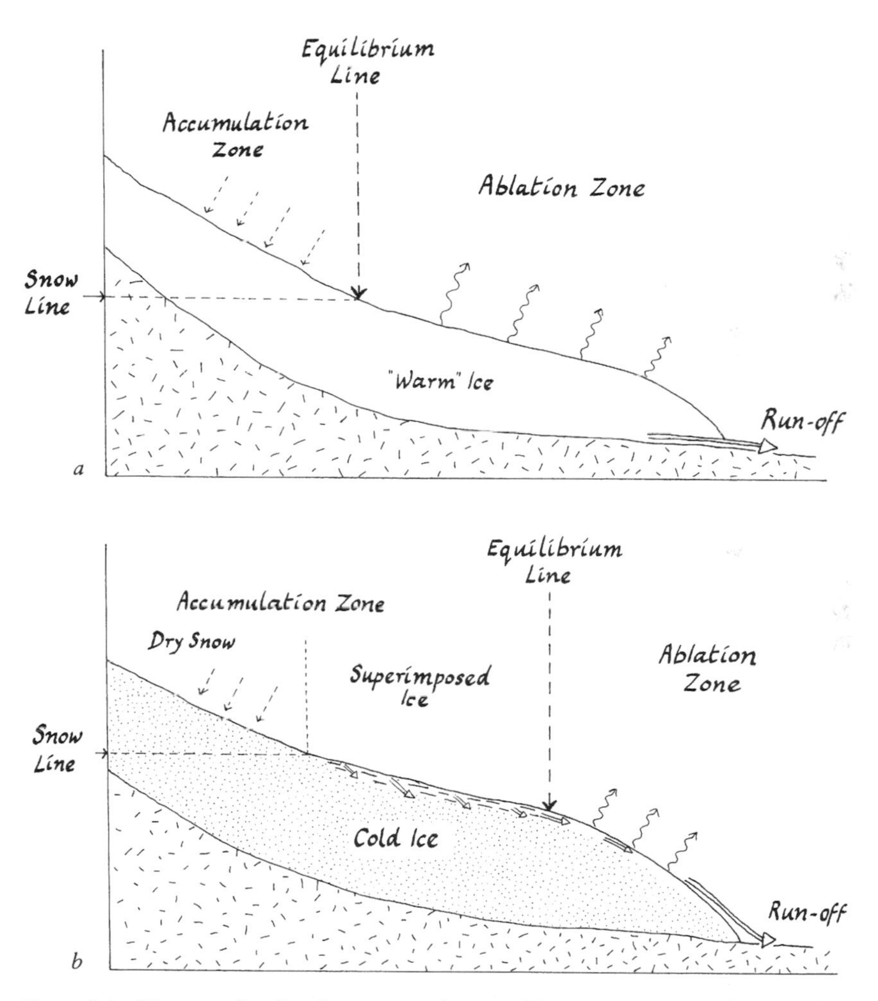

Figure 3.1. Diagrams showing the contrasts between (a) a temperate zone glacier, and (b) an arctic glacier. Dashed arrows show snow falling at all seasons. Wavy arrows show loss of ice through ablation. Hollow arrows show flowing water. Cold ice (much below freezing) is stippled; comparatively "warm" ice (near freezing) is white. In (b), the horizontal dashes show refrozen, superimposed ice.

of transverse cracks and crevasses, give a glacier the intricate surface patterns to be seen from an aircraft window. Melt-streams are often deep and swift. Crossing them can be dangerous; to fall in is to be swept helplessly away over a glass-smooth surface that offers no purchase. The streams meander sinuously with a pattern that is a miniature version of a meandering river's pattern, but far more precisely regular (fig. 3.2). Glaciologists disagree on why these meanders form.

Some melt-streams are large enough to be called melt-rivers. Flowing in the same channels summer after summer, they last long enough to be shown on topographic maps as permanent rivers flowing on ice. Also in summer, rivers gush down "moats" on either side of a glacier. In warm weather, the roar of meltwater torrents on and beside the ice can be heard from afar. Meltwater lakes form too, wherever a glacier dams a river in a tributary valley. The existence of an unfrozen lake in contact with a glacier shows convincingly that the glacier must be melting and must owe its continued existence to the ice cap that constantly feeds it.

3.2 Permafrost

The arctic could well be called "permafrost country." *Permafrost*—the name denotes permanently frozen ground—is the salient characteristic of all arctic lands, and governs the shape and texture of the ground surface everywhere. You can only appreciate arctic scenery to the full by being conscious of permafrost, and its sometimes astonishing effects.

Below a certain depth, which depends on the climate, arctic soil and subsoil remain continuously frozen year after year, if not forever, then for centuries or millennia. The upper surface of the frozen ground, known as the *permafrost table,* can be anywhere from 3.5 m below ground surface in the subarctic to a mere 20 cm in the High Arctic. The thickness of the permafrost layer varies too, from zero at its southern limit to 500 or 600 m in the High Arctic. The ground below the permafrost, known as *talik,* is unfrozen. The ground above it, between the permafrost table and the surface, is called the *active layer*. This is the layer that thaws in summer; all plant roots are confined to it.

Temperature fluctuates with the season even within the permafrost, though only in its uppermost layer (fig. 3.3); here the frozen ground is warmer in summer than in winter, while remaining below freezing point all year. These seasonal fluctuations become less and less with increasing depth down to the *level of zero amplitude,* which is the level at which the effect of surface changes disappears. This happens at a depth of about 20 m below the permafrost table, where the temperature may be 3° or 4° C higher than its annual average at the surface; these figures are

Figure 3.2. An ouflow glacier flowing from an ice cap. Note the miniature meander patterns in the run-off streams near the toe of the glacier.

representative, they are not the same everywhere. Below the level of zero amplitude, the temperature of the permafrost rises steadily, because it is warmed from below by the earth's internal heat. On average, it rises at about 1° C per 50 m of increased depth. At the level where the temperature reaches the melting point, permafrost stops and the unfrozen ground from there on down is talik.

Permafrost exists wherever the average temperature is low enough; it forms a continuous underground layer throughout the whole of the tundra-covered region, except under lakes and rivers big enough and deep enough not to freeze to the bottom in winter. Liquid water (in sufficient quantity) stores enough warmth to prevent the ground below it from becoming permafrost. South of the tundra zone permafrost is present in patches rather than as a continuous layer; the warmer the climate, the smaller, thinner, and more widely spaced these patches become, until they finally peter out altogether. Near its southernmost limits, permafrost forms only where conditions are especially suitable, for example, on north-facing slopes with vegetation dense enough to insulate the soil below so that it doesn't warm up in summer. The map (fig.

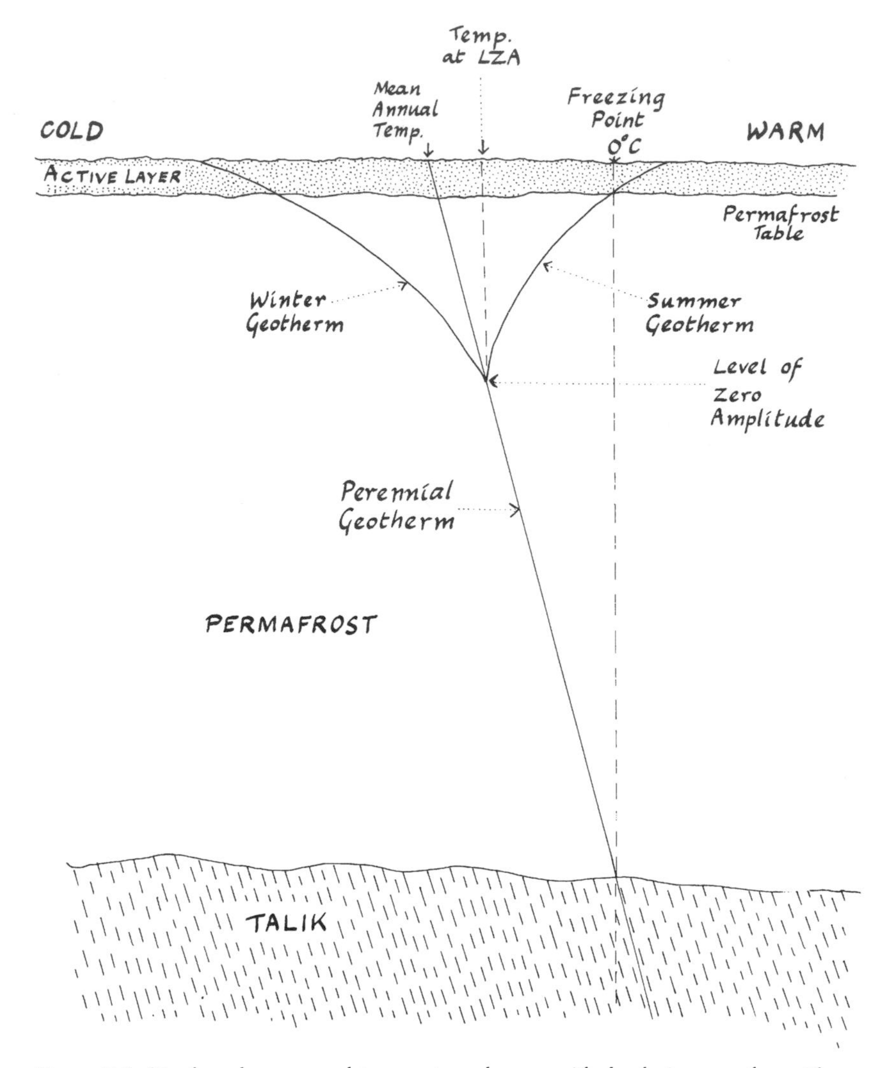

Figure 3.3. To show how ground temperature changes with depth, in permafrost. The unshaded layer is permanently frozen; the stippled layer at the top (the active layer) thaws in summer; the cross-hatched layer at the bottom (talik) is permanently unfrozen.

The curved Y is the geotherm, showing how temperature (increasing from left to right) changes with depth (increasing from top to bottom). The geotherm forks at the Level of Zero Amplitude (LZA). Below the LZA, ground temperature is unaffected by seasonal temperature changes at the surface; above it, temperature fluctuates seasonally between the summer and winter limits shown. The active layer is the layer in which the summer geotherm lies to the right of 0° C. At the bottom of the diagram, permafrost gives way to talik where the perennial geotherm crosses to the right of 0° C.

(*Note:* The diagram is not to scale; it magnifies the detail in the upper layers.)

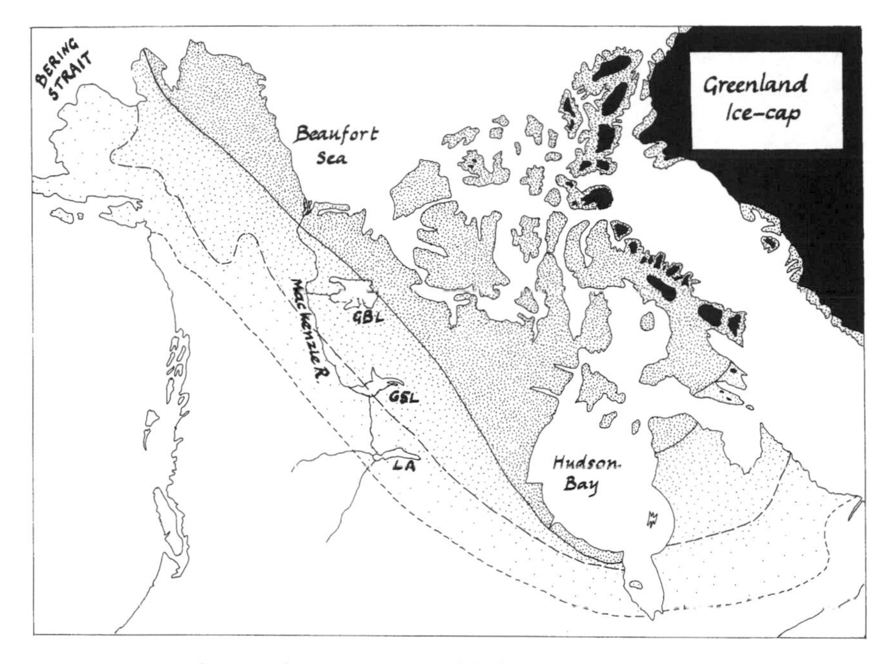

Figure 3.4. Map showing the arctic ice caps (black), and the three permafrost zones: from north to south, they are the continuous zone, the discontinuous zone, and the patchy zone. GBL, Great Bear Lake. GSL, Great Slave Lake. LA, Lake Athabasca.

3.4) shows the boundaries of zones in which permafrost is continuous, discontinuous but widespread, and patchy. Of course, the boundaries are not nearly as clearcut on the ground as they look on a map; the changes are gradual.

The most striking feature of the map is the way the boundaries dip southward as you go from west to east. Tundra-covered permafrost country—indeed, the whole "Arctic"—is much more extensive around Hudson Bay than it is farther west, where it forms a comparatively narrow strip along the shore of the Beaufort Sea. The cause is Hudson Bay itself. Near large bodies of water, such as the ocean or the Great Lakes, the climate is usually more moderate (with cooler summers and milder winters) than it is in the continental interior. This is because water has great heat-holding capacity; it is slow to warm in spring and slow to cool in the fall; this evens out summer and winter temperatures, reducing the contrast between them. The process fails in Hudson Bay because the ice pack covering its shallow waters persists until midsummer. Once the spring sun has melted the snow on shore, the dark ground

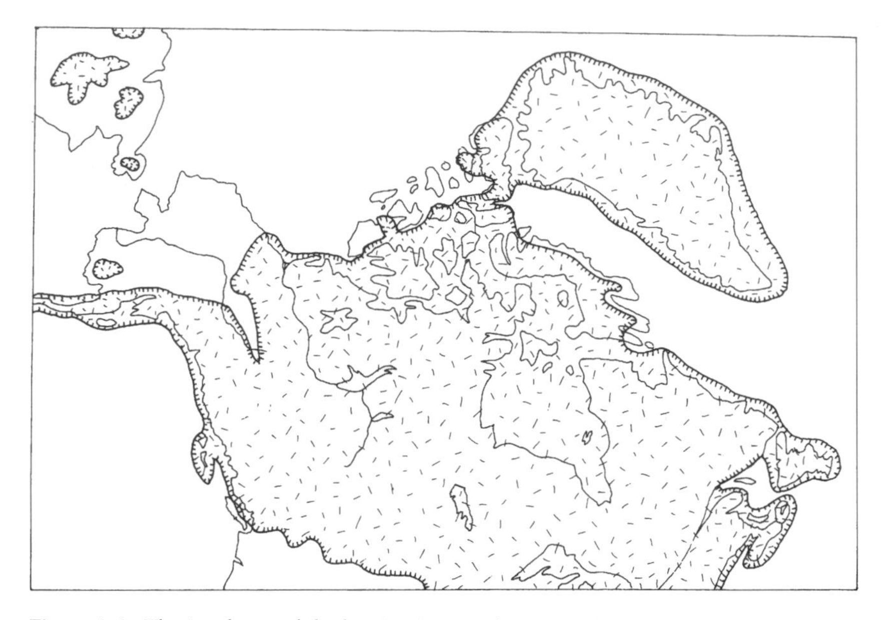

Figure 3.5. The ice sheets of the last Ice Age, as they were about 18,000 years ago.

absorbs the sun's rays and warms up. Meanwhile, out on the Bay, the dense, white pack ice continues to reflect the rays instead of absorbing them and the melt is greatly delayed. Spring warming of the water, instead of being merely slow, is nonexistent. By the time the ice has all melted, it is too late for the water to warm up appreciably before fall. The chilly waters of the Bay therefore keep summers cooler than they would otherwise be, so that seasonal temperatures fluctuate around a very low average. Permafrost is the result.

Permafrost has existed unmelted in the Arctic for tens of thousands of years. The evidence could hardly be more convincing: undecayed woolly mammoth carcasses have been found imbedded in it. It is believed that the permafrost of the arctic coastal plain has been frozen at least since the beginning of the most recent ice age, about 100,000 years ago. It could conceivably have existed through several earlier ice ages, without once melting in the short (ca. 20,000 years) gaps that separated them, going back for a million years or more.

Studies are in progress on how climatic warming is likely to affect permafrost. A rise in surface temperature takes time to penetrate down through the permafrost, and the effect—the "signal"—weakens as it descends. The fate of a "pulse" of surface warmth, starting abruptly and lasting for a known length of time, can be theoretically predicted. The pulse would travel slowly down and, some considerable time later, at any given depth the temperature of the frozen ground would gradually rise to a maximum and then gradually fall; the response becomes slighter, later, and more drawn out as it gets progressively deeper. For example, 3 years after a pulse of one year's duration at the surface, the maximum temperature response would be 10 m down in the permafrost, and its strength—the magnitude of the temperature change—would be only 10 percent of the strength of the original pulse. After 2 more years, it would have advanced to a depth of 17 m, and its strength would have dwindled to only 5 percent of the original.

The earth is said to have a "thermal memory" for events on its surface and some of the "memories" have been inspected, with surprising results; several drill holes were bored down into the permafrost of the arctic coastal plain of Alaska, and ground temperature measured at a sequence of depths. The results have led researchers to conclude that surface temperature has risen by about 2° C during the past several decades; the change may have taken as much as a century. The cause of the change is not certainly known, but global warming resulting from the greenhouse effect is a likely possibility.

3.3 Traces of the Ice Age

Not all arctic landforms are the result of present-day permafrost. There are, as well, many traces of the past: landforms that owe their existence to the last ice age, when the whole of northern Canada, like modern Antarctica, was buried under huge ice sheets. Figure 3.5 shows the extent of these ice sheets when they were at their maximum, about 18,000 years ago. The only ice-free lands then were parts of Alaska, Yukon, and some of the Arctic Islands.

Three very noticeable traces of the ice age that arouse the interest of all arctic travelers are *glacial fluting, raised beaches,* and *eskers.*

Glacial fluting is particularly common in the Barrenlands of the mainland Canadian Arctic. It is most easily seen from the air (fig. 3.6). The land has been fluted, or corrugated, by the heavy ice sheets that slowly spread over it, and the direction of the long, narrow, precisely parallel ridges and valleys shows the direction of ice flow. The flutings are partly the result of erosion—the valleys were scraped out by the ice—and partly the result of deposition—the moving ice plastered the ground with *till* (the mixture of mud, gravel, and rocks that adheres to the bottom of an ice sheet), molding it into elongate ridges in the process. Such ridges are known as *drumlins.*

Raised beaches are to be seen inland—sometimes a long way inland—

Figure 3.6. Glacial fluting, as seen from the air.

of many arctic shorelines. They are old sea beaches above the modern beach, often forming a series of terraces (fig. 3.7). Clam shells can sometimes be found on these old beaches, and even parts of the skeletons of whales stranded long ago (see fig. 5.6). The beaches show that the land must have been rising, relative to the sea, over the past several thousand years.

The resultant fall in the level of the sea relative to the land seems paradoxical in light of the fact that, because of the melting of the ice sheets, the volume of water in the world's oceans has increased greatly since the end of the last ice age. However, the lands that were once ice-covered have been rising because of rebound: when the ice sheets were present, their enormous weight pressed down the earth's crust beneath them, and when melting started, the crust began rebounding, like the side of an oilcan when pressure is removed. Crustal rebound is now in the process of compensating for the increased volume of the ocean, and creates the raised beaches.

Although the last of the mainland ice disappeared over 6000 years ago, crustal rebound still continues. For example, the floor of Hudson Bay is at present rebounding at the rate of 15 mm per year.

Eskers are the third large-scale relics of the ice age. They are long, narrow, sinuous, steep-sided ridges of gravel and sand, anywhere be-

Figure 3.7. Raised beaches, as seen from the air.

tween 10 and 50 m high, wandering across the landscape in an arbitrary fashion that bears no relation to other landscape features. From the air, they look like giant worm casts (fig. 3.8). They are the sediments deposited by rivers flowing through tunnels in and under the ice at a time when the ice sheets were melting. So in a sense they really are casts, the casts of those vanished tunnels. The melt-rivers, with their big loads of sediment, flowed through and under "warm" ice sheets, in which the ice at all levels had a temperature close to the melting point. Eskers were produced independently of, and later than, the "flutings" described above, which were formed by spreading ice sheets when the ice age climate was still cold. Therefore eskers lie on top of the ridges and valleys of glacial fluting, and wind across them in directions unrelated to the direction of the fluting.

3.4 Tundra Polygons

The ice age isn't over: although the great ice sheets have disappeared leaving only a few ice caps as remnants, the land everywhere in the Arctic is still in the grip of permafrost, and will probably stay that way until the next ice age overtakes us, unless present-day global warming wholly alters the future course of nature. Permafrost affects the shape

Figure 3.8. An esker, as seen from the air.

of the land in many ways and on many scales; this section describes one of the most distinctive permafrost landscapes to be seen from an airplane window—tundra polygons.

Tundra polygons are such a common feature of the arctic landscape that a special symbol is used to represent them on topographic maps (fig. 3.9). Commonly, each polygon is anywhere from 5 to 50 meters in diameter. As their shape suggests, they were formed when the ground cracked as it froze, in the same way (though on a larger scale) that the muddy bottom of a puddle cracks when the water dries. The network of cracks in the ground that outline tundra polygons take decades or centuries to develop; they are held open and repeatedly enlarged by the growth of wedges of ice within them (fig. 3.9).

The process starts (in places with appropriate soil) when the ground is exposed to extreme winter cold. At temperatures of −15° C or lower, the ground becomes brittle enough to crack as it contracts from the cold. The following spring, meltwater seeps into the cracks and freezes, forming vertical seams of ice known as ice wedges. The ice wedges form lines of weakness which crack open again when the temperature falls in the next winter; then the next spring's meltwater drains into the re-opened cracks where it freezes. And so on, year after year for centuries and millennia, progressively enlarging the ice wedges, making them

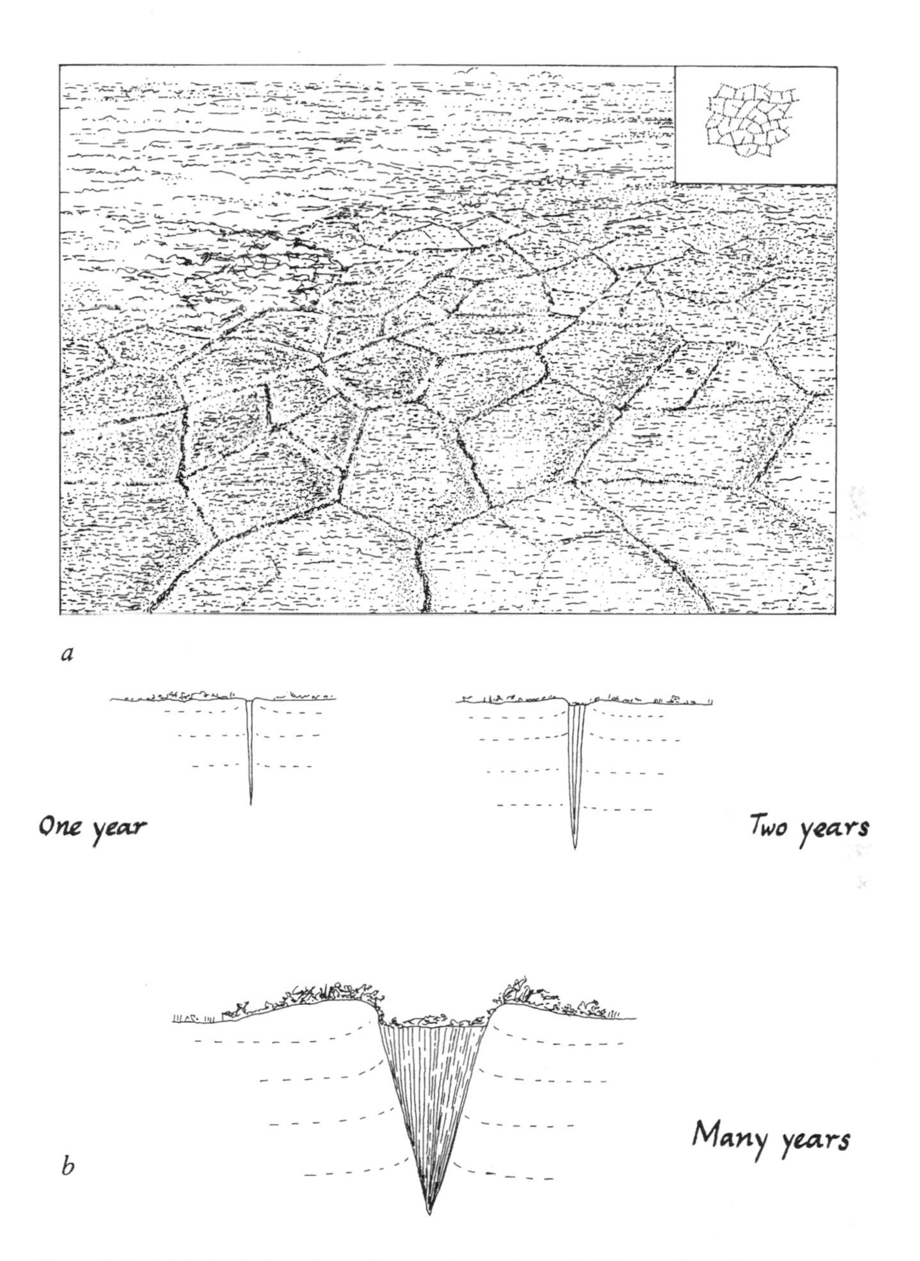

Figure 3.9. (a) A field of tundra polygons. *Inset:* the symbol for tundra polygons used on topographic maps. (b) Diagrams to show how ice wedges form.

thicker and deeper. As an ice wedge thickens, it forces up the soil on each side of it into a low shoulder. Consequently, at this stage (there is more to come), each polygon is slightly lower at its center than at its rim, and is called a *low-center polygon*. It functions as a shallow basin in which a pool of meltwater often collects in summer. A network of such tundra pools can form an astonishingly regular array (fig. 3.10) when seen fom the air. Sometimes the polygons are many-sided, sometimes they are rectangles.

After seeing tundra polygons from a distance, it is interesting to examine on the ground the cracks that outline them. The cracks appear as long trenches, about 1 or 2 m wide, bordered by low ridges; the ice wedges, their tops concealed by soil and vegetation, lie beneath the troughs. The troughs (which are sheltered) and the ridges (which are exposed) provide dissimilar habitats for plants; they support markedly distinct plant communities, making the polygonal pattern visible from afar. The centers of the polygons, which are waterlogged even when they are not submerged, constitute still another plant habitat. The growth of these various plant communities can gradually change the landscape, causing low-center polygons to develop, over centuries, into *high-center polygons* (fig. 3.10).

It happens in this way. To begin with, the wet centers of the polygons are covered with sedges (including cottongrass) and moisture-loving mosses, while the surrounding ridges support communities of "dry land" plants—grasses, lichens, dwarf willow, and in the Low Arctic, dwarf birch, crowberry, bearberry, and cloudberry. As the plants die, their remains accumulate; at the same time, plants both living and dead intercept dust-laden winds. These two processes gradually raise the ground level, making it drier. The plants on the ridges can then spread to fill the polygon centers, while the aquatics and wetland plants that had flourished there die out. A stage comes when the polygon surfaces are level and entirely covered by dry-land plants. But change still goes on. The remains of dead plants continue to accumulate in the form of peat (for more on peat, see sec. 3.6), raising the polygon surfaces yet higher; then, as each year's spring meltwater drains from the peat polygons into the troughs separating them, the flowing water deepens the troughs and erodes the margins of the polygons leaving them slightly domed. High-center polygons (fig. 3.10) are the result. (Low-center polygons don't invariably evolve into high-center ones; the sequence of events is sometimes entirely different; see sec. 3.10.)

Scientists have found that the ice wedges separating high-center polygons are wider, on average, than those separating low-center polygons. This is what we should expect: the high-center polygons are older, and

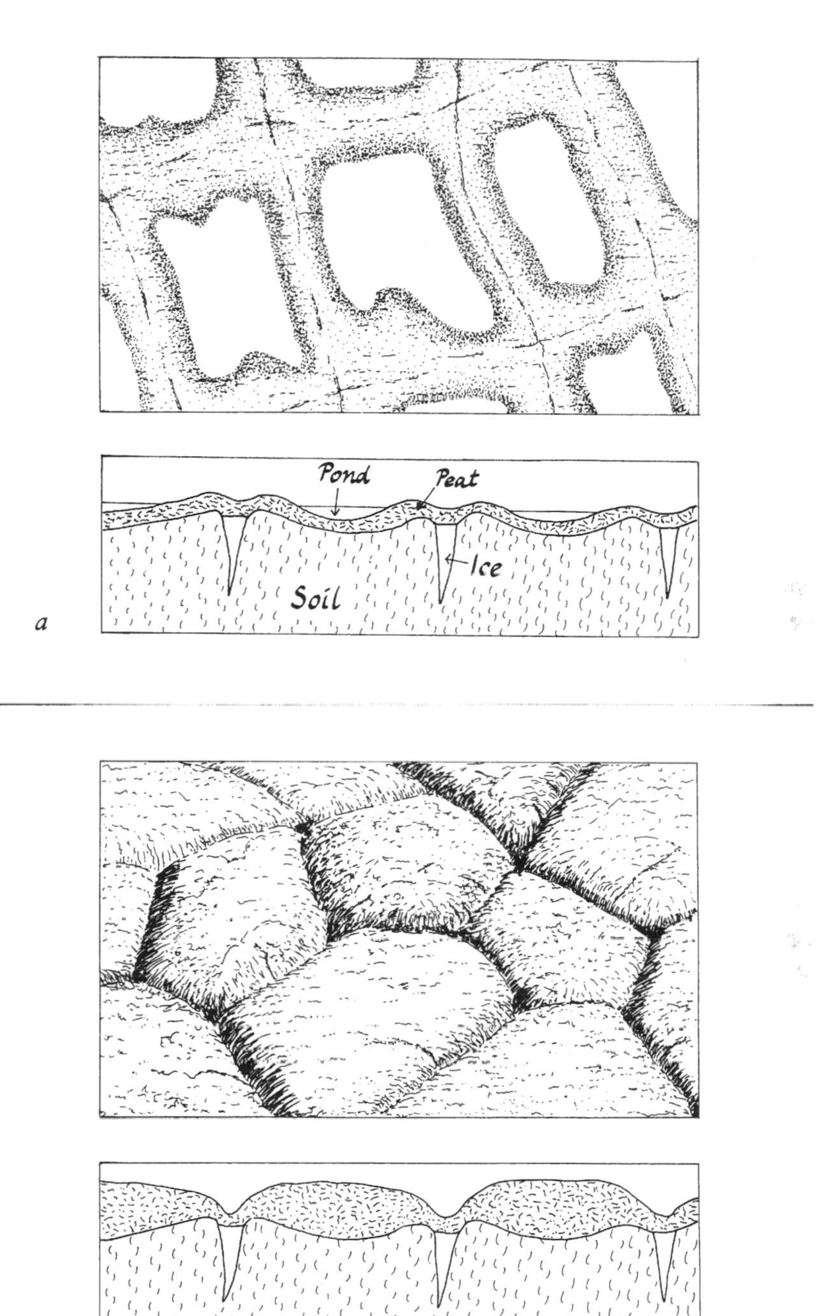

Figure 3.10. Views from the air, and diagrammatic soil sections showing: (a) Low-centered polygons, with a tundra pond in each polygon. (b) High-centered polygons.

therefore the ice wedges between them have had time to grow bigger. Some fields of high-center polygons in the Central Arctic are known to be 10,000 years old; they must have begun to form as soon as the shrinking ice sheets left the ground bare, a few thousand years after the time of maximum ice (fig. 3.5). On the Alaskan North Slope, which was not covered by ice sheets during the last ice age, 14,000-year-old ice wedges have been found buried under the soil, the biggest 10 m wide at the top.

Where the climate is cold enough—in the continuous permafrost zone (see fig. 3.4)—the ice wedges bordering tundra polygons are still actively growing, thickening at a rate of about 1 or 2 millimeters per year. Tundra polygons found in the warmer climate of the discontinuous permafrost zone are "fossil," that is, the ice wedges are now inactive (no longer growing).

3.5 Pingos

Pingos are another arctic landform showing the power of permafrost to shape the land. A pingo is a conical hill with a core of clear ice. A big one may be as much as 75 m high and 500 m across, making it a

Figure 3.11. Two pingos near Tuktoyaktuk. The biggest pingos are about one-half as tall as the largest Egyptian pyramid.

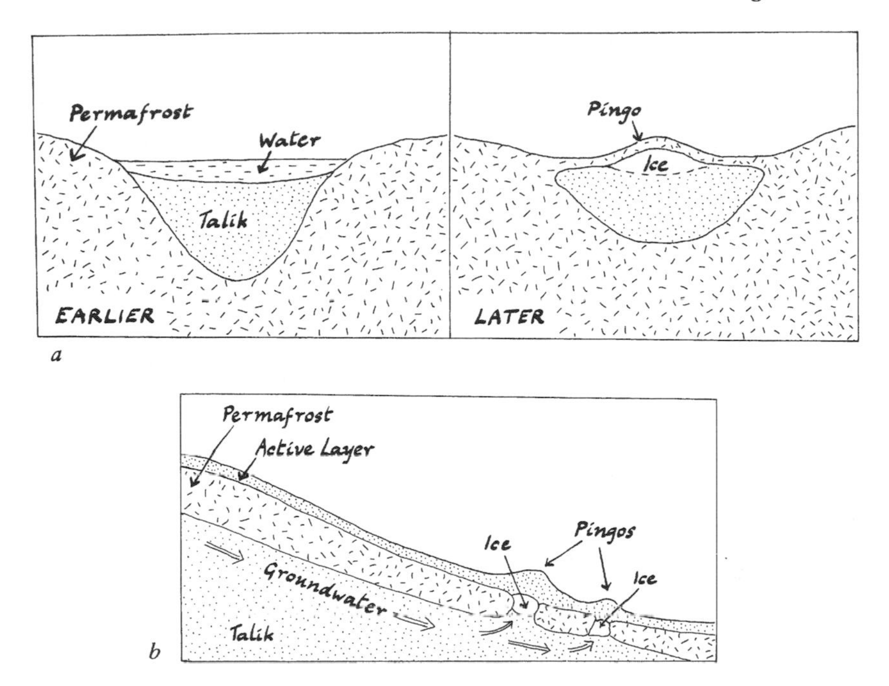

Figure 3.12. Stages in the growth of (a) a closed system pingo, and (b) open system pingos. In (a), the active layer is too thin to show in the diagram. More details in text.

conspicuous feature of the landscape (fig. 3.11). Sometimes, as in the figure, the summit has a crater at the top, making the whole pingo look like a little volcano.

Pingos exist in two varieties, *closed-system pingos* and *open-system pingos;* in the former, the water that freezes to form the pingo's core is in an enclosed space surrounded by impermeable frozen soil, whereas in the latter the water flows in from an outside source.

Closed-system pingos are found only in the continuous permafrost region. They are most numerous in the western Arctic, in the coastal plain extending westward from the Mackenzie Delta into Alaska (the Alaskan North Slope); this is because they develop best in the unfrozen silty sediments that are emerging above sea level along this coast as the earth's crust rebounds. They are believed to form as shown in the diagram in figure 3.12. The process begins when a tundra pond becomes drained—ponds drain naturally quite often, as explained below (sec. 3.10). The drained pond must have been deep enough not to freeze to the bottom in winter; then, because water is a good insulator, the soil beneath the lake will have remained permanently unfrozen to a considerable depth; in other words, there will be a thick layer of unfrozen talik, saturated with water, between the lake bottom and the permafrost table.

When such a lake drains, the insulation it had provided suddenly disappears. The lake floor, newly exposed to intensely cold air, freezes into a new permafrost layer; at the same time, the existing permafrost table below the lake floor starts to rise. Thus two permafrost layers converge on each other from above and below, trapping a pocket of unfrozen waterlogged soil between them. The pocket is squeezed like a sponge, and the water squeezed out of it freezes. In doing so, it expands in volume (water expands by 9 percent on freezing), forcing up the overlying permafrost layer into a dome—a newborn pingo—which goes on growing as the squeezing progresses. The finished pingo consists of a cone-shaped plug of pure ice entirely encased in permafrosted soil.

The growth of many pingos was probably triggered by the onset of the climatic cooling that started several thousand years ago, when the warming trend which ended the last ice age reversed itself. (Note: the post–ice age "climatic optimum" is now long past). The largest pingos may have grown for 1000 years before exhausting their internal water supplies. Some—probably a minority—of those now to be seen began growth more recently and are still growing; and more will develop in the future. But the majority existing at any one time are in the process of decay. As a pingo ages, the soil covering its summit inevitably erodes until the icy core is exposed. Thereupon the core melts rapidly and finally the outer wall collapses, leaving recognizable, donut-shaped "ruins" that last for a long time.

The same process that produces closed-system pingos occasionally produces temporary ice-cored mounds ("minipingos") in the centers of low-center polygons; but they don't last long, sometimes for only one year.

Open-system pingos form quite differently. They don't become as large as closed-system pingos. They are found in valleys, often in groups, in the discontinuous permafrost zone. Figure 3.12 shows how they develop: a thin layer of permafrost, with gaps here and there, lies below the active layer on the sides and floor of a valley. Groundwater flows downhill in the unfrozen talik below the impermeable permafrost. Being under pressure, this water is forced toward the surface wherever it can get through, that is, via the holes in the permafrost. If the temperatures of the water and of the permafrost surrounding the holes happen to be just right, the upward-seeping water will freeze before reaching the surface, forming a rising plug which forces up the overlying soil; the resultant mound is a newborn pingo. All this can happen only if conditions are "just so." Therefore it is probably an incomplete explanation of exactly what goes on in open-system pingos; scientist believe that other, more complicated glaciological happenings must be in progress as well.

3.6 Peat: Peat Mounds, Palsas, and Peat Plateaus

Peat, shaped into mounds and hillocks and slabs, is the material forming many arctic and subarctic landscape features. It is an interesting substance, owing its very existence to the climate in which it develops. The cause of its formation deserves a couple of paragraphs.

Consider what happens to plants when they die. There are three possibilities. If the climate is warm and moist, dead plants quickly decompose or decay (the words mean the same thing) and become part of the soil. Their tissues are consumed by vast numbers of microbes: bacteria, fungi, insects, mites, and an assortment of tiny invertebrate animals. This consumption *is* the process of decay, and it cannot happen unless conditions are suitable—warm enough, moist enough, and sufficiently well aerated—for the organisms that bring it about. If the climate is dry (either warm or cold), dead plant material can't decay, and there is not much of it in any case, because of the water shortage. Such plants as do grow wither and crumble when they die, whereupon wind scatters the dustlike fragments far and wide, and they mix unnoticed with the soil.

If the climate is cool and moist, and the ground cold, soggy, and airless, dead plants can neither crumble nor readily decay. Their remains simply accumulate in wet masses; this is peat. Very slow decay is sometimes possible (caused by organisms that don't require air), but in any case, undecayed or only partly decayed peat persists for thousands of years. That is why some peat is still to be found in the High Arctic, where modern conditions are desertlike; it is old peat, perhaps between 5000 and 10,000 years old, formed in the milder climate that followed the end of the ice age. In the Low Arctic, peat formation still goes on, though slowly. With the climate as it is at present, peat is now accumulating most rapidly in the subarctic boreal forests.

The flatness of arctic tundra is often relieved by blocks and masses of peat of various shapes. In the High Arctic, the commonest peat landforms are low *peat mounds,* seldom more than 1 m tall. Going southward, the mounds become progressively taller until, in the Low Arctic, they reach a height of 6 or 7 m, and are given the name *palsas* (fig. 3.13). These are not, as might at first be thought, young pingos; the core of a pingo is clear ice, that of a palsa frozen peat, with occasional thin layers of ice at most. Palsas are found only in the discontinuous permafrost zone (see map, fig. 3.4). Near treeline, on the open tundra and among scattered stunted spruces and tamarack, one also finds *peat pla-*

Figure 3.13. Palsas.

teaus, extensive flat tracts of peat, slightly raised above the level of the surrounding countryside.

Even when these landforms don't have an eye-catching outline (which is especially true of peat plateaus), they attract attention with their distinctive vegetation. The water in peat is poor in mineral nutrients and slightly acid. This is because it is not continuous with the relatively mineral-rich groundwater that flows slowly through the mineral soil below; rather, the water in a body of peat is a "private" supply, isolated from groundwater and replenished only by water from the sky—rain and snow. These conditions favor distinctive plant communities: mosses on high arctic peat mounds and such typical bog plants as cloudberry, Labrador tea, bog rosemary, shrub birch, and a variety of lichens on low arctic and subarctic peat plateaus and palsas. Besides tolerating the low nutrient levels of peat, these bog plants also require the slightly better drainage that the raised peat surfaces ensure; the surrounding low-lying land is usually waterlogged with groundwater and covered with sedge meadows. The color contrast between the sedges and the bog plants is what makes peatlands stand out in the scenery.

Landforms constructed of peat don't last forever; they eventually erode and disappear. Indeed mounds, palsas and plateaus all go through a definite cycle: they are "born," and they "die." In simple cases the sequence of events is this.

A mat of sphagnum (also called peat moss, see sec. 5.15) starts to grow on a waterlogged sedge meadow. When it is thick enough, the central layer of the sphagnum fails to thaw in summer, becoming an isolated flat slab of permafrost. The ground surface overlying the permafrost slab starts to rise, at first because the water held in the peat expands on freezing, and subsequently because additional water migrates toward the expanding slab. These expansions cause a raised dome to appear; sometimes the doming is accelerated because marsh gas (methane), produced when the underlying sedge meadow vegetation decays without oxygen, pushes up the confining "lid" of permafrost on top of it.

Once doming has started, a feedback effect makes it continue. The newly formed dome is more windswept than the surrounding lowland and is therefore covered in winter with only a thin layer of hard-packed snow, which is a much less effective insulator than the comparatively soft, deep snow elsewhere. Moreover, the frozen peat of winter conducts heat far more readily than the wet peat of summer, so that more heat is lost in winter than is gained in summer. Because of all this, the core of the dome keeps cooling, the permafrost block keeps expanding, and the dome keeps rising. But not indefinitely. As it rises, the dome becomes better drained and starts to dry out in summer. Its surface cracks; rain and warm air reach its frozen core and before long the center thaws and the whole structure erodes and collapses.

Scientists have deduced this "life cycle" by studying peat landforms in different stages of growth and decay, which are often found side by side. The sequence of events is inferred by taking a core and examining the peat at different levels. Peat is not all the same: the dead plants of which it consists are recognizable. Thus in a full-grown palsa, for example, peat formed from woody bog plants typically forms the top layer; sphagnum peat, with which growth of the palsa started, lies below; and sedge peat, remaining from the meadow on which the sphagnum mat once grew, comes at the bottom.

3.7 Tundra Hummocks

Fields of hummocks (fig. 3.14) are the commonest of the characteristically arctic terrains; unlike such comparative rarities as pingos and palsas, they occur everywhere. In a well-developed hummock field, all the hummocks are roughly the same size, and all have matching vegetation (one kind of plant on the hummock tops, another in the crevices between); the result is a pattern almost as regular as textured wallpaper. Different hummock fields are not necessarily like one another. The individual hummocks tend to be smaller in high arctic than in low arctic

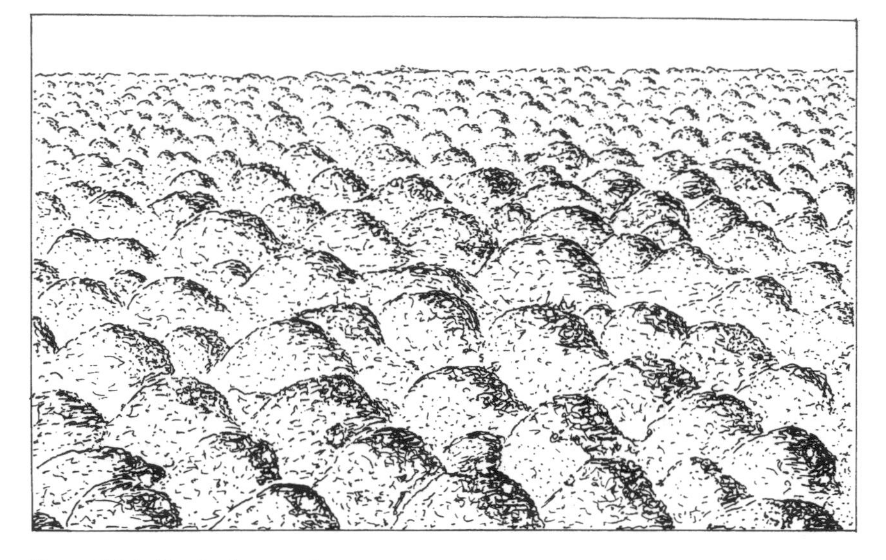

Figure 3.14. Tundra hummocks.

fields; in the High Arctic, they are typically about the size of a human head. Sometimes, but not always, the hummocks have cores of ice. And the plants growing on the hummocks are not necessarily the same in different hummock fields; although the commonest hummock-top plant is arctic dryad, on wetter ground it may be purple saxifrage; the crevices have different plants, or sometimes none at all. Hummock fields of pure clay, with no plant cover, are occasionally found. Hummock fields are not confined to flat areas; they can also develop on fairly steep slopes.

Many theories have been devised to explain how hummocks first develop. Here are three of them:

1. Winter freezing causes a polygonal network of cracks to form in the active layer of the soil. (These polygons are many times smaller than the big tundra polygons described in an earlier section, which are bordered by cracks descending far into the permafrost.) Trickles of meltwater flowing in the cracks in summer erode and enlarge them, rounding the polygons into hummocks with troughs between.

2. If the active layer is very muddy and wet, scattered "nuclei" of ice will form when the layer freezes; soil-water in contact with each ice nucleus will freeze onto its surface, and more water will be sucked in toward it, as by a dry sponge, from the surrounding mud. This process (known as *segregation*) causes each small nucleus to grow bigger; it will also expand because of the way water expands on freezing. Each swelling

lump of ice forces up the soil above it into an ice-cored hummock, and a field of hummocks is the result.

3. On damp ground covered by cushions of absorbent moss, each cushion is an incipient hummock. Because a moss cushion acts as an insulator, the cylinder of soil directly below it freezes later in the year than the exposed soil between cushions. As the exposed soil freezes, it expands, and the sideways pressure it exerts forces up the cylinders of soft, unfrozen soil below the moss cushions, like toothpaste from a tube.

Hummocks are not static. However they are formed—perhaps by one of the mechanisms outlined above, or a combination of them—events are happening that will change them. Some processes increase "hummockiness" regardless of how it originated: flowing water deepens the channels among the hummocks; wind-borne dust, captured by the plants on hummock tops, adds to their height. Other processes reduce hummockiness: if, for any reason, the hummocks dry out and the plants growing on them die, they are soon "sandpapered" flat by wind erosion. The stumps of eroded hummocks, appearing as concentric rings, can sometimes be found on dry ground.

3.8 Patterned Ground of Other Kinds

Tundra polygons, pingos, palsas, peat mounds, and tundra hummocks are all known as "patterned ground." Other ground patternings are found here and there as well, some of them even more striking.

All are caused by alternate freezing and thawing, and they develop best in ground where vegetation is sparse or absent. Their interesting designs make up for the scarcity of colorful flowers in the polar desert. And they illustrate, convincingly, that absence of life does *not* imply absence of activity. A lifeless ground cover of clay, silt, sand, pebbles, and boulders is not inert: energy from the sun activates it. When the sun shines, rocks are heated and ice is melted; when the sun disappears (behind clouds or below the horizon), rocks cool and water freezes. These repeated changes cause continual small movements, which accumulate to produce some extraordinary patterns.

The commonest are *frost boils,* more or less circular patches of smooth, bare clay. They form where an underground pocket of wet, unfrozen clay or silt (*fines,* to use the geologists' term), oozes up through a weak spot in the surface. This happens when the sediment enclosing the pocket freezes, squeezing it; as oozing proceeds, the top of the boil rises. It may be as much as 20 cm above the surrounding ground in winter; then it subsides in summer. In an actively forming frost boil the fines may be as soft as quicksand (as you discover when you step on

Figure 3.15. An array of frost boils.

one), but in an old, dried-out boil they are firm and hard. Big tracts of tundra are often dotted with numerous pale, circular boils (fig 3.15); such a tract could well be called "spotty ground."

The thick layer of broken rock covering the ground in much of the High Arctic is subject to repeated frost-heaving. The effect is to sort the rocks by sizes, since the largest are "heaved" the most. In time, the material becomes stratified, with big stones at the top, then smaller stones, pebbles, grit, sand, and fines forming successively deeper layers. The top layer, at the ground surface, is often *felsenmeer* (German for a sea of rocks), a huge expanse of broken, angular rocks where no plants grow—or, at best, a few tiny cushion plants rooted in the windblown dust concealed in occasional crevices.

When the oozing of buried fines produces frostboils in felsenmeer, extraordinary patterns can form. Sometimes a whole array of boils forms; if widely spaced, they remain circular; if crowded, they abut on one another and are forced into polygonal shapes. In either case, the fines ooze up in low mounds, causing the overlying rocks to roll to the outer edge of each boil. Large rocks roll farthest; successively smaller ones roll successively shorter distances, until each boil is outlined by concentric rings (or polygons) of rocks, graded by size. The finished pattern is called a network of *sorted circles* or *sorted polygons* as the case may be (fig. 3.16).

Earth scientists have also come up with another possible explanation for the development of sorted circles and polygons. (The theories outnumber the phenomena they are meant to explain!) According to the other theory, freezing of the ground starts at a number of nodes and spreads outward from them. Consequently, the freezing front—the isothermal surface separating frozen from unfrozen ground—shifts side-

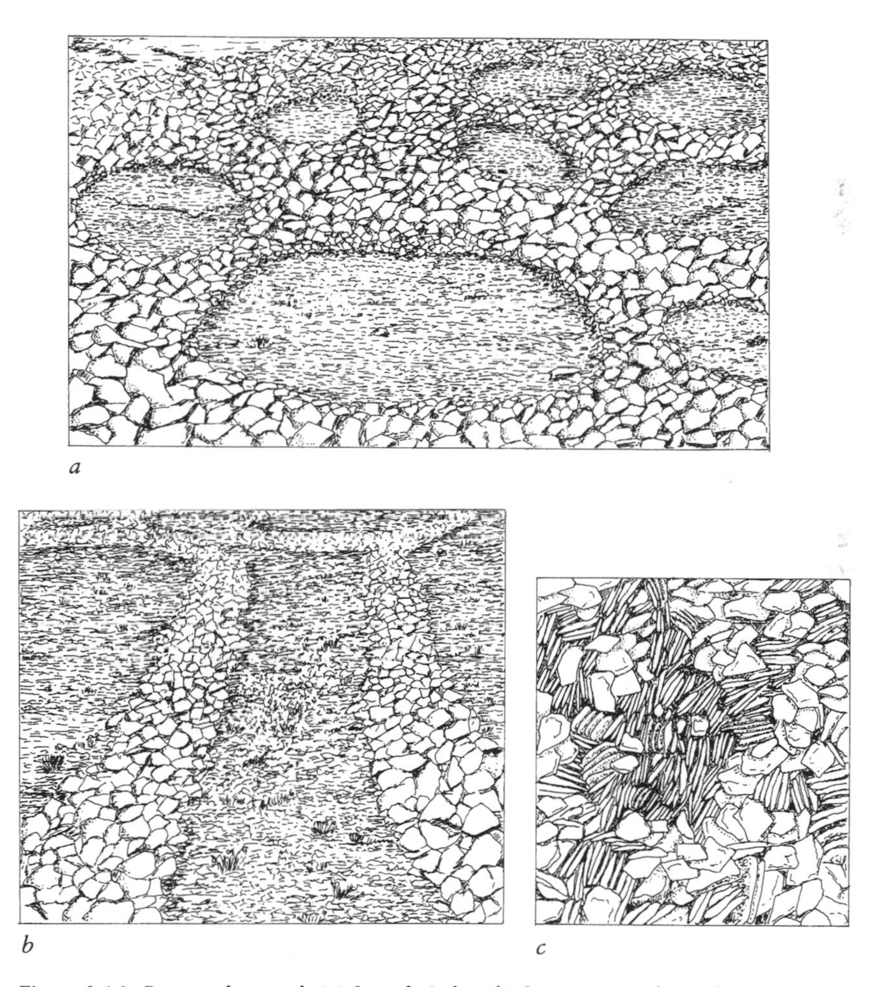

Figure 3.16. Patterned ground: (a) Sorted circles. (b) Stone stripes, descending a gentle slope away from the viewer, toward stone circles on level ground below. (c) Frost-tilted rocks.

ways as well as in an up-and-down direction, producing sideways frost-heaving and a sideways sorting of rocks; the larger rocks, which are heaved farther, end up beside (instead of above) the smaller ones. The finished pattern is (again) a network of sorted polygons or circles, with large rocks outside and small rocks inside.

Good networks of circles or polygons form only on level ground. If the processes that produce them take place on a gentle slope, the whole surface slides downslope. The slide is imperceptibly slow—only a few centimetres per year—and it produces a pattern of parallel stone stripes (fig. 3.16): stripes of smooth fines alternating with stripes of coarse rocks. The fines, which move slightly faster than the coarse rocks, act like parallel "rivers," each one sweeping the coarse rocks aside, causing the latter to collect in stripes of their own.

The patterns just described are known as *sorted patterns,* because they show fines and larger rocks separated or sorted out. Corresponding but much less spectacular *unsorted patterns* can be discerned where unfrozen fines have oozed in ground devoid of pebbles and rocks. In unsorted circles and polygons, the undisturbed ground between the frostboils usually supports a modicum of vegetation. In unsorted stripes, the actively moving stripes alternate with motionless, vegetated stripes.

A final pattern to look for in rock fields is that produced by frost-tilting. Alternate freezing and thawing does more than heave rocks bodily upward through the fines: if the rocks happen to be thin and flat they are also tilted on edge (fig. 3.16).

For a detailed, scientific account of research on patterned ground, see *Geocryology: A Survey of Periglacial Environments,* by A. L. Washburn (London: Edward Arnold, 1979).

3.9 Underground Ice

Away from ice caps and glaciers, the only ice to be seen on the surface of the ground during the Arctic summer are occasional patches of *aufeis* or frozen river water (see sec. 3.11) and, along the coast, temporarily stranded ice floes left on shore by a falling tide. Underground ice, by contrast, is probably widespread, but because it is not visible its total extent is unknown. Its presence is sometimes revealed naturally by the slumping of overlying scree or soil on a hillside and sometimes by exploratory drilling.

These hidden layers, or blocks, of pure, clear ice owe their existence to various causes. In some cases, water froze underground and the resultant underground ice has never seen the light of day. In other cases, what was once surface ice has become buried.

Examples of ice masses that have always been concealed below the ground are ice wedges (sec. 3.4), and the cores of pingos (sec. 3.5). Much more common (probably) is so-called *segregation ice,* which occurs as buried layers sometimes 40 m thick and more. These layers must originate from thin sheets of ice in the soil, which then slowly thicken as new ice freezes onto the ice already present. Layers many meters thick must have been growing for centuries. The necessary water comes via the soil, where it is temporarily held in the tiny pores among the soil grains. In fine-grained soil, the pore water moves upward because of capillary action (the process by which water rises in a sponge) and freezes onto the bottom of the growing ice layer. In coarse-grained, sandy soil, the pores are too big for capillary action to occur. Instead, the water freezes in the pores, expanding as it freezes. Consequently its volume becomes too great for the pores to hold all of it and some is squeezed up and out of the sand and freezes onto the overlying ice layer.

Some of these ice layers in the western arctic coastal plain near the Mackenzie Delta are known to be at least 40,000 years old. They lie sandwiched between beds of sediment that have been deformed—folded or faulted—by the shoving action of moving glaciers. The beds of sediment and ice are parallel, showing that they were all deformed together and proving that the ice beds must have remained frozen continuously, without any thawing and refreezing, ever since the folding and faulting happened.

Buried remnants of the huge ice sheets that covered much of the Arctic in the last ice age (sec. 3.3) are also known, for example in northwest Victoria Island and northern Ellesmere Island. As the climate warmed, expanses of melting ice sometimes became covered by windblown dust, by sediments carried in meltwater streams, or by mudflows; once the ice was covered, plants grew on the overburden, died, and formed peat (sec. 3.6). The old ice was kept cold by this insulating blanket and has persisted, unmelted, to the present day. Some underground ice masses are probably the remains of frozen lakes and ponds that became buried in the same way.

The cores of ice-cored moraines are another kind of buried ice. Outflow glaciers (sec. 3.1) from ice caps in the eastern Arctic Islands usually carry a tremendous load of broken rock on their margins, rocks that have rolled onto the glacier from the cliffs on either side. Where the toes of the glaciers have retreated, owing to the climatic warming that followed the cold period of the "Little Ice Age" (roughly 1450 to 1900), this rubble has been left behind as *lateral* and *terminal moraines,* ridges of loose rock beside, and at the end of, the glaciers' old beds. These moraines often have cores of ice, ice that failed to melt when the exposed glacier beside it melted because it was insulated by the overlying rock rubble. Ice-cored moraines can be found in glaciated mountains in many parts of the world, but they are best developed, and last longest, in the Arctic, where very little ice melts in the short, cool summers. Ice-cored moraines look no different from other moraines; only drilling through the overlying rocks can show, indisputably, whether or not a moraine has a core of ice.

3.10 Thawing Landscapes

A great variety of landforms come into being when frozen ground thaws. The most widespread are those caused by the annual melting of the active layer of soil above the permafrost, which happens everywhere.

If it is of the right consistency, and on a slope, the thawed active layer oozes downhill like melted wax down a candle, building up *solifluction lobes* (fig. 3.17a). Solifluction means soil flow, from any cause; sticklers for precision prefer the term *gelifluction* (the flow of seasonally frozen soil when it thaws), or even *congelifluction* (applied to gelifluction when the surface below the flow is permafrost). Sagging, U-shaped lobes festooned across a hillside are obvious solifluction lobes. But sometimes the lobes are simply low banks that escape the notice of anybody not consciously on the lookout for interesting landforms.

Given the right soil (cohesive peat that bends without breaking) and the right topography (a wide, gently sloping valley opening onto an expanse of flat ground), solifluction can also produce a string bog (*strangmoor*). A thawing peat layer slithers slowly down the valley and, on coming to a stop, piles into a series of transverse ridges as would a heavy rug being pushed along a surface on which it couldn't slide freely. The parallel ridges are the "strings" of the bog; the valleys between, which fill with water to become long, narrow ponds parallel with the contours, are *flarks* (fig. 3.17b).

Solifluction is brought on by the annual thawing of the active layer. Other landforms are produced by the thawing of permafrost itself. Landforms produced in this way are collectively called *thermokarst.* Where permafrost or underground ice melts, the very ground loses its solidity and the overlying soil surface collapses. The process resembles what happens in limestone country where caverns, deep in the bedrock, are hollowed out and enlarged over the centuries by percolating rainwater (because rainwater is a weak acid, it dissolves limestone); every time a cavern caves in, the ground above subsides leaving a new hollow or valley in the landscape. Such terrain is known as *karst.* Hence the name thermokarst for terrain similarly produced, but by the melting of permafrost rather than by the dissolution of limestone. Of course, the processes producing thermokarst happen much faster, and, because of the warmth required, they happen only in the Low Arctic.

Thaw slumps are common in thermokarst terrain (fig. 3.18). Slumping happens most often on coastal cliffs, lake shores, and river banks, when

a

Figure 3.17. (a) Solifluction lobes. (b) A string bog, or *strangmoor*.

Figure 3.18. Thaw slumps on coastal cliffs.

they are undercut by waves or currents. Cliffs, shores, and banks are worn away more rapidly in permafrost country than elsewhere because erosion is enormously speeded by melting. When an undercut has become so pronounced that the overhang above it collapses, the result is a thaw slump. Masses of soft, wet soil slide down, leaving a headwall of permafrosted ground suddenly deprived of insulation and exposed to the summer sun. Because of this, further melting, and further slumping occur. As the process continues the slump recedes farther and farther back from its starting point. The speed at which this happens depends on many things; as an example, a thaw slump in Yukon, observed over the years, has eaten its way back into the land at about 1 km per decade.

Human activities are another cause of slumping. Vehicles destroy the vegetation and peat that insulate the permafrost. The resulting meltwater runs off in the little gullies formed by vehicle tracks; these gullies become eroded into larger and larger gullies, whose walls slump, leaving the tundra damaged.

Thawing can also start where gaps in the insulating layer are opened up by frost-heaving. And sometimes, of course, meltwater doesn't flow away but collects to form tundra ponds.

Vast numbers of ponds dot the low arctic tundra: swarms of them occupy the hollows of low-center polygons, created by networks of buried ice wedges (see sec. 3.4). If the soil doesn't soon become stabilized by the roots of growing plants, these small ponds tend to grow larger;

even the tiny ripples on the smallest ponds can erode their shores appreciably, because this is thermokarst erosion, boosted by permafrost melting. Then, as each pond expands, the wind raises ever larger waves, speeding the erosion. In time, several neighboring ponds coalesce to form quite a large *thaw lake;* from the air, the pattern of the original polygons can often be seen through the shallow water covering the lake floor. In regions with a marked prevailing wind, wave erosion lengthens all the thaw lakes in the same direction; an array of elliptical lakes, aligned in parallel, develops.

Landscapes never stop evolving. Thaw lakes don't last forever. Some become overgrown by sedges and ultimately dry up. Others drain away; this happens when a tundra stream, eroding its course headward, breaches a lake's low banks. Once a lake is emptied, tundra polygons form anew on its dry bed. If the lake was shallow, the network of ice wedges under its floor will have survived and will become reactivated, reproducing the old polygonal pattern exactly as before. The ice wedges under a deeper lake—one that didn't freeze to the bottom in winter—will have melted, however, and a new polygonal pattern of wedges then develops. In either case the cycle begins again, with a new batch of small ponds that gradually expand. A single cycle probably takes thousands of years to complete.

In theory, the cycles don't depend on changes of climate. In fact, the rate at which permafrost melts is obviously governed by the prevailing warmth, which never stays constant for long. A warm period (warmer than now) followed the melting of the great ice sheets of the last ice age (see sec. 3.6). Another warm period seems to have begun about 100 years ago bringing an end to the "Little Ice Age," a period of roughly 500 years during which the world was much colder than it is now. Trends are difficult to detect when you are in the midst of them and minor details obscure the larger picture, but a warming trend seems to be in progress (see sec. 3.2), perhaps caused, at least in part, by human activities. Permafrost melting and thermokarst erosion seem likely to increase.

3.11 Arctic Rivers

All wilderness travelers pay close attention to rivers. Backpackers wonder whether they will be able to ford the rivers lying across their routes; kayakers, canoeists, and rafters wonder whether they will be able to negotiate the rapids. They have to consider some of the unexpected ways rivers behave in the arctic environment.

Arctic rivers flow from a variety of sources. Some are fed by melting ice caps and glaciers, some by the melting of temporary snowfields that, in most years, disappear by the end of summer, and some flow from lakes. The way a river flows, through the summer season and even through each day, depends on the kind of source it flows from.

Many of the rivers in Ellesmere, Devon, and Baffin Islands are fed by meltwater from ice caps and glaciers. The flow of such a river varies from month to month and from hour to hour, in response to changing temperatures. A stream that is only a few scattered pools of ice in winter, can become a raging torrent in July. Not only that, in summer its flow varies between the cool of the "night" (when the sun is low in the north) and the heat of the "day" (when the sun is comparatively high in the south). Campers can hear the difference: a river that roars deafeningly all afternoon becomes quieter and quieter at night, as its flow diminishes. In the coolest hours of the 24, anywhere from 4 A.M. to 6 A.M. sun time, the flow—and the noise—reach a minimum; this is the safest time to attempt fording such a river. The contrast between day and night is least at very high latitudes, where the sun's elevation changes comparatively little between noon and midnight. It becomes progressively more pronounced the lower the latitude.

Rivers flowing from melting snowfields behave in much the same way, at the beginning of the warm season. But with advancing summer, the snowfields shrink and the water supply dwindles; by season's end, the rivers have dried up.

The third kind of river encountered by arctic travelers are those flowing from lakes. They are particularly numerous in that part of the Arctic lying on the Canadian Shield, that is, nearly all of the mainland Arctic east of about 125° W longitude, and most of Baffin Island. The Shield itself is a huge expanse of hard, ancient rock, that has been called North America's geological heartland. Half a billion years of erosion have smoothed it, so that it lacks dramatic mountain ranges; but the ice sheets of the last Ice Age gouged out innumerable hollows in the bedrock, which are now occupied by lakes. Indeed, the land is peppered with these lakes. Unlike the ever-changing thaw lakes and thaw ponds described in section 3.10, these deeper, rock-bottomed, shield lakes have changed scarcely at all in the few thousand years since the ice sheets melted, because the hard rock underlying them is highly resistant to erosion.

One of the noteworthy things about these lakes is that they are all at different levels. Look around from a ridge top, and you can often see a dozen or more lakes each at a slightly different elevation. Numerous steep streams, full of rapids and falls, drain higher lakes into lower ones; in summer, the air is filled with the sound of rushing water. An area like this is said (by geologists) to have *deranged drainage.* If no geological changes occur, and the streams continue to flow, after tens (perhaps hundreds) of millions of years, the streams will have had time to erode their rocky beds down, draining their source lakes in the process, until no rapids and falls are left. The stream beds will become less steep, until in time (eons of it), they will lose all their erosive power, and equilibrium will be reached. The lakes will all have gone, drained to the sea, and the rivers will be silent. In other words, the "derangement" will have been corrected!

To return to the present: the attractiveness of shield terrain, with its numerous deep lakes and swift rivers, is due to the fact that the ancient land surface has recently been reshaped; only a short time has passed (as geologists measure time) since it was covered by the ice sheets of the ice age. The rivers do not behave like those flowing from ice caps or snowfields; their flow in summer doesn't vary between day and night. Ice covers the lakes for 9 or 10 months of the year, to a depth of between 2 and 3 m on average, but water continues to drain away under the ice, at least in early winter; if a lake's outflow channel is steep enough, water may keep flowing all winter long. The loss of water from a lake leaves the surface ice unsupported, so that it sags in the middle. The odd result is a lake whose frozen surface, instead of being level, is a shallow bowl.

Another consequence of the continued flow of water from lakes too deep to freeze to the bottom is the formation of *aufeis.* Aufeis is a sheet of ice filling a river valley until late summer. Except in the High Arctic or in the mountains, it is the only pure ice to be seen above ground in summer and is visible from a long way off against the brown of the tundra. Aufeis forms in winter, when water flowing in a deep stretch of river is dammed by ice filling a shallower stretch. The dammed water overflows the river banks, floods the valley, and freezes. This happens again and again; the frozen floodwaters build up layer upon layer, until by spring there is an extensive sheet of aufeis 2 or 3 m thick. Because of its thickness, it takes a long time to melt when summer comes. A sheet of aufeis is a surprising find in hot weather; it is a reminder that, in the Arctic, hot weather never lasts long. The tundra close to aufeis is a good place to look for flowers that are over elsewhere; the cold air close to the ice often delays flowering for weeks.

The quality of a river's water depends on the river's source. Rivers from lakes are usually clean and clear, those from glaciers and ice caps are dirty and opaque. But while the water from an ice-fed river may be unattractive to drink, it is interesting to observe. The fast-flowing water of a river in spate will carry a tremendous load of material—rocks, pebbles, sand, silt, and clay—provided the material is there to be picked up. Beneath a glacier, there is abundant material. Rocks of all sizes, from boulders bigger than houses down to the finest "rock flour," become embedded in the bottom of the moving ice; and all are released into the meltwater, when they reach the toe (downstream end) of the glacier. Swift rivers can shift large boulders: you can hear them grating and

a

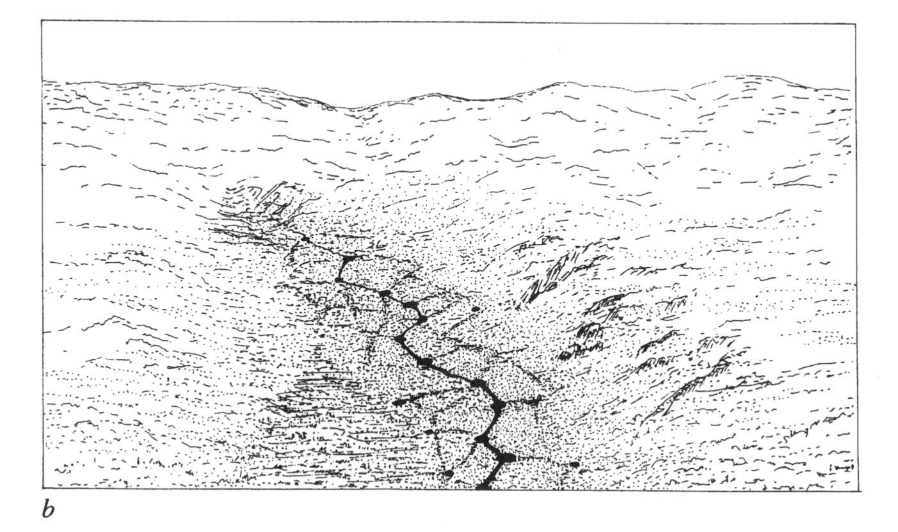

b

Figure 3.19. (a) A braided river. (b) A beaded stream.

banging and tumbling, concealed below water made opaque by rock flour and silt. The name Rollrock River (a river in northern Ellesmere Island) describes matters concisely. Rolling boulders are an obvious danger to hikers fording a river—another reason for fording in the coolest part of the day. This is also the best time for filling water bottles: rivers become noticeably less turbid when the flow weakens.

A river in spate in high summer, bearing a big load of sediment, has enormous erosive power. Given a wide valley—perhaps carved out by a now-vanished glacier—the river rapidly undercuts its banks. Undercutting happens especially fast in permafrost country, because the flowing water melts the riverbanks as well as eroding them. As a result, rivers fed by glaciers and ice caps make wide channels for themselves. Then, when temperatures drop and melting slows, the river slows too and dumps its load; large rocks come to rest first, then cobbles, pebbles, gravel, sand, silt, and clay. As the water level falls, the sediments are left as islands, splitting the river into numerous small streams that diverge and then rejoin each other in intricate patterns. Such a *braided* river channel (fig. 3.19) is a typically arctic scene.

Turning from large rivers to small streams, another typically arctic scene is *beaded drainage* (fig. 3.19). Look for it where tundra polygons have formed. A beaded stream is one that follows a zigzag course with little pools at the angles. The reason for this distinctive pattern is often obvious, especially from the air. The stream is following the ice wedge cracks that delimit the polygons (see sec. 3.4), and it widens into pools where the cracks meet.

4 Seas

4.1 Icebergs

Images of icebergs and pack ice are what first come to mind when the Arctic Ocean is mentioned. They are entirely different things. Pack ice is what forms when the sea freezes. The ice of icebergs (or bergs, for short) is freshwater ice; it never was sea water. A berg is a mass of ice that has broken away from the margin of an ice cap or glacier, where it meets the sea.

The great majority of bergs, including all the big ones, come from the Greenland ice cap; the ice caps on Ellesmere, Devon, and Baffin Islands also contribute to the flock of 10,000 bergs or more adrift (and sometimes aground) in eastern arctic waters at any one time. Once the toe of an advancing outflow glacier (see sec. 3.1) has been pushed forward into the sea, by the pressure of the ice behind, it will float; it will move up and down with the tides and be buffeted by waves. Under these stresses, the ice eventually fractures, and a newborn berg is *calved* from the end of the glacier, usually with a tremendous roar as it topples, before it turns and settles into the equilibrium position that allows it to float steadily.

Because of their source, and the currents that shift them, bergs are found only in the eastern Arctic; they are particularly abundant in Baffin Bay and Davis Strait, and seldom stray far into the channels leading west from these waters (see fig. 4.1). Anticlockwise currents carry the majority of them from the place where they were calved, north around the head of Baffin Bay, and then south past Baffin Island and the coast of Labrador. In winter they are firmly frozen into the pack ice, but all the time they are afloat in open water they are gradually melting and wasting away. Only the large ones last long enough—mostly between 2 and 3 years—to get as far south as the Grand Banks of Newfoundland; when disaster struck the *Titanic,* in 1912, she was south of 42° N latitude, closer to the equator than the pole.

Bergs come in vastly different sizes. A big one can weigh 10 million tons. Because the density of ice is only slightly less than that of water, most of a berg is submerged, and the tip that protrudes above the surface amounts to only one-eighth of its total bulk. Even so, the visible part of a berg is sometimes 60 or 70 m tall, and the tallest one so far reported towered to more than 200 m. Not surprisingly, the underwater parts project to considerable depths; this explains why sea currents, not winds, control the movement of bergs. It also explains why big bergs often run aground, remaining stalled at one spot for months at a time. Their jagged bases can also gouge deep grooves in the ocean floor. The risk of *ice scour,* as it is called, precludes the use of undersea cables and pipelines in many places.

Of course, not all icebergs are big at birth; or even if they start out big, they shrink to small size before ultimately melting away. Bergs between 1 and 5 m tall are called *bergy bits;* those less than 1 m are *growlers.*

As a berg floats through the sea, its underwater parts continuously melt away, making it top-heavy. Its balance becomes more and more precarious until, without warning, it suddenly overturns, with a thunderous roar; the waves raised by the upset spread for kilometers in a calm sea, while the berg, rocking gently now, settles to a new position of rest.

Adjustments like these recur at intervals as the berg slowly melts. Sometimes, instead of overturning completely, a berg will merely tilt at a new angle. If it has spent time aground, unable to rise and fall with the waves, its walls will have been undercut at the waterline; when the berg tilts to a new angle, the old waterline will be tilted too (fig. 4.1). The part of the berg that was underwater will have become smoothly rounded, while the parts that have always been above water remain craggy. Sharp crags and cliffs form anew if a berg breaks.

Some bergs are "winged": their above-water parts consist of more than one turret, rising up separately from the submerged "body" of the berg. The underwater link between the turrets shows blue-green through the shallow water covering it. Likewise, bergs that slope gently down to the waterline and below it are surrounded by a rim of the same intense blue-green, where the water is shallow enough for the submerged ice to "glow" through it. The color of submerged ice contrasts with the bluer shadows in above-water ice. In sunny weather, bergs are brilliantly colorful. Occasional bergs contain layers of "dirty" ice, picked up when they were parts of moving glaciers.

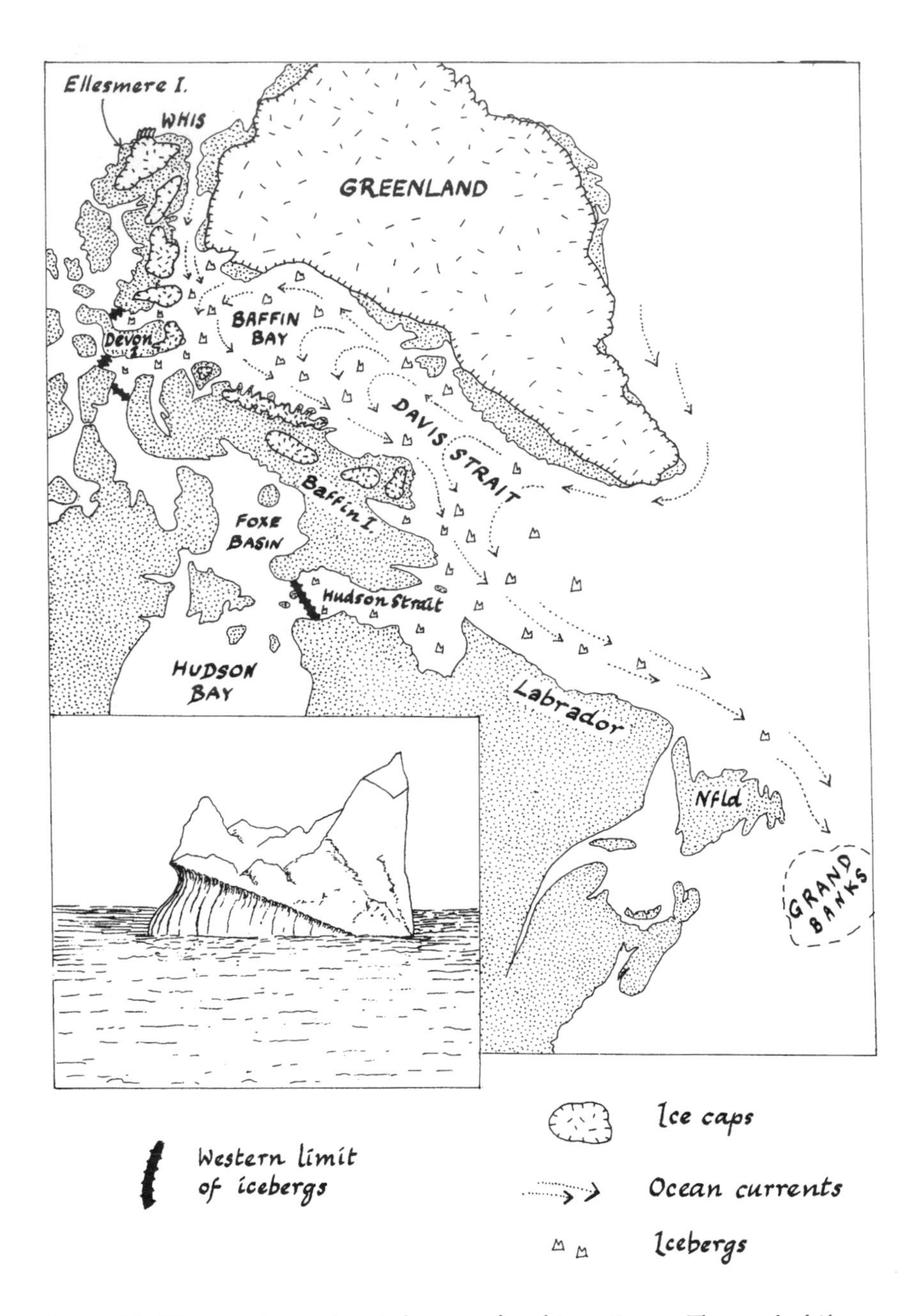

Figure 4.1. The map shows where icebergs are found in arctic seas. They rarely drift west of the limits shown. *Abbreviations:* Nfld, Newfoundland. WHIS, Ward Hunt Ice Shelf (see sec. 4.3). *Inset:* A typical iceberg. It has spent time aground, and while it was unable to rise and fall with the waves it became undercut at the waterline. It has since floated free and settled to a new equilibrium position, leaving the old waterline tilted.

Many of the events in the life of a berg can be inferred by examining it carefully. Even the color of the ice—opaque white rather than clear—is informative. It shows that the ice was once part of an ice cap or glacier, and was formed by the packing of snow that was originally fluffy and full of trapped air. As each year's snow layer gradually turned to ice, under the ever-growing weight of centuries of further snowfalls, the millions of tiny air bubbles remained trapped; they give the ice its whiteness. And, if berg ice is used as the ice cube to add to a drink, it fizzes energetically as the compressed bubbles are released.

For a narrative account of the many-faceted natural history of a big iceberg, see *Voyage of the Iceberg* by Richard Brown (Toronto: James Lorimer and Co., 1983).

4.2 Sea Ice: The Pack

Almost as soon as the arctic summer is over, the sea begins to freeze. Sea water, because of the salt it contains, freezes at a lower temperature than fresh water.

Another difference between sea water and fresh water is the way the density changes as the water cools. As fresh water cools, its density increases until its temperature falls to 4° C; with further cooling, its density *decreases*. The coldest water, that just at 0° C (the freezing point of pure water), therefore forms a thin skin floating at the surface, where it freezes to form a paper-thin sheet of ice, clear, brittle, and glasslike. This is how ice forms on a calm lake.

Sea water freezes in an entirely different manner. Its density steadily increases as it cools, so that water on the verge of freezing (which it does at about −2° C, depending on the saltiness) sinks instead of floating at the surface. A comparatively thick layer of water becomes ready to freeze all at once. Millions of tiny, separate, free-floating ice crystals form throughout the layer, creating a "soup" known as *frazil ice*. As the temperature drops, the soup thickens and congeals. So-called *grease ice* may form for a while, a scum of ice strong enough to support a standing gull, but not rigid. It undulates gently as the swell passes beneath it, and, when pushed by a breeze, forms ripples like those on cooling fat in a tilted frypan.

Soon the ice solidifies. The pack ice has formed. The more southerly pack is present only in winter: it is the *winter ice pack*, consisting of *annual ice*, which melts in spring and freezes anew in fall. The extent of the ice varies from year to year, depending on the temperature.

Farther north lies *multiyear ice*. It occurs in two zones: in the central, island-free part of the Arctic Ocean is the *permanent polar pack*, a great area of ice that, though not rigid, remains as one piece and moves as a unit (see sec. 4.3). Around it, plugging many of the channels among the Arctic Islands, is the more mobile *summer ice pack*. It breaks up to some extent in summer, and the floes drift free. The ice at the head of narrow inlets also melts in summer, even at high latitudes; it is thawed by meltwater draining from the land. The average limits of the different zones of ice are shown in figure 4.2.

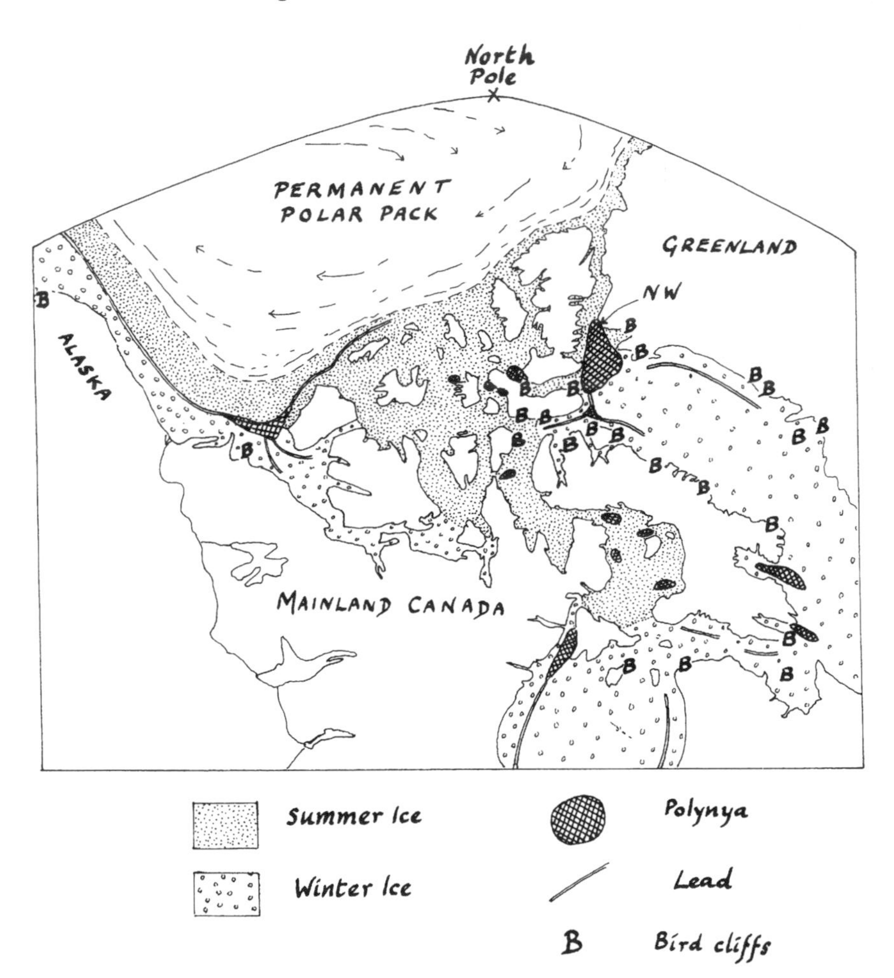

Figure 4.2. Map showing the average extent of summer (multiyear) and winter (annual) ice. Also shown are: the permanent polar pack, which slowly rotates in the direction shown by the arrows (see sec. 4.3); the locations of recurrent polynyas and leads; and the locations of "bird cliffs" where colonially nesting seabirds nest in vast numbers. The North Water is marked NW.

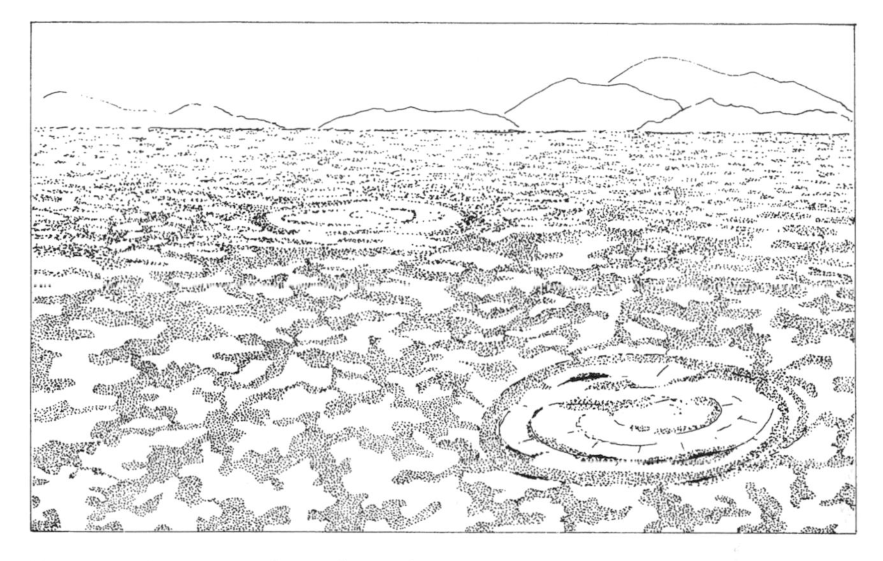

Figure 4.3. An expanse of annual ice in late spring, an interlocking pattern of ice and shallow meltwater, which is blue on a sunny day. The two circular patterns are frozen over polynyas (see text).

Annual ice seldom becomes thicker than 2 to 2.5 m. It is often smooth and level, and when it starts to melt in summer, it becomes covered with a myriad shallow blue pools of meltwater, as elaborately shaped as the pieces of a jigsaw puzzle (fig. 4.3). Summer rain adds to the pools. Further melting causes the ice to break up into separate floes.

Multiyear ice remains frozen year after year, but this does not mean that the ice it consists of is unchanging. Rather, it goes through a cycle. In summer, multiyear ice loses its topmost layer: the warmth melts the preceding winter's snowfall and also some of the upper surface. The following winter, the ice gains a layer because, as it cools, the water below freezes onto the bottom of it. This water, which lies just under the ice, is almost fresh and freezes easily (salt water has a lower freezing point than fresh); the fresh water comes from the preceding summer's meltwater pools, which have drained away through holes in the ice, and come to rest floating on the underlying, denser, salt water.

Thus as the seasons alternate, a layer of "old" ice is lost from the top of multiyear ice, and a layer of "new" ice is added at the bottom. When the process is in perfect equilibrium, the amount of ice added each year exactly balances the amount lost. Each layer migrates upward; the time it takes for a layer to travel from bottom to top—in other words, the lifetime of the layer—is probably about 12 years. Because of this continuous loss and replacement, multiyear ice doesn't thicken indefinitely. The annual layers composing it add up to a thickness ranging from about 3 m or less near its southern boundary to as much as 8 m in the far north.

These figures are very approximate, of course, and the cycle described above doesn't always work like clockwork. Unseasonably mild winters or cool summers can interfere with the cycle, as can irregularities in the thickness of the ice.

The pack everywhere, whether made up of annual or of multiyear ice, is floating and breakable. It cracks in places, and the cracks widen into *leads* separating individual floes. Drifting floes collide; the edge of one floe rides up over another, breaking into thousands of pieces poised at all angles. The result is a high *pressure ridge* of jumbled ice blocks. Because the whole pack is afloat, these thickened ridges sink somewhat, forming inverted ridges ("keels") on the bottom of the pack, corresponding with the visible ridge on top.

The ridges in annual ice are not nearly as high as those in multiyear ice. This makes it possible to see, from the air, where the annual ice ends and the multiyear ice begins; the latter has a rougher surface, and therefore looks darker, than the annual ice of the winter pack. Moreover, multiyear ice is dirtier than annual ice because windblown dust from the land accumulates on it year after year; annual ice, lasting for only one season, doesn't persist long enough to collect much dust.

Another difference to note is that between *landfast ice,* the part of the pack firmly frozen to the shore, and freely floating pack. The two are usually separated by a conspicuous lead.

The saltiness of pack ice deserves mention. On average, it is only about one-tenth as salty as sea water—about 3 parts per thousand (ppt) as against 35 ppt in sea water far from ice; sea water close to melting ice is diluted by meltwater, which (so long as the sea remains calm) floats at the top.

As salt water freezes, the ice crystals incorporate only a small fraction of the salt in the water; consequently, the diminishing quantity of water remaining unfrozen becomes steadily saltier until fairly concentrated pockets of liquid brine remain in the lower part of the ice. When this brine leaks into the sea from ruptured pockets, it sinks (the more concentrated the salt, the greater the density of the brine). Multiyear ice is quite fresh at the top, where the ice is oldest; only the youngest layers, at the bottom, contain a small quantity of salt.

Pack ice is a unique environment, neither land nor sea. In winter, it bridges all the interisland channels, allowing large animals (especially polar bears) to island-hop at will. Most of its enormous area is desert:

no plants grow, and no nourishment is available. But there are exceptional spots, openings in the ice where the sea remains unfrozen. These are *polynyas;* big, long-lasting polynyas are vitally important in the lives of marine mammals and seabirds. They are described in section 4.4.

Other polynyas are merely small, temporary holes in the pack, likely to arouse the interest of anybody flying over them, and worth watching for. They form where upwelling currents of comparatively warm water make a warm spot in the sea that melts through the floating ice directly above it. Such a polynya is unlikely to last long, because a shift in the pack shifts the polynya with it, so that it no longer overlies the warm spot that caused it; it then freezes over. The freezing takes place from the rim inwards. If the hole in the ice was circular, which is often the case, the new ice takes the form of concentric rings, making what looks like a blister on the surface of the pack (fig. 4.3).

4.3 Ice Islands

In addition to icebergs and the various kinds of pack ice, the arctic seas also contain ice in a third form, *ice islands.* These are large, flat masses of ice, up to 80 m thick, calved from the Ward Hunt Ice Shelf on the northern shore of Ellesmere Island (fig. 4.1).

Probably 30 or 40 ice islands, with diameters of half a kilometer or more, are afloat in the Arctic Ocean at any one time. A few large ones are several kilometers across and hundreds of square kilometers in area. The islands become frozen into the permanent polar pack, which, driven by currents and winds, slowly rotates; it turns clockwise, around a center about halfway between Point Barrow (the northernmost point in Alaska) and the North Pole. The whole mass takes about 7 to 10 years to complete a single rotation.

The Ward Hunt Ice Shelf, like the ice shelves of the Antarctic, is the margin of a land-based, freshwater ice cap floating out over the sea. The ice forming it is called *shelf ice.* Whenever a piece of the Ward Hunt Shelf breaks away to form an ice island, multiyear landfast sea ice (MLSI for short) forms in the sea to fill its place; MLSI also grows on the outer margin of the shelf. As a result, what looks like a single expanse of ice is really a patchwork of two different kinds of ice, and the contrast between them can be seen from the air. Both kinds have undulating surfaces, with long, parallel "rolls" (rounded ridges). Between the rolls are long valleys, filled with meltwater lakes in summer. The scene looks rather like a string bog (see fig. 3.17) constructed of ice. The rolls on shelf ice are much higher than those on MLSI, however, and are separated by valleys ten times as wide. The contrast makes them recognizable. Cores drilled through the two kinds of ice show, also, that shelf ice is about five times as thick as MLSI. The ice islands have rolls and valleys that match those of shelf ice, showing that they originated as shelf ice.

Some of the large ice islands have been used as floating bases for scientific research stations. One of them, called T-3, which was calved from the ice shelf in 1946, carried a research station for 20 years before it finally broke up. Another, calved in 1982, was named Hobson's Ice Island, after the former director of Canada's Polar Continental Shelf Project; it carried a research station from 1984 to 1992, before it unexpectedly broke into three pieces and veered southward from its hoped for route around the Arctic Ocean, drifting, instead, into the channels among the Queen Elizabeth Islands (the islands north of Parry Channel).

4.4 Open Water: Polynyas and Leads

The Arctic is home to numerous warm-blooded animals that breathe air and feed at sea: whales, seals, walruses, and polar bears among mammals; fulmars, murres, guillemots, dovekies, and some of the gulls among birds. If it were not for the existence of patches of open water in the winter pack ice, none of them could survive. The chief open-water bodies are polynyas, which form where warm, upwelling sea currents prevent the surface from freezing. Many polynyas—those described in section 4.2—are small and temporary. Others, the ecologically more important ones, are much bigger and remain open all winter, even though they shrink in size as winter advances. They appear at the same place year after year and are known as *recurrent polynyas* (see fig. 4.2). By far the biggest of them is the North Water, at the head of Baffin Bay. It is a saltwater "inland sea" surrounded by ice, and is sometimes as big as Lake Superior; it has been known to whalers since its discovery in 1616.

The recurrent polynyas are a dependable part of the environment for many years at a stretch. But not unfailingly: in exceptionally cold years they can freeze over, making breeding impossible that year for all the birds that normally rely on them.

Less important to sea life though still useful, are leads, long, open cracks in the pack ice. Leads can open up suddenly, almost anywhere where the pack is unconsolidated; they can also disappear as suddenly, when the pack on either side closes in again. In a few places, however, the wide leads that form between shorefast ice and the floating pack are predictable and long-lasting; they are also shown in figure 4.2.

What makes polynyas so valuable is that they have more to offer than merely open water; they are rich feeding grounds for all manner of animals because they are an ideal habitat for the microscopic organisms at the bottom of the food chain. Although they are in the Arctic, the environment is the very opposite of harsh. Unfrozen sea water cannot be colder than its freezing point of about −2° C. This is wonderfully mild compared with the bitter winter temperatures on land. And polynyas contain an above-average supply of nutrients, especially phosphorus and nitrogen, brought up from the sea bottom by the upwelling currents that keep the surface warm enough not to freeze.

The conditions are therefore ideal for the growth of *ice algae,* microscopic one-celled plants that live and breed in pores in the underside of the ice surrounding polynyas. The ice algae, which are so abundant they turn the bottom few centimeters of the ice a yellowish brown color, are the lowest link in the marine food chain. Next above them in the chain come minute animals of the plankton, including countless amphipods (tiny, shrimplike crustaceans), then larger crustaceans, squid, fish, and finally the warm-blooded animals.

Ice algae develop early in spring, long before the weather starts to warm up. The low temperature (compared with that of temperate seas) does not prevent their growth, which begins as soon as the spring sun rises high enough to give them enough light for photosynthesis. Indeed, the growing season in the sea starts while vapor from the open water still condenses into clouds of steam as it rises into the wintry air. This early start to the season is important to seabirds; they need time to build up their strength before they can lay and tend their eggs. And since seabirds can only hunt for food in open water, they are entirely dependent on polynyas. The large birds capture fish, the smaller ones scoop up plankton organisms. The nesting cliffs of those seabirds that nest in densely crowded colonies are all within flying distance of polynyas; the sites are shown on the map in figure 4.2. (For further details about seabirds, see chap. 6.)

Seals and walruses are not so dependent on polynyas as the seabirds are, since they are capable of spending most of their time under water, making only quick visits, at fairly long intervals, to their breathing holes through the ice. In spite of this, they tend to congregate at polynyas, no doubt because of the richness of the food supplies. Whales do the same (see chap. 7).

Food is certainly not unobtainable away from polynyas, however, since ice algae grow wherever sufficient light penetrates the ice. These algae make ice-covered seas unique. At all latitudes, shallow seas (shallow enough for sunlight to reach the sea floor) have a community of organisms living on the bottom; the bottom community, which includes a great variety of small and microscopic organisms, is known as the *benthos.* Ice-covered seas have a "ceiling" as well as a bottom, a ceiling of ice that supports what could be called an inverted or overhead benthos. Polar cod, one of the fish species that graze on this overhead benthos is specially adapted to do so; it has a jutting lower jaw, enabling it to graze easily on the "ceiling."

The living community on the undersurface of the ice, which flourishes exceedingly where nutrient-rich sea currents well up, makes the ice edge an unusually productive environment. All the animals that find their food in the sea come together at the ice edge in spring, to make it a scene busy with life of all kinds.

The sea floor, as well as the sea "ceiling," supports a varied community too. Some of its members are available for inspection by anyone traveling over the sea ice in early summer. What happens is this: numerous small sea-bottom invertebrates live among blades of kelp—brown seaweeds that grow as long, flat strips—and blades of "Devil's apron," a seaweed of big, flat, brown sheets riddled with small holes (burned by sparks from the devil's forge?). These seaweeds grow attached to rocks, but are often torn loose by storms, after which they drift at the surface with numerous small animals trapped among them. When winter comes, they are frozen into the ice. The following spring, as the ice surface melts, the dark seaweeds are uncovered; because they are dark, they absorb the sun's heat more readily than the white ice around them, which reflects the sun's rays instead of absorbing them. Soon the ice surface is pocked with small, deep pools of clear, fresh meltwater, each lined with a blade of dark seaweed accompanied, often, by the undamaged remains of a number of sea-bottom organisms, such as whelks, clams, mussels, sea urchins, seastars, brittle stars, and the like. Sometimes you will find the hollow shells of *euphausids,* the long-limbed shrimps known as *krill,* which abound in arctic seas and form the food without which bowhead whales could not exist.

For more on the life in polynyas, see "Arctic Seas That Never Freeze," by Maxwell J. Dunbar, *Natural History,* April 1987. For more on ice algae, see "Ice Algae" by Anya Waite, *Canadian Geographic,* Oct.–Nov. 1985.

4.5 Ice and Beaches

The beaches surrounding icy seas are very different from those of warmer latitudes. Not surprisingly, the intertidal zone is almost devoid of life, at any rate on the surface. No living thing can survive constant grinding and battering by sea ice.

Patches of rockweed manage to grow in the lee of protecting boulders, and sometimes a beach will be found strewn with stranded pinkish-red jellyfish, of the species *Cyanea arctica.* Their giant size, as well as their color, make these jellyfish spectacular. They are sometimes over 2 m across, with tentacles 60 m long. Note that they are almost certainly poisonous, and anybody handling one risks being severely stung; reports of stings are rare or nonexistent—the species does not live in waters where people swim—but swimmers off Atlantic beaches are often painfully stung by a closely related jellyfish of temperate seas, which is seldom more than 30 cm across. So if toxicity is related to size, the arctic jellyfish is an animal to fear.

Arctic beaches are interesting less for the life to be found on them than for the way sea ice shapes them. The effects depend greatly on the tide range (the difference between the water level at low and high tides). Tide ranges are low in the western Arctic and high in the eastern Arctic (fig. 4.4), and therefore different beach forms are to be looked for in west and east.

Where the tide range is slight, and a beach consists of deep beds of rough rock fragments or shingle (smooth, waterworn pebbles), *ice-push ridges* form (fig. 4.4). When pack ice is driven against the beach by the wind, it pushes the shingle up into a ridge, much as a road grader would. This cannot happen where a large tide range means that the floating pack ice never stays long at one level; push ridges should be looked for where the tide range is small, and the ups and downs of the pack almost negligible.

The ridges nearest the shore are the youngest; they are formed anew each season. Those farther back are the result of unusually strong winds forcing the ice onto the land: the higher the ridge, the more exceptional the winds; a really powerful storm can force the pack ice, with a ridge of shingle ahead of it, so far up a sloping beach that the top of the ridge is 10 m (vertically) above mean sea level. Therefore, higher beaches are likely to be older beaches; this can be seen by searching for plants in the troughs between ridges; the higher the trough, the more the vegetation.

Sometimes, push ridges extend so far up the sloping ground at the back of a beach that it seems impossible for them to have been made by ice floating on the sea. It would indeed have been impossible if sea level had stayed unchanged for thousands of years. In fact, the land has been rising, and the uppermost in a long series of push ridges will have been formed closer to the sea than it now is, when the land was lower. The highest "modern" push ridges, made by storms of once-in-a-century magnitude, merge imperceptibly into "fossil" ridges now beyond the reach of any storm. These fossil ridges are raised beaches (see sec. 3.3). The most conspicuous raised beaches are those with steep, narrow push

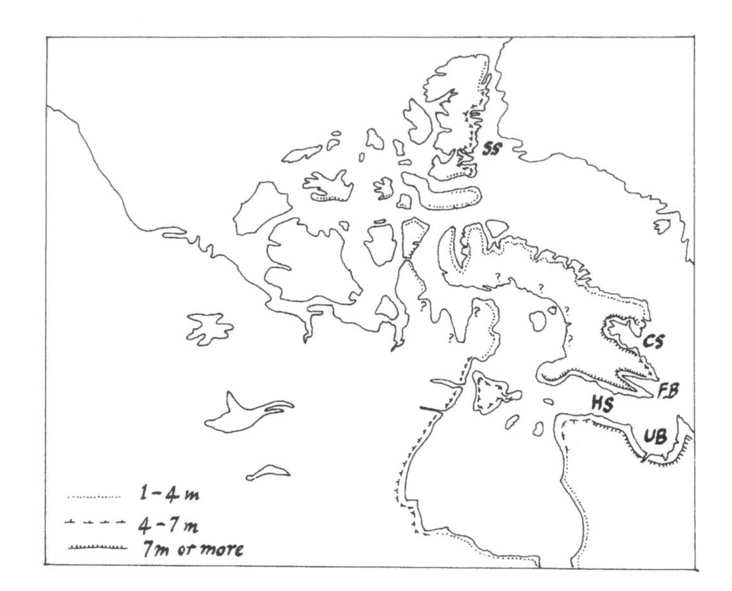

Figure 4.4. *Above:* Map to show approximate tide ranges along arctic shores. (The numbers give "large" tide ranges, i.e., the maximum range at spring tides.) Note that tide recording stations are widely spaced in the Arctic, so the map cannot be exact. Most shores left unmarked have a tide range of less than 1 m (this may not be true of eastern arctic shores lacking tide recording stations, and marked "?"). *Abbreviations:* SS, Smith Sound. CS, Cumberland Sound. FB, Frobisher Bay. UB, Ungava Bay. HS, Hudson Strait. *Below:* A beach backed by a long series of ice push ridges. Compare with fig. 3.7.

ridges, as in figure 3.7. Raised beaches lacking push ridges, as in figure 5.6, are much less obvious.

Where the tide range is big, the interaction between the sea ice and the land is quite different. The pack rises and falls twice a day as the tide goes in and out. At every low tide, broken blocks of ice are stranded in the intertidal zone: some of them land on a level part of the beach, some are perched on boulders, and some are let down where they cannot balance; they topple and break. At every high tide the blocks are refloated and rearranged. All the while, if it is winter, the quantity of ice is steadily increasing. By winter's end, the ice nearest the shore is a wild jumble of ice blocks, of all shapes and sizes. The strip of rough, hard-to-cross ice, between the shore and the comparatively smooth pack farther out, may be well over a kilometer wide.

The biggest ice blocks last well into summer; some even survive into the following winter. They rise and fall with every tide, and are dragged this way and that by winds and currents the minute they are even partly afloat. The result is a gouged beach, that looks as though it has been bulldozed at random, into a disorderly array of pits and hillocks.

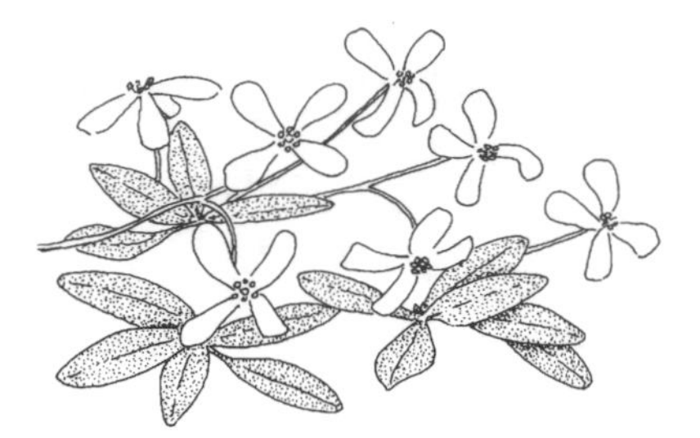

5 *Plant Life*

5.1 Why Trees Can't Grow in the Far North

Somewhere between the dense evergreen forests of northern Canada and the open, treeless tundra of the Arctic lies "the treeline." Defining it is difficult; the very word has two meanings. It is sometimes defined as the northern boundary of the forest-tundra zone beyond the dense forests; this is the zone where forest with patches of tundra (treeless barrens) merges into tundra with patches of forest, and it hasn't, of course, a distinct boundary (the map in fig. 5.1 shows an approximation to it). Sometimes it is defined as the northern limit of the growth of trees 5 m or more in height, another line that is obviously difficult to map precisely. Progressively smaller trees are able to endure at scattered points beyond the line, the most northerly being spruce *krummholz,* the name for gnarled, dwarfed spruce "trees" that grow so slowly because of the cold and are so continually bent by icy winds that they never grow taller than low shrubs.

The treeline on a map is therefore an artificial construction, an attempt to represent sharply something that in reality is fuzzy. What anybody flying poleward from forested, temperate latitudes into the Arctic notices is not a visible treeline but merely that the trees below become smaller and smaller, and at the same time more and more widely scattered, until none are left. Evidently the arctic climate is unsuitable for trees, even though the southernmost stretches of arctic tundra are densely carpeted with a rich variety of smaller plants. If they can stand the climate, why not trees too? Cold, of itself, doesn't kill trees; the coldest winters in the northern hemisphere occur in forested country, not in the tundra, and laboratory experiments show that the trees of the far north

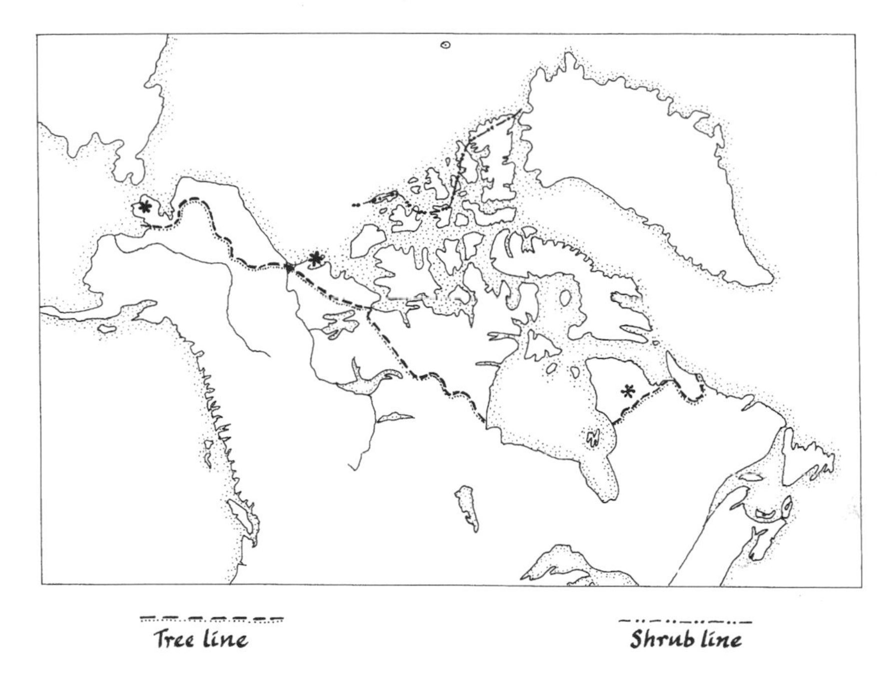

Figure 5.1. To show the treeline, the shrub-line (see sec. 5.5), and some sites (*) north of treeline where semi-fossilized spruce logs have been found.

can endure temperatures lower than any ever experienced in the real world.

It is lack of summer warmth that sets a northern limit to tree growth. Sunlight—the "energizer" for photosynthesis—is plentiful in the arctic summer, but the *photochemical* reactions of photosynthesis cannot happen unless plant tissues are warm enough. With the climate as it is nowadays, the arctic growing season is short. The growing season is the period in the year during which temperatures are high enough for photosynthesis, the production of new plant tissue, to occur. Where new tissues are formed too slowly, a plant cannot "afford" to grow tall. Nearly all its annual growth must be devoted to producing the essentials for life: the leaves without which it would starve, and the roots and other underground parts in which it stores the food made in summer to tide it over the rest of the year. There is nothing left to convert into wood, which is what gives a tree the strength and rigidity to stand upright. In a nutshell, leaves and roots are a necessity, wood is a luxury.

The need for summer warmth explains why the treeline does not lie along a parallel of latitude. As is clear from the map, it trends southeastward all across the Central Arctic, coinciding, roughly, with the southern limit of continuous permafrost (see fig. 3.4). Both the treeline and the permafrost boundary are controlled by summer temperatures, which in turn are controlled by the presence of Hudson Bay, as explained in section 3.2.

Permafrost also affects trees directly. Where the permafrost table is close to the surface, the active layer is too thin to accommodate their roots. Some trees, especially spruces, can get by with exceedingly shallow root systems. Even so, a tree's height is inevitably limited by the depth and holding power of the soil it is rooted in. This is another reason, besides lack of warmth, why trees become shorter and shorter as you travel from forest to tundra.

5.2 The Northernmost Tree Species

By far the most important trees at treeline are two spruces: Black spruce (*Picea mariana*) and white spruce (*Picea glauca*). Both species are adapted to the harsh conditions, black even better than white. Its root system is so shallow that it can grow in an active layer only 25 cm thick; it does well on heavy clay soils and also in wet peat bogs, saturated with rather acid soil water. White spruce requires better-drained, less-acid soil.

The two species can always be told apart by their twigs (if the needles are dense, bend them back to expose the twig surface). Those of black spruce bear a fuzz of dark brown hairs; those of white spruce are light-colored (tan or pale green) and hairless. The trees also differ in shape as seen from a distance (fig. 5.2); a clublike top is a sure indicator of a black spruce, but its absence does not necessarily indicate a white spruce, as not all black spruces have it. The cones also differ; black spruce has small egg-shaped cones, white spruce has elongate ones. Often no cones will be found, as the trees may be growing in too cold a climate for cones to develop (see below).

These two dark, evergreen spruces dominate the northernmost woods immediately south of the tundra, but they are not the only trees. Another fairly common conifer is larch or tamarack (*Larix laricina*), with deciduous needles that are apple green in spring, turn gold in late summer, and drop in early fall.

Besides the three conifers, there are three subarctic hardwoods. The most northern is balsam poplar (*Populus balsamifera*), the only broad-leaved tree in the region having deeply furrowed bark (fig. 5.2); in the Western Arctic it grows on the gravelly shores of lakes and of rivers draining into the Beaufort Sea, which makes it the most northerly growing tree species on the continent. Its close relative trembling aspen (*Populus tremuloides*) does not grow quite so far north; it has smooth, chalky bark, and leaves that flutter in the slightest breeze because the leaf stalks are pliable flat strips.

The third subarctic hardwood is paper birch (*Betula papyrifera*). It has the peeling, papery bark typical of birch trees, but the trunks are often brown or cream-colored rather than a conspicuous white like the paper birches of temperate latitudes. For this reason, the tree is often called Alaska birch, and some botanists classify it as a separate species, *Betula neoalaskana.*

Figure 5.2. (a) Black Spruce. (b) White Spruce. (c) Balsam Poplar.

In sum, the subarctic trees are easy to recognize. But they sometimes seem unfamiliar at first because, compared with members of their species growing in temperate forests, they are often small and stunted.

5.3 The Wavering, Shifting Treeline

The position of the treeline is not fixed. Over the centuries it shifts to and fro, from south to north and back again, in response to the fluctuating climate. Since the peak of the last ice age the climate has at times been considerably warmer than it is now. Temperatures peaked about 9000 years ago in the Western Arctic and 5000 or 6000 years later in the east; indeed, a "wave" of warmth, lasting around 2000 years at any one spot, progressed from west to east across the land.

At the height of the warm period, the treeline was between 150 km and 250 km north of its present position. This is known from evidence left behind: tree pollen from ancient forests is preserved in the sediments of tundra lakes. And semifossilized spruce logs are sometimes found far beyond the modern treeline, lying on the tundra or buried in peat bogs; the map in figure 5.1 shows the sites of three such finds.

Living reminders of historic climates exist too, in the form of spruces growing at places where, given today's climate, they cannot produce seeds (seed production requires a longer warm season than mere growth). Instead, the trees propagate by "layering"; this happens when a low branch droops to touch the ground and develops a new root system at the point of contact. Then a new trunk grows above these roots, resulting in a new, independent tree once the original linking branch has died. Gardeners force this process to happen when they propagate shrubs such as forsythia by artificial layering. It works with many plants but by no means all. Of the subarctic conifers, black spruce layers easily, white spruce and larch less so.

A group of trees formed by layering is a clone: all its members are genetically identical, since no seeds were required to produce them. A clone can expand for centuries, never producing cones and seeds; it can go on spreading in a cooling climate long after seed production has become impossible, although its original ancestor must necessarily have started life as a seedling, which it can only have done when the climate was warmer. Clones like these usually hang on until a fire destroys them. No new trees can grow to replace those destroyed, because there is no source of seed within range; therefore burned patches remain treeless—another factor that makes the treeline ragged and diffuse.

Many segments of the present treeline are historic, in the sense that the trees forming it are survivors from a past warm period, unable, because of the cold, to reproduce by seed. If the climate warms because of the greenhouse effect, the treeline won't migrate immediately. Migration can't start before the warmth is great enough to allow the currently northernmost trees to produce seeds; then some of these seeds will be blown northward into the tundra, seedlings will begin to grow north of their parents, and a northward drift of the treeline will be under way.

5.4 Trees and Permafrost

Vegetation, especially forest, is obviously a heat insulator—which raises the question: does it prevent the ground beneath it from warming up or from cooling down? The answer isn't obvious. Studies have shown that in the High Arctic, a cover of vegetation keeps the ground warmer in winter than it would otherwise be. Farther south, vegetation keeps the ground cool in summer by insulating it from sunlight and warm air.

This means that in the subarctic there are surprising feedbacks between plants and permafrost. Under bare, open, sunlit soil, the permafrost table (the upper surface of the permafrost) lies deep, leaving the surface a hospitable habitat for trees. Conditions are good enough for white spruce, balsam poplar, and paper birch to flourish. When the trees grow big enough to form a closed canopy, the ground becomes shaded and is soon covered with the dead leaves of poplar and birch, and with fallen white spruce needles; in time, a carpet of moss develops as well, which is an especially good insulator. As it becomes increasingly insulated from the sun's warmth, the soil gradually cools, and the permafrost table rises until the ground is no longer suitable for the trees now growing on it. As they die they are replaced by stunted black spruces and larch, which can grow on a thinner active layer and on soil made acid by peat (undecomposed plant remains). The new trees are too widely spaced to shade the ground, so such tundra plants as Labrador tea, cottongrass, and blueberries can grow among them. Dense carpets of lichen often develop. The sequence is a perfect example of ecological succession.

The question arises: how do tracts of bare ground originate, to allow the sequence to begin? One cause is fire. Another is the presence of a body of water. Under lakes and rivers, permafrost is forced to great depths, or is absent completely, because water has such tremendous heat-retaining capacity. Trees and shrubs growing on lake shores are taller than elsewhere, not because the plants benefit directly from the water, but because the permafrost table is farther below the surface, and the active layer thicker, than at some distance back from the water's edge.

A meandering river in forested permafrost country gives you a snapshot view of vegetational succession in all its stages. This is because the bends in the river—the meanders—do not remain in one place but are forever shifting slowly downstream. The course of events is as follows (fig. 5.3): In a meandering river, the water flows fast on the outsides of the curves, eroding the bank and adding to the load of sediment it carries. It flows much slower on the insides of the curves and consequently deposits some of the sediment—especially the coarse gravel—that it was carrying downriver from farther upstream. The deposited gravel settles out on the downstream margin of each lobe of land. The result of all this is that the lobes migrate downstream as each is continuously eroded on its upstream side and continuously augmented by added sediment on its downstream side. The deposited sediment consists of bare, well-drained, sun-warmed gravel, with permafrost far below or absent. It provides an ideal site for ecological succession to start; various willows are the first-comers, then balsam poplar, and then white spruce (fig. 5.2). By the time the succession has reached its end, with black spruce and larch on a thin, peaty, active layer over near-surface permafrost, the land it occupies is on the upstream side of the migrating lobe, soon to be eroded and destroyed. As the undercut banks give way, the trees on them fall into the river. The cycle continues endlessly.

Permafrost and individual trees interact in different ways in different places. Usually the crown of a tree functions as an umbrella in winter, causing the snow just around the tree to be shallower than it is elsewhere. Snow—especially the soft, fluffy snow in sheltered woodlands—is an effective insulator which, where it is deep enough, prevents the soil from freezing. If, under trees' crowns, it isn't deep enough, then each individual tree can sometimes develop its own private permafrost "pedestal," which slowly expands upward as the ice accumulates, forming a mound with the tree on top.

This doesn't happen everywhere. Trees on the Mackenzie Delta, and perhaps in other places where spring floods deposit little hillocks of sediment around each tree, have basin-shaped depressions in the permafrost table below them. The reason is that the sandy hillocks hold less moisture than surrounding soil, allowing summer warmth to penetrate downward; also, the sediment prevents the growth of an insulating carpet of moss which would keep in the cold. Thus, depending on circumstances, a tree can cause new permafrost to form directly below it or can cause the top of existing permafrost below it to melt.

The hummocky ground that occurs throughout the permafrost country of the arctic and subarctic regions (see sec. 3.7) is often covered by a woodland of small spruce trees in its southernmost tracts. The frost-heaving that produces the hummocks has an odd effect on the trees;

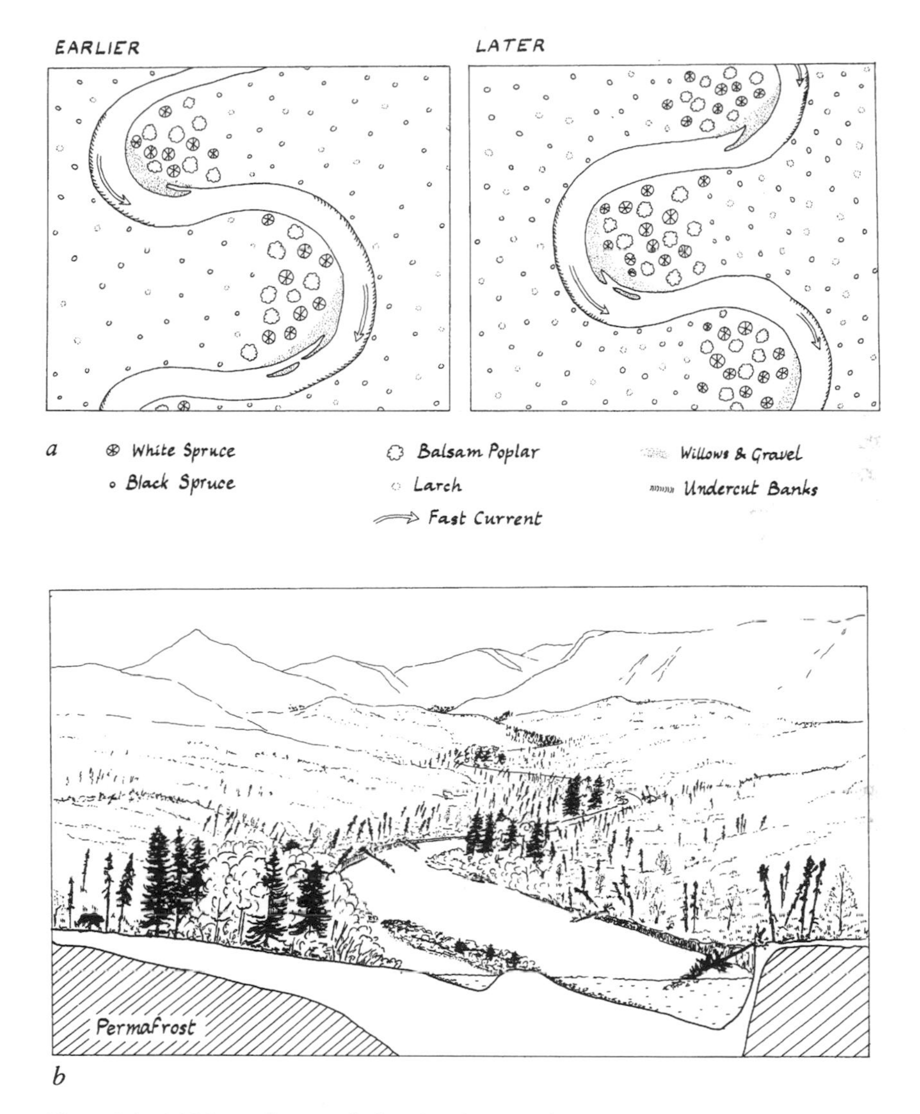

Figure 5.3. (a) Maps of a meandering river in permafrost country, at an earlier (*left*) and later (*right*) time. The later map shows how the meanders have shifted downstream (southward) over the course of many decades. The permafrost table is close below the surface everywhere except under the river itself, and under the "lobes," which consist of sediments recently deposited on the river bottom. (b) Oblique aerial view of the same scene.

it tilts them this way and that so that they lean in all directions, and the result is aptly known as a drunken forest.

5.5 Tundra and Polar Desert

Beyond the northernmost woodlands, the tundra proper begins. The low-growing plants—shrubs, herbs (nonwoody plants), mosses, and lichens—don't alter at the treeline: they continue, for some distance, as before. Only the trees show a change: they become smaller and sparser until they finally disappear. But this is not to say that the tundra is uniform; far from it. The southernmost tundra is comparatively luxuriant; plants of one kind or another cover the ground everywhere, and dense thickets of tall shrubs are common. There are many different species of plants.

The farther north you go, however, the more austere the vegetation becomes. With increasing latitude the growing season becomes progressively shorter, and late-lying snowbanks lie longer. Species disappear one after another, until only the hardiest remain. About 400 different species of vascular plants (this excludes lichens and mosses) are to be found in the southernmost tundra, which has a mean July temperature of 11° C; the number of species falls off in proportion to this temperature, so that in the most northwesterly of the Arctic Islands, where the July temperature is 4° C or less, there are no more than 50 species.

As the climate becomes harsher, the plants become shorter and more widely spaced. Patches of bare ground appear, and become larger. The ever sparser vegetation becomes steadily less effective as a windbreak. Plants crowded together protect each other; the fewer and farther between they become, the less well can they shelter their neighbors from the rigors of a cold, windy climate.

The increasing cold means, too, that there is less time in summer for the growth of wood; just as there is a treeline, there is also a shrub-line, north of which shrubs cannot grow. It is shown in figure 5.1.

Water is in short supply, as well. Over much of the Arctic Islands, annual precipitation (rain and snow) is less than 10 cm per annum. Much of the land is covered by immense fields of sharp rock fragments with only occasional tiny plants tucked among them; it is aptly described as "polar desert." In this desert the drought is most intense on windswept ridges. Few are the places where mosses can grow well enough to form continuous moist carpets, as happens in the wet tundra farther south where moss carpets form water-retaining seed beds for other plants. Among the few sites in the polar desert with a continuous water supply are the margins of snowbanks that shrink in summer without disappearing; but here the late-lying snow means that the growing season starts even later than elsewhere. In unusually cold years, some of these snowbanks may fail to shrink, leaving plants that normally experience a short growing season with no growing season at all. Only a few plants—mountain sorrel is one of them—can survive under snow for several years in succession, and this is one of the reasons why even relatively moist areas in the polar desert don't have as diversified a plant cover as the wet tundra of the Low Arctic.

This is not to say that the High Arctic is a scene of uninterrupted bleakness, however. The arctic desert has its share of oases, formally described as High Arctic oases. These are especially favored places where, by a lucky accident of topography, water doesn't drain away and cold winds don't intrude. A good soil (by arctic standards) can develop, supporting an atypically lush and varied vegetation, which, in turn, supports muskox and caribou herds and their predators. One of the best-known High Arctic oases is the valley of Lake Hazen in northern Ellesmere Island, at 82° N, where more than 110 species of flowering plants have been found.

5.6 Carbon in the Tundra Ecosystem

Away from the extreme rigors of polar desert, wet tundra, with an unbroken cover of plants, is the typical vegetation of much of the Arctic. It owes its ample moisture supply less to rain and snow than to permafrost, which prevents spring meltwater from draining away. Stagnant pools form in great numbers on flat ground and in the basins formed by low-centered tundra polygons (see sec. 3.4). Cottongrasses and other sedges form extensive meadows in the shallows. The abundant flowering plants for which the Arctic is justly famous grow as cushions and tufts and mats wherever the ground is a little less soggy.

Tundra owes its awe-inspiring spaciousness to the shortness of its plants, but in spite of appearances, the plants are not particularly small. In wet tundra, only 5 percent of the plants' biomass (living material) is above ground and visible; the remaining 95 percent is underground, in the form of roots and *rhizomes*. A rhizome (root stock) is one of the horizontally growing underground stems by which many plants spread; as a rhizome grows, new tufts of roots grow downwards, and new tufts of leafy shoots grow upward, at intervals. (Any gardener who has battled with couch grass knows all about rhizomes even if unaware of the name.) In these big underground organs the plant stores the carbohydrates it will need to begin growth at the beginning of next year's spring.

Because of the cold, photosynthesis, hence growth, is slow. The cold also slows decomposition; dead plants and plant parts, instead of decaying rapidly as they would in a warm climate, accumulate as peat. Because of this, tundra is a *carbon sink*. This means that tundra absorbs more carbon than it gives off, bringing about a net reduction in the amount of carbon dioxide in the atmosphere. All ecosystems exchange carbon dioxide with the atmosphere. Green plants absorb it as they carry out photosynthesis; and all living organisms (including green plants) give it off as they breathe, among them the bacteria and fungi which decompose (in other words, eat) dead plants and animals. In low latitudes, the exchange goes on at a brisk pace because warmth speeds up the activities of the decomposers, and dead material doesn't accumulate. In the cold of high latitudes, dead material—peat—does accumulate, and large quantities of carbon, withdrawn from the atmosphere by green plants, are locked within it. Indeed, nearly all the carbon in wet tundra is stored, dead, in peat; only a tiny fraction of it is in living organisms. Slow decomposition also has the effect of reducing the amount of mineral nutrients available in the soil; like carbon, the nutrients are locked in the peat. Nitrogen, especially, is in short supply. This is another reason for slow plant growth in the Arctic.

Ecologists have predicted how the tundra ecosystem will be affected when (or if) the climate is appreciably warmed by the greenhouse effect. Their conclusions are disturbing. They expect that the peat will become warmer and drier and begin to decompose. Decomposition yields carbon dioxide and methane, both of them greenhouse gases given off by the organisms (fungi and bacteria) doing the decomposing. Methane soon oxidizes to carbon dioxide and water vapor, but before this happens its "greenhouse effect" is considerably stronger than that of carbon dioxide. The extra carbon dioxide is unlikely to be absorbed by photosynthesizing plants; they cannot grow much faster than they are growing now, owing to the infertile soil. Consequently, instead of being a carbon sink, the tundra ecosystem will become a carbon source, a system giving off more carbon (as carbon dioxide and methane) than it absorbs. This, of course, will enhance the greenhouse effect, and an unwelcome positive feedback will come into play.

5.7 Plant Adaptations to Arctic Conditions

Cold is only one of the hardships of arctic life that a plant must be adapted to endure. The others are the short growing season, drought, the effects of frost-heaving, the strong winds, and the infertile soil.

Cold itself is the first consideration. Arctic plants have to be adapted to it in two quite different ways: First, they must survive the deep cold of winter, when they are dormant. Second, they must survive spells of freezing weather in summer; summer cold spells are not nearly as cold as winter, of course, but they come at a time when the plants are actively growing rather than dormant. The chemical processes that constitute a plant's life—in a word, its physiology—are not the same in the dormant season as they are in the growing season; therefore, the physiology of arctic plants must be adapted to cope in both seasons.

The growing season is short as well as cool, and plants must make the most of what warmth there is. Most arctic plants are low-growing, with their leaves close to the ground. As soon as winter snow has melted, the dark ground surface absorbs the heat of the spring sun; this warms the plants and they can quickly resume active life after a winter of dormancy. Even in snowbanks, some plants manage to get an early start. This happens where patches of snow partially thaw in daytime, and then refreeze at night. If this happens repeatedly, the refrozen slush sometimes becomes a "window" of clear ice, about a handspan above ground level; plants below the window then find themselves in a small, natural greenhouse, in whose warmth they can begin growth days ahead of less well protected plants.

Another adaptation by many plants to the short growing season is *wintergreen,* or semi-evergreen, leaves. These are leaves that develop late in the summer and survive through winter without withering. They remain green and can start photosynthesis as soon as the weather is warm enough in spring, before there has been time for the new season's leaves to expand and start functioning. They finally wither after the new leaves have taken over. (Wintergreen leaves are not limited to the Arctic; many plants of the northern forests have them too.) Among common arctic plants with wintergreen leaves are arctic poppy, thrift, alpine saxifrage, and several kinds of chickweeds and starworts.

The plants least adapted to a short growing season are annuals. These are plants that go through their whole life cycle, from germination to seed production, in one growing season and die before winter sets in; they exist in the cold months only as seeds. Very few plants can compress a complete life cycle into the short time interval provided by an arctic growing season. Probably no more than half a dozen plant species are able to manage it, and they tend to be tiny, inconspicuous, and rare.

Another small group of arctic plants is annual or biennial, according to circumstances. They flower and set seed only once in their lives, dying immediately afterward; this is the hallmark of both groups. In the Low Arctic, with its relatively long summer, they can behave as annuals since there is time enough for them to grow and flower in a single season. But where summers are shorter, their life cycles occupy two years. In the first summer they grow leaves but no flowers; in the second they flower, set seed, and then die; in other words, they behave as biennials.

Except for these comparative oddities, all arctic plants are perennials; that is, they live for several years, flowering each summer. Perennials have long-lived root systems, and many have rhizomes as well, branching and radiating extensively through the soil. These underground parts are at risk from frost-heaving, which, not surprisingly, is common in the Arctic. Not only does it control the pattern of the ground over large areas (see chap. 3), it also damages plants, whose roots and rhizomes, especially when young, can be injured or broken by the shifting soil.

The plants best adapted to invade bare, frost-heaved soil are those that quickly grow long roots, binding the soil. By stabilizing the soil in this way, the pioneers create a safe habitat for the seedlings of other plants as well as for themselves. Moss-campion, a compact cushion plant with a deep taproot, is an example of one such pioneer.

Now we come to the wind, the fierce, cold, relentless, drying wind typical of the arctic climate. It affects plants profoundly. It dries them out in summer and, much worse, it "sandblasts" them in winter with wind-driven snow particles. The snow carried by the wind isn't soft and fluffy; it is hard and abrasive. It consists of millions of sharp snow particles driven by gale-force winds just above the surface, the material of blizzards. Snow abrasion is a far greater threat than low temperatures to a plant's well-being in the arctic winter.

The commonness of cushion plants in the Arctic is a response to the wind as much as to the cold. Their dead and withered leaves do not drop off; instead, they accumulate, making the plant even more cushiony. The dead leaves shelter the living leaves from snow abrasion in winter and from drying out in summer. In summer, too, they absorb the sun's warmth and then trap the warmed air so that the temperature inside a cushion rises several degrees above that of the outside air. Therefore, the abundant dead leaves in, for example, cushions of prickly saxifrage and arctic dryad are not simply waste material, they are an indispensable protection.

Plants that cannot endure the wind can survive only in sheltered spots. In the High Arctic, near the limit of woody growth (the shrub-line; see sec. 5.5), the two northernmost woody plants show an interesting contrast in their adaptations to the harsh conditions. The plants are arctic willow and arctic white heather. The heather is a short, upright subshrub that cannot endure the wind. It is found only in sheltered hollows where, because the wind is suddenly slowed, deep snowdrifts form; the snow acts as an additional protection, and heather can tolerate the shortened growing season which is the price a plant must pay for a roof of snow in winter. Arctic willow has a different survival strategy. It is a prostrate shrub tough enough to withstand snow abrasion in winter and dessication in summer; therefore, it can grow on exposed ridges where heather would quickly succumb. Away from the desertlike conditions of very high latitudes, arctic white heather is less restricted in its habitat; in the better-vegetated tundra of the Low Arctic, it gets the protection it needs from plants growing around it.

The final hardship endured by arctic plants is the poor soil. It is low in nutrients because dead plants fail to decompose; as explained above, they accumulate as peat. Arctic plants are adapted to stay alive without much mineral nutrition, but lack of nutrients certainly slows their growth. This has been shown by ecological experiments in the High Arctic; plants in an artificially fertilized experimental plot grew much bigger than those outside the plot. The infertility of most soils also becomes obvious when you see how surprisingly large and flourishing a normally small plant can become if it happens to grow in a naturally fertilized spot such as below a bird perch, or beside a caribou carcass.

A cushion plant can compensate to some extent for the soils' infertility by intercepting wind-blown dust, which helps renew its own patch of soil; the dead leaves held in the cushion provide some nutrients too. Plants of wet sites that form big tussocks use the same strategy; they collect and conserve nutrient sources, living and dead, around themselves. Various sedges, including cottongrass, are examples.

5.8 Arctic Flowers

The beauty of its abundant flowers is one of the glories of the arctic tundra in summer. A subsequent section (sec. 5.13) tells how to identify the common ones. Here we consider them from the ecological point of view.

As an adaptation to the shortness of the warm season, nearly all arctic plants grow their flower buds in late summer. These buds overwinter and are ready to go—to open and start functioning—as soon as temperatures are high enough in the following spring. This ensures that every flower will have as long a warm season as possible; no time is wasted on preliminaries at the beginning of the season. Overwintering flower buds on arctic plants don't catch the eye of summer visitors, of course; to see an example in temperate latitudes in winter, look at a twig of common Labrador tea (its arctic relative is very similar). The compact little spherical bud at the center of the radiating leaves is a preformed flower bud, from the summer before, ready for action as soon as spring arrives.

The purpose of conspicuous flowers, as all naturalists know, is to attract insects so that they can carry pollen from one flower to another. This is the first step (and only the first step) in the production of viable seeds. Flowers attract insects by their colors and scents and reward their insect visitors with nectar and pollen to feed on, and sometimes with a warm place to bask in as well.

Most insect pollination in the Arctic is done by flies of several kinds and by bumblebees. "Flies" means more than merely houseflies and insects that look like them; it means any insect with two wings (as opposed to four, or none) and includes midges, crane flies, hover flies, mosquitoes, gnats, and many more (see secs. 9.3 and 9.4).

Flowers of different colors attract different kinds of insects. Flies are attracted by whites and yellows, bumblebees by blues and purples. Purple saxifrage, for instance, is a bumblebee flower. Flower shape makes a difference, too. Flowers like louseworts and milk-vetches, whose pollen-bearing stamens are surrounded by enclosing petals, are accessible only to bumblebees; flies are not strong enough to force their way in. Flowers adapted to flies are open and often bowl-shaped.

Some of the light-colored bowl-shaped flowers that attract flies have more to offer than nectar and pollen; they also provide warmth. An insect visiting such a flower often stays there simply basking, long after it has finished feeding. In two very common flowers—arctic dryad and arctic poppy—the air becomes several degrees warmer inside the flower than outside whenever the sun shines. This happens because of the flowers' shapes. They act like parabolic reflectors, focusing the sun's rays, and warmth, onto the center of the flower and making it an attractive spot for flies to settle on. An added advantage is that the "working parts" of the flower—where pollen develops and seeds are fertilized and ripen—are warmed too; warmth speeds these processes and thus improves a plant's chances of setting seed before cold weather comes.

The heating mechanism only works, of course, when the sun shines directly into a flower. For it to work effectively, the flower must keep turning to face the sun as it moves across the sky. This is, in fact, what happens. Arctic dryad flowers follow the sun except during a few hours "at night," stopping only when the sun is low in the north. Poppy flowers follow the sun continuously (so long as it is shining) all through the 24 hours of daylight.

Few people ask the obvious question: how can this happen without the flower stalk becoming more and more twisted? In fact, there is no twisting. This is what happens: a flower stalk grows continuously, and it needn't grow at the same speed on all sides. Growth results from elongation of the microscopic cells of which the stalk is composed, and if the cells on one side elongate more rapidly than those on the other the flower bends toward the slow-growing side instead of facing straight up. This raises another question: what causes the growth rate to be slower on the sunlit side, as the process requires? The answer is, hormones. Plant growth hormones are produced in the cells at the tip of the stalk and seep downward to stimulate elongation in the cells below them; sunlight inhibits the formation of the hormones, with the result that growth is slowed on the sunlit side.

Bowl-shaped flowers are not the only ones to warm up to temperatures greater than that of the surrounding air. Willow catkins warm up too, presumably because their fairly dark color enables them to absorb the sun's warmth; it is even believed that the hairs covering them may function like greenhouse windows. (For more on arctic willows and their catkins, see sec. 5.13.)

Nectar and pollen to feed on, and warmth to loaf in, make a "fly-flower" an attractive spot for a fly in the High Arctic. The cold does limit flowers' capacities to some extent, however. The quantity of nectar they secrete is greatly affected by the warmth they receive from the sun. They yield far less in the High Arctic than in the Low Arctic, and flowers growing on south-facing slopes, the best place for catching the sun's rays, are better nectar producers than those at less-favored sites.

Because a flower is visited by insects that come away dusted with pollen, it doesn't necessarily follow that the flowers depend on the insects to perform cross-pollination for them (cross-pollination is pollination by pollen from another plant). Only by experimentally preventing cross-pollination can botanists discover whether a particular plant species must be cross-pollinated, or whether it can set viable seed by pollinating itself (self-pollination). Examples of self-pollinating plants are arctic poppy and arctic bladderpod. Cross-pollination is believed to be necessary for arctic willow and capitate lousewort. Arctic dryad and purple saxifrage seem to do best with cross-pollination although both manage to ripen some viable seeds without it.

5.9 Seeds and Fruits: The Way Plants Spread

If all goes well (and in many years it doesn't: see the next section), pollination starts a process that ends with the production of ripe seeds. At this stage the fierce gales of the High Arctic become a blessing rather than a curse. They blow seeds for enormous distances over the slippery, crusted snow, ensuring that every plant's offspring will be dispersed over a wide area. Because tundra vegetation is low (and polar desert vegetation even lower), not many twigs protrude above the winter snow to impede the wind or stop the sliding seeds. The frozen sea is no barrier, either. In winter the Arctic Islands are no longer islands; the whole Arctic Archipelago is an extension of the mainland.

Certain high arctic plants are adapted to make the most of the wind. In some, the flower stalks grow much taller once the flowers are over, so that the pods or capsules containing the seeds will be high enough to protrude through the snow. The stalks are short while the flowers are functioning, keeping them low down in the still, warm air close to the ground where conditions are best for insect visits, pollination, and seed ripening. But when it needs the wind, each flower stalk, now a "seed-capsule stalk," grows up to intercept it. This strategy is well developed in plants of the mustard family, for example, Pallas's wallflower (fig. 5.4), and some of the louseworts.

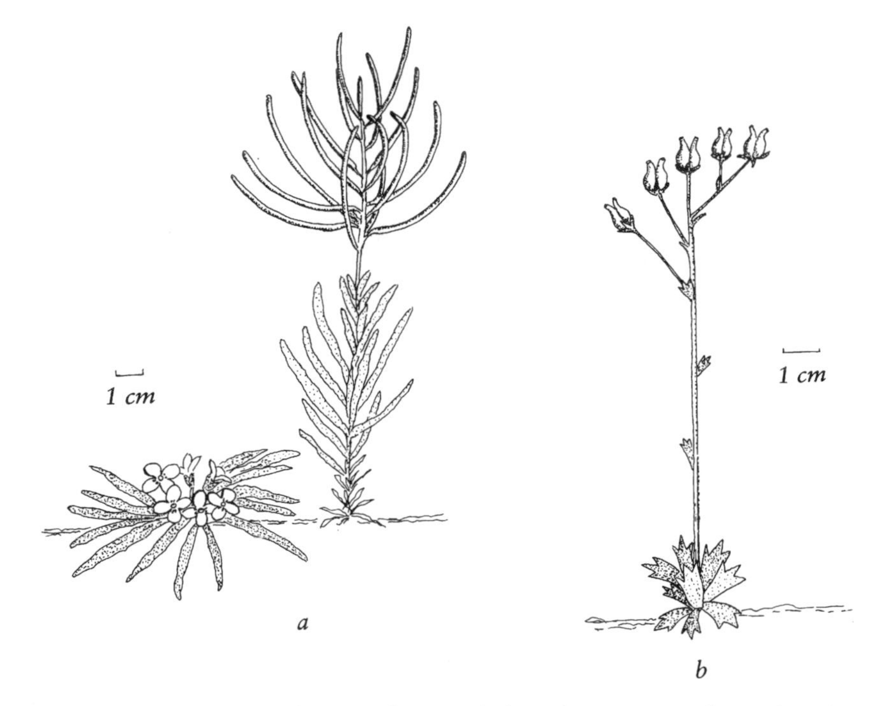

Figure 5.4. (a) Pallas's wallflower, in flower and after it has gone to seed; note how the stalk lengthens to bring the seed pods higher above the ground than the flowers were. (b) Prickly saxifrage with ripe capsules; note that they are upright, with openings only at the top.

In many of the saxifrages with seed capsules atop tall stalks, the adaptation is further refined. The stalks are stiff and the capsules held rigidly upright. Each capsule consists of two "horns" (fig. 5.4), each with a small opening at the top through which the seeds ultimately escape, but they do not escape easily, for they cannot simply fall out. They are held in the capsule until a winter gale sets the rigid stalk rapidly vibrating; then the seeds are violently shaken out onto the smooth snow, and into a wind strong enough to carry them far away.

This mechanism ensures that seeds leave the parent plant only during a strong wind; it also makes it likely that the seeds will not all be released at the same time. If they were, they would all be blown in the same direction, and might all chance to end up at an unsuitable destination—on the sea ice, for example. An adaptation that causes seeds to be shed in installments, all through the winter and into spring, is obviously valuable; it reduces the risk that all the seeds will "go astray" in one unfortunate accident. Some members of the mustard family achieve the same result by having long seed pods that open by degrees, releasing the seeds inside one at a time.

Seeds spread by gale-force winds are usually round and smooth, like tiny peas. Their journey over ice and smooth snow usually comes to an end in a snowdrift, where wind-blown dust—the makings of a new pocket of soil—is deposited with them and where, when spring comes, the melting snow provides a steady supply of water for the seedlings. But not all plants with wind-borne seeds rely on strong winds to disperse them. Several plants have plumed seeds that are carried by gentle winds. Some examples are arctic dryad, the fireweeds, and the cottongrasses.

As in warmer latitudes, many seeds are spread by animals. In the Arctic seeds are dispersed by lemmings, voles, ground squirrels, and a multitude of seed-eating birds. Some animals prefer dry seeds; others are attracted to berries, of which five varieties are available: blueberries, cranberries, cloudberries, bearberries, and crowberries. These berry-bearing plants are common only in the Low Arctic, however. The birds and mammals they rely on to disperse their seeds are not abundant in the High Arctic and "berry plants" become rarer as you go northward.

5.10 When Seeds Fail

In spite of all the marvelous adaptations plants have evolved to maximize their chances of producing, and dispersing, viable seeds, they often fail to achieve their object. In the High Arctic especially, the warmth of summer is often cut short before seeds are ripe. And even when a good crop of viable seeds is produced, unless the subsequent spring and summer are warm enough they will not germinate. Seeds of most plants can remain dormant through cool growing seasons for many years, while they await conditions suitable for growth. Some can last almost indefinitely, probably for thousands of years, if they are buried in frozen peat.

Because of the hazards of reproduction by seed, most plants, most of the time, reproduce vegetatively. To emphasize the distinction, it is better to say that they *multiply* vegetatively, since the word reproduction usually implies sexual reproduction, and vegetative multiplication is completely asexual (it is one of several forms of so-called asexual reproduction). All that happens is that part of a plant becomes detached and establishes itself as a new plant without a seed being involved.

It happens in many ways. Low, spreading plants with branches trailing across the ground are common in the Arctic, and in nearly all of them roots can develop along the undersides of the branches wherever they touch the ground. When such a branch breaks, the broken-off part continues life on its own: it is a new plant. Other plants spread by *rhizomes* (horizontal underground stems that root at intervals), or *stolons* (often called runners: above-ground branches that grow new plantlets at their tips, as in strawberries). An example of a high arctic plant with conspicuous stolons (they are bright red) is spiderplant, one of the saxifrages.

Another means of vegetative multiplication is by *bulblets* (also called *bulbils*). These are tiny buds that form in place of some (or all) of a plant's flowers. They easily become detached, and blow over the ground almost as readily as seeds. When a bulblet comes to rest at a suitable spot, it sprouts and grows roots, developing into a new plant. Its capacity for long-distance travel therefore gives it some of the advantages of a seed, with the added bonus that its production doesn't depend on successful pollination and seed ripening; however, a bulblet cannot survive years of dormancy, as a seed can. Three examples of plants that produce copious bulblets are nodding saxifrage, grained saxifrage, and viviparous knotweed.

Vegetative multiplication has a disadvantage, however, in spite of the fact that it enables a plant to multiply in years when seed production fails. The drawback is that all the new plants derived from one "parent" (plus the parent itself) form a clone, a group of genetically identical plants. If the process were to continue year after year, without the exchange of genes brought about by sexual reproduction, any species would gradually lose its genetic diversity and hence its chances of adapting to changing environmental conditions.

Figure 5.5. An expanse of cottongrass tussocks. The species is *Eriophorum vaginatum*.

5.11 Some Plant Communities

Summer temperature is the overriding factor determining what plants can grow in any given region in the Arctic. That is why the vegetation is zoned in the way already described, with the numbers of species of flowering plants falling off steadily as the cold increases (sec. 5.5).

Other environmental factors—soil moisture and soil fertility being the most important—control the vegetation within each temperature zone. The contrast between moist sites and dry sites makes for the most conspicuous differences. A much wider variety of plant species, and most of the colorful flowers, are to be found on moderately moist ground, while sedge meadows—smooth tracts of uniform bright green—occupy many of the wet areas. Slightly drier areas are sometimes covered by seemingly endless expanses of cottongrass tussocks (fig. 5.5).

The wetness of the soil is controlled by the microtopography of the land much more than by annual precipitation. Rain and spring meltwater are prevented from soaking into the ground by the impermeable permafrost not far below the surface. Water on or in the active layer doesn't disappear, because evaporation, and the absorption of water by plants, are slowed by the low temperatures. Therefore, water simply collects in every little hollow, however small, and on every tract of low ground, where it remains in the form of wet marshes dotted by innumerable tundra ponds.

The soil's wet-and-dry pattern is revealed by a matching pattern in the plant communities growing on it. Variations in soil fertility also have visible effects on plants, as does the way the soil is disturbed, by freezing and thawing and by the activities of animals (including humans). A number of distinctive plant communities and community patterns are easy to recognize and worth watching for. Some of the more noteworthy ones follow (all plants mentioned by name are described individually in secs. 5.13, 5.14, and 5.15).

Communities Controlled by Microtopography. As described in chapter 3, the annual alternation of freezing and thawing of the soil often produces remarkably regular ground patterns. These microtopographic patterns range in size from arrays of big tundra polygons (each up to 50 m or more across) to fields of small (less than 1 m) hummocks. As already explained, plants respond to these patterns, besides contributing to their development.

Plants that can endure drought and exposure to the wind grow on hummock tops and on the ridges surrounding low-center polygons; such plants are also the final colonists on high-center polygons when the ground has dried out. The commonest flowering plant in these sites is arctic dryad, often accompanied by drought-adapted lichens such as mane lichen. The comparatively damp troughs between polygons and hummocks are habitats for a great variety of moisture-loving plants, for instance, viviparous knotweed and sulphur buttercup; these troughs also give arctic white heather the shelter it needs from strong winds. The shallow pools and waterlogged areas occupying the centers of low-center polygons are usually dominated by sedges, sometimes mixed with cottongrass. The varied vegetation highlights the topographic structure that underlies it.

"Fertilized" Minicommunities. These are very typical of the Arctic. For the most part, arctic soil is poor in nutrients, but here and there a patch of soil is lavishly fertilized. It may be the last resting place of an animal carcass, perhaps of a caribou or a muskox. Or it may be manured by bird droppings; droppings accumulate in quantity at the foot of nesting cliffs, and also below the perches—usually big rocks—used by carnivorous birds to survey their hunting territory. In flat tundra, such high

Figure 5.6. A long-tailed jaeger at its lookout perch on an old bowhead whale skull on a raised beach. The unusually tall arctic poppies and nodding saxifrage growing around the skull are fertilized by droppings.

spots are few, and are well patronized. Figure 5.6 shows an example, sketched on Devon Island: the perch is the skull of a bowhead whale, stranded hundreds or thousands of years ago on a beach now left far inland as a raised beach (see sec. 3.3); the bird is a long-tailed jaeger; the luxuriant plants below the perching spot are chiefly arctic poppies and nodding saxifrage. All plants grow more luxuriantly at such sites than elsewhere. Some species typical of these sites are Pallas's wallflower, northern water carpet, and roseroot, and, below sea-birds' rookeries, the shore plant scurvy-grass.

Communities on a Thick Active Layer. In places where water can drain away quickly, such as river bars, sand dunes and eskers (see sec. 3.3), the permafrost table is deep underground and the active layer thick. The surface soil is warm, loose, and dry; as a plant habitat it contrasts strongly with cold, wet tundra, and is therefore home to some distinctive plants. Some of these places are also home to burrowing animals such as ground squirrels, who make conditions even better by contributing manure, and it may be hard to tell which of the merits of a site, ample manure or warmth, is more important from the plants' point of view. Some plants typical of these sites (whether manured or not) are river beauty, pasque flower, both species of liquorice-root, rock-jasmine, and Jacob's ladder.

Beach and Saltmarsh Communities. As in other latitudes, sea beaches and salt or brackish marshes provide environments intolerable to many plants but ideal for those that are especially adapted. Some common plants of arctic sea beaches are seashore starwort (on sheltered beaches), scurvy-grass, sea lungwort (on shingle or cobble beaches), and sea-purslane (on sandy beaches). Plants to look for near, rather than on, the seashore are seaside buttercup, Pallas's buttercup, mastodon flower, thrift, seashore chamomile (in the Low Arctic), and northern primrose (only in the west).

Snowpatch Communities. These are found on ground where winter snowdrifts are deep, so that the snow lasts long into spring; in cool years, some snow may remain all through summer. To survive in such places, plants must be fast growers, able to flower and set seed in a growing season unusually short even by arctic standards; if a series of cold years prevents the snow from melting, they have to survive several missed growing seasons, in the cold, damp, dark conditions under the snow. To compensate, the snow protects them from winter wind, and provides trickles of water all through the drought of summer. Snowpatch communities are most noticeable in the polar desert; often they are the only patches of ground where, because of the constant meltwater supply, carpets of mosses can grow. These provide a moist habitat for other plants, for example, mountain sorrel, snow buttercup, dwarf buttercup, arctic cinquefoil, sibbaldia, and finger lichen. In the Low Arctic, sites where snow lies late are usually occupied by the same community as surrounding land, but the plants' development is delayed. In high summer, such spots are worth exploring. Though they appear brown and lifeless from afar, you are likely to find flowers at their prime that are "over" everywhere else; they shine out among last year's flattened grass and beneath the gray, leafless twigs of birches and willows whose winter has scarcely ended.

Peat Communities. The way peat forms, and the way it functions as a habitat for plants, is described in section 3.6. To repeat, characteristic plants are cloudberry, Labrador tea, bog rosemary, and dwarf birch. The way in which peat and its ecology can be pigeonholed equally well with landforms or with plant ecology is an excellent illustration of the way the various topics of arctic natural history mesh with each other: none can be treated in isolation.

5.12 Where Have They Come From?

In the last ice age all but a small fraction of the Arctic was buried under enormous ice sheets and devoid of life (see fig. 3.5). Therefore, all plants living in the Barrenlands and most of the Arctic Islands today are the descendants of immigrants; their ancestors must have lived somewhere else 18,000 years ago. This prompts the question: where did their ancestors come from? While the ice age was at its height, these ancestors must have survived in ice-free regions where the environment was hospitable to life.

Several different *refugia,* as they are called, provided sanctuary for the plants of the North American Arctic. The biggest refugium, but also the most distant, was North America south of the ice sheets. An equally important refugium, because of its size and proximity, was *Beringia.* This is the name given to unglaciated Yukon and Alaska, plus northeastern Siberia and huge expanses of additional land forming a wide land bridge between Alaska and Asia; this land is now beneath the sea, but was not submerged during the ice age when sea level was about 100 m lower than it is at present. Even some small areas within the Arctic—islands and parts of islands—were not covered by ice and no doubt functioned as refugia for a few plants, but conditions for life must have been harsh in the extreme; the cold was intense, and it was simply lack of snow that kept the land (the westernmost Arctic Islands and easternmost Baffin Island) ice free. All the ice sheets were formed by the accumulation, year after year and century after century, of snow that failed to melt in the cool summers and turned to ice under its own enormous weight.

For many of the plant species you see in the Arctic today, the refugium they came from is known, if not with certainty, then with a high degree of probability. It adds to the enjoyment of arctic flowers to know something of their history, so here are short lists of some of the species from the three sources (every species listed here is illustrated in sec. 5.13 or in fig. 5.7).

From south of the ice sheets came least willow, viviparous knotweed, mountain sorrel, moss-campion, arctic dryad, yellow mountain saxifrage, and alpine blueberry. The reason for supposing that these species migrated north from a refugium in the south, as the ice sheets melted, is that they are known to have lived at middle latitudes south of the ice when the ice age was at its height. Their fossil remains (leaves, fruits, or pollen) have been found buried in sediments at the bottom of mid-latitude lakes and have been dated by radioactive carbon dating. It is

Figure 5.7. The map shows: Stippled: areas that were not covered by ice sheets in the last ice age. Outlined with a dashed line: the area containing all known occurrences of the three plant endemics pictured below. Of the three, only Vahl's cinquefoil grows throughout the area, including Greenland; the other two species are confined to arctic North America. (a) Arctic parrya. (b) Silvery oxytrope. (c) Vahl's cinquefoil.

likely (but of course not certain) that these fossils were among the ancestors of modern arctic plants.

For Beringia, two lists of plants can be compiled, one for species that survived in Beringia and have remained there without spreading into the central and eastern Arctic when the land became free of ice, and another list for species that have spread. Into the first list could be put nearly all the plants recorded in the plant field guide (sec. 5.13) as having range [A], that is, the Alaskan Arctic slope and Arctic Canada as far as the Mackenzie Delta, for example, mountain heliotrope, lagotis, arctic forget-me-not, Siberian phlox, northern shooting star, Alaska spiraea, and Alaska boykinia. All but the last of these are known as *Amphi-Beringian* plants, for they are found on both sides of Bering Strait. The last, Alaska boykinia, is *endemic* to the Alaska-Yukon region, i.e., it is found nowhere else in the world. The other, much shorter, "Beringia" list is of Amphi-Beringian species whose ranges do extend eastward. Judging from their present distributions, it seems reasonable to assume that Beringia was the ice age refugium of saussurea, rock-jasmine, arctic dock, northern Jacob's ladder, and arctic groundsel, and that they migrated after the ice melted.

It is remarkable how few Beringian refugees have expanded their ranges eastward. The Mackenzie Valley evidently acts as an ecological barrier to the eastward spread of Beringian plants. Crossing is not impossible, of course, merely improbable. It is a matter of chance: for any given plant species, the chance of crossing the barrier is small. Consequently, successful crossings of the barrier do happen, but only very occasionally, at intervals of (perhaps) many centuries, and there has been time for only a few crossings since the ice melted.

Last come species that, to judge from their present ranges, may have spread from refugia in the unglaciated Arctic Islands. They are not numerous. Three worth watching for are silvery oxytrope, Vahl's cinquefoil, and arctic parrya. They are endemics, found nowhere outside the ranges shown on the map in figure 5.7.

For more on plant migrations following the ice age, see *After the Ice Age,* by E. C. Pielou (Chicago: University of Chicago Press, 1991). For a detailed technical account, see *Postglacial Vegetation of Canada,* by J. C. Ritchie (Cambridge: Cambridge University Press, 1987).

5.13 A Field Guide to Arctic Flowering Plants

PRELIMINARY NOTES

The following pages describe the majority of conspicuous arctic flowering plants. Rarities unlikely to be found and hard-to-identify plants

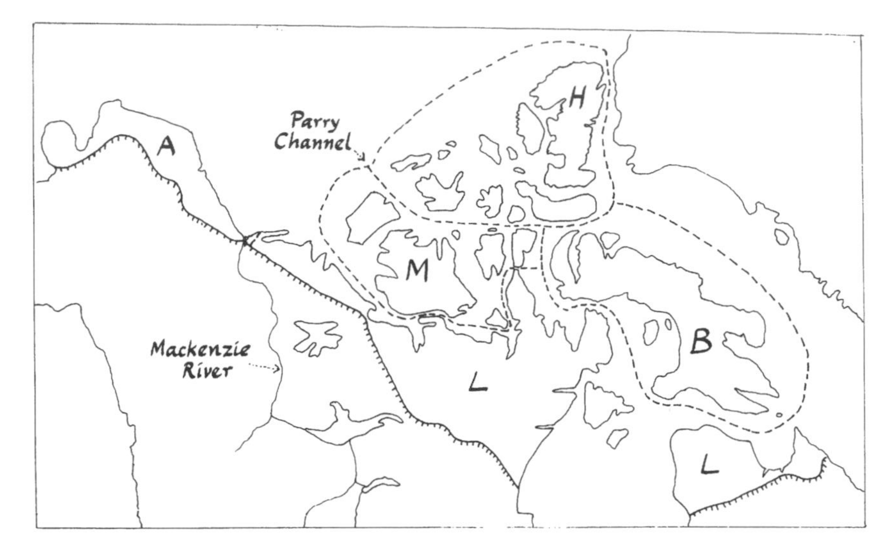

Figure 5.8. To show the boundaries (dashed lines) of the regions used for specifying a species' geographic range. (Note: these are not ecological boundaries.) The hatched line is the treeline, the southern boundary of regions A and L. The regions are: A: land west of the Mackenzie Delta. H: the High Arctic islands (the Queen Elizabeth Islands) north of Parry Channel. M: the Mid Arctic, i.e. the islands (excluding Baffin Island) south of Parry Channel. L: the Low Arctic, i.e., all the mainland Arctic east of region A. B: Baffin Island. See text for further details.

with inconspicuous flowers (this includes most grasses, sedges, and rushes) are excluded. Ferns and their relatives are described in section 5.14, and mosses and lichens in section 5.15.

Besides its English name, every plant species has a two-word scientific (Latin) name; the first word gives the plant's *genus* (plural *genera*), the second its *species* (plural also *species*) within the genus. The genus is the more inclusive class (several species can belong to one genus). When several species of the same genus are mentioned together, the generic name is given in full at the first mention only; after that, its initial is used. Genera themselves are collected into *families*. Plants are grouped by families in the descriptions below. To identify a specimen, consult the charts after first reading the instructions.

For each species described, its English name is given first, followed by its geographical range in square brackets, and then its scientific name in italics. The geographical range is given in the letter code shown on the map in figure 5.8 and described in the legend. A capital letter means that the species is common and/or widespread in the region concerned, a lowercase letter that the species is uncommon or only locally common. Keep in mind that there is always a chance you may find a species outside its range as known at present.

Note that the ease with which you can identify a plant varies greatly from family to family. Thus all arctic species of the Lily Family (there are only 5) are easily distinguished from one another. In the Pea Family, some species are instantly recognizable, but others belong to big genera containing numerous rather similar species that are hard to distinguish. And in the Mustard Family, even the genera are sometimes difficult to recognize, and the best an amateur can do is to name the family correctly. Because of this variability, different families are given different treatment in the descriptions that follow. For a nonspecialist, it makes more sense to identify a particular plant's genus (or even family) correctly and let it go at that, than to make an uncertain, and possibly mistaken, identification of its species.

For "difficult" groups, identification of all the separate species is impracticable except for specialists, who may need to use a high-powered magnifier and may need specimens in fruit as well as in flower. They would also need to consult an exhaustive technical Flora of the Arctic, a book giving exact scientific descriptions of all the species. References are given below. For those who want to defer consulting a Flora (often a bulky book!) until they get home from a trip, notes are given, where appropriate, of the characters of a plant to observe with particular care; for example, to identify many members of the Mustard Family, you must know the shape of the seed pod and what sort of hairs (single or forked) grow where, on stems and leaves.

Plant identification is made much easier if a checklist of local plants exists for the area you visit. Naturalists' lodges often have checklists; remember that they are inevitably incomplete, and you may be able to add new species. But don't assume that a plant that is not easily identifiable is a "discovery." Plants of a single species can vary considerably, in size, showiness, and general robustness, depending on the environment they grow in. Unexpectedly large specimens of a species may be found in moist, or sheltered, or exceptionally mineral-rich sites; conversely, specimens in harsh environments may be few-flowered and dwarfed.

References

TECHNICAL FLORAS

Vascular Plants of Continental Northwest Territories, Canada, by A. E. Porsild and W. J. Cody (Ottawa: National Museums of Canada, 1980).

Illustrated Flora of the Canadian Arctic Archipelago, by A. E. Porsild (Ottawa: National Museums of Canada, 1957).
Flora of Alaska and Neighboring Territories, by Eric Hultén (Stanford, CA: Stanford University Press, 1968).

POPULAR FLOWER GUIDES

Eastern North America as Seen by a Botanist, vol. 1: *The Arctic region,* by In-Cho Chung. Available from the author at 1308 Laurel Drive, Daytona Beach, Florida 32017.
Barrenland Beauties: Showy Plants of the Arctic Coast, by Page Burt (Yellowknife, NWT: Outcrop Publishers, 1991).
Wildflowers of the Yukon, Alaska, and Northwestern Canada, by John G. Trelawney (Victoria, BC: Sono Nis Press, 1988).
The Alaska-Yukon Wild Flowers Guide, by Helen A. White and Maxcine Williams, eds. (Anchorage: Alaska Northwest Publishing Co., 1974).
The Wildflowers of Churchill, by Karen L. Johnson (Winnipeg: Manitoba Museum of Man and Nature, 1987).

The Identification Charts

INSTRUCTIONS

Charts 1 through 4 are for use with plants in flower and allow identification of all plants described in the field guide; chart 5 is for identifying plants with conspicuous berries, which may attract your attention after their flowers are over.

Plants in flower are divided into four separate groups, each with a chart of its own, so that once the correct chart has been chosen any specimen can be tracked down without turning a page.

Plants whose flowers are inconspicuous or green	Chart 1
Plants whose small flowers are crowded into dense heads, spikes, or flat sprays	Chart 2
Plants whose flowers are widely spaced and/or large and in which the petals are separate	Chart 3
Plants whose flowers are widely spaced and/or large and in which the petals are joined	Chart 4

Having picked the appropriate chart, start at the top and follow the arrows downward, choosing between the alternatives at each step. You will finally reach a "destination," which is usually the name (in capital letters) of your plant's family. Sometimes the chart identifies your specimen more precisely, naming (in lowercase) its genus or species. The more precise names have been given for plants that differ in some im-

Chart 1

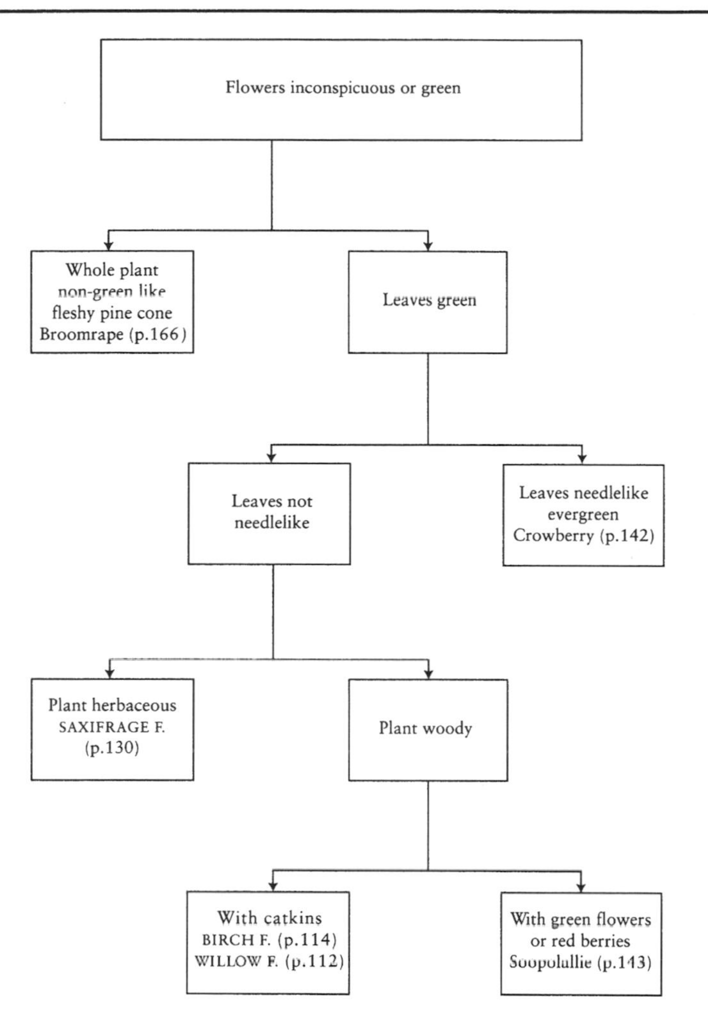
Flowers inconspicuous or green
Whole plant non-green like fleshy pine cone Broomrape (p.166)
Leaves green
Leaves not needlelike
Leaves needlelike evergreen Crowberry (p.142)
Plant herbaceous SAXIFRAGE F. (p.130)
Plant woody
With catkins BIRCH F. (p.114) WILLOW F. (p.112)
With green flowers or red berries Soopolullie (p.143)

Chart 2

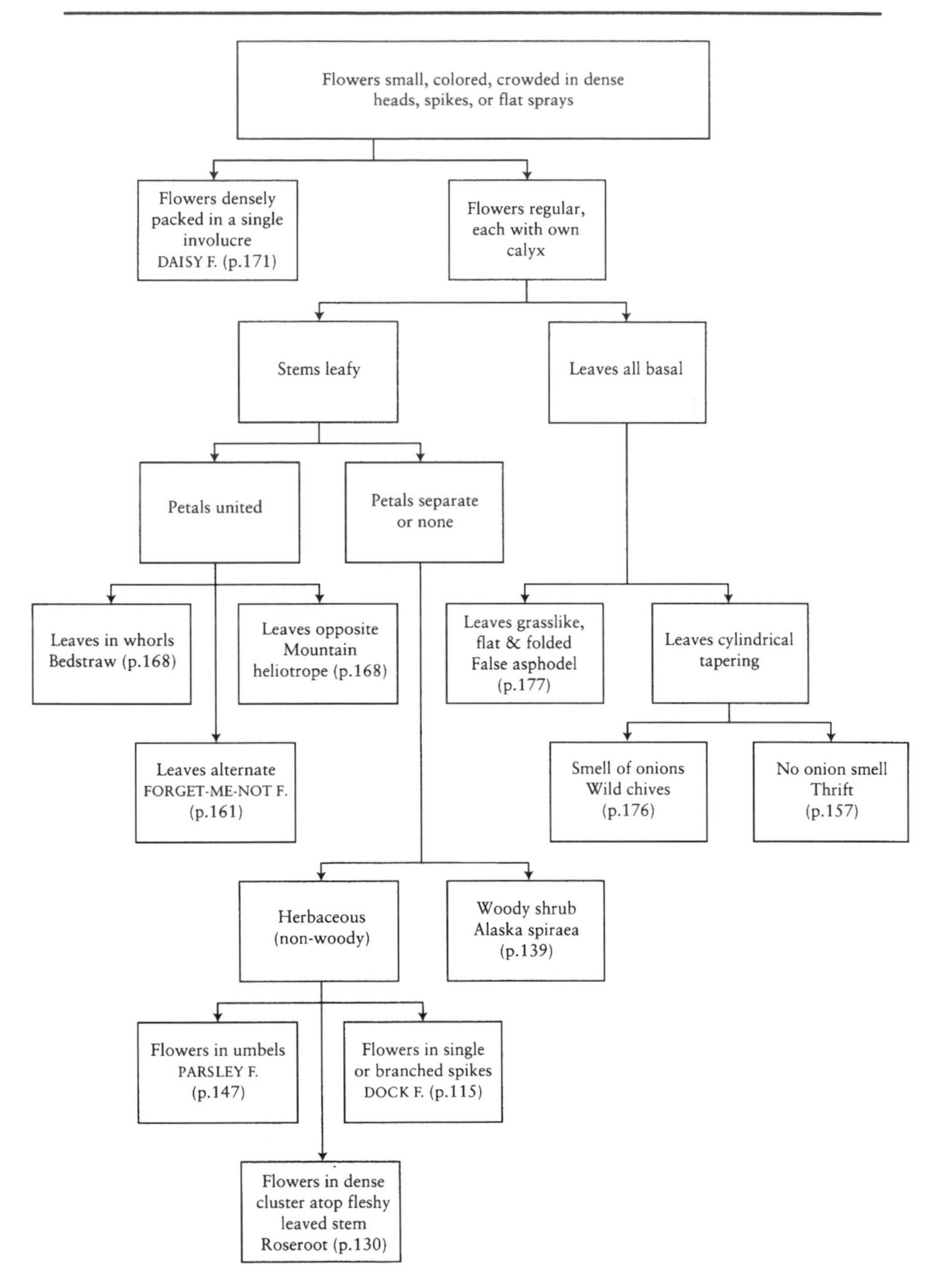

Flowers small, colored, crowded in dense heads, spikes, or flat sprays
Flowers densely packed in a single involucre DAISY F. (p.171)
Flowers regular, each with own calyx
Stems leafy
Leaves all basal
Petals united
Petals separate or none
Leaves in whorls Bedstraw (p.168)
Leaves opposite Mountain heliotrope (p.168)
Leaves grasslike, flat & folded False asphodel (p.177)
Leaves cylindrical tapering
Leaves alternate FORGET-ME-NOT F. (p.161)
Smell of onions Wild chives (p.176)
No onion smell Thrift (p.157)
Herbaceous (non-woody)
Woody shrub Alaska spiraea (p.139)
Flowers in umbels PARSLEY F. (p.147)
Flowers in single or branched spikes DOCK F. (p.115)
Flowers in dense cluster atop fleshy leaved stem Roseroot (p.130)

Chart 3

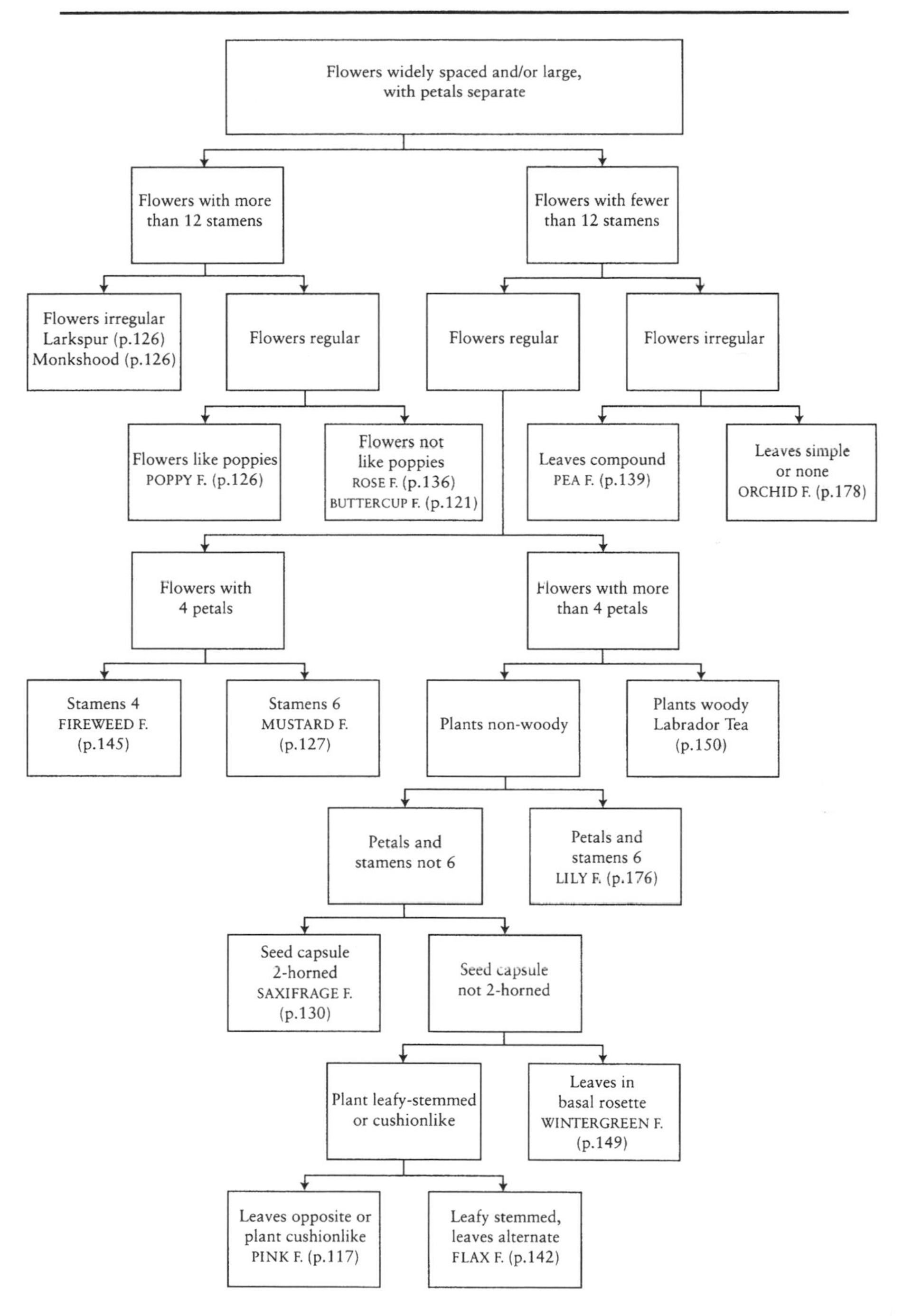

Flowers widely spaced and/or large, with petals separate
Flowers with more than 12 stamens
Flowers with fewer than 12 stamens
Flowers irregular
Larkspur (p.126)
Monkshood (p.126)
Flowers regular
Flowers regular
Flowers irregular
Flowers like poppies
POPPY F. (p.126)
Flowers not like poppies
ROSE F. (p.136)
BUTTERCUP F. (p.121)
Leaves compound
PEA F. (p.139)
Leaves simple or none
ORCHID F. (p.178)
Flowers with 4 petals
Flowers with more than 4 petals
Stamens 4
FIREWEED F. (p.145)
Stamens 6
MUSTARD F. (p.127)
Plants non-woody
Plants woody
Labrador Tea (p.150)
Petals and stamens not 6
Petals and stamens 6
LILY F. (p.176)
Seed capsule 2-horned
SAXIFRAGE F. (p.130)
Seed capsule not 2-horned
Plant leafy-stemmed or cushionlike
Leaves in basal rosette
WINTERGREEN F. (p.149)
Leaves opposite or plant cushionlike
PINK F. (p.117)
Leafy stemmed, leaves alternate
FLAX F. (p.142)

Chart 4

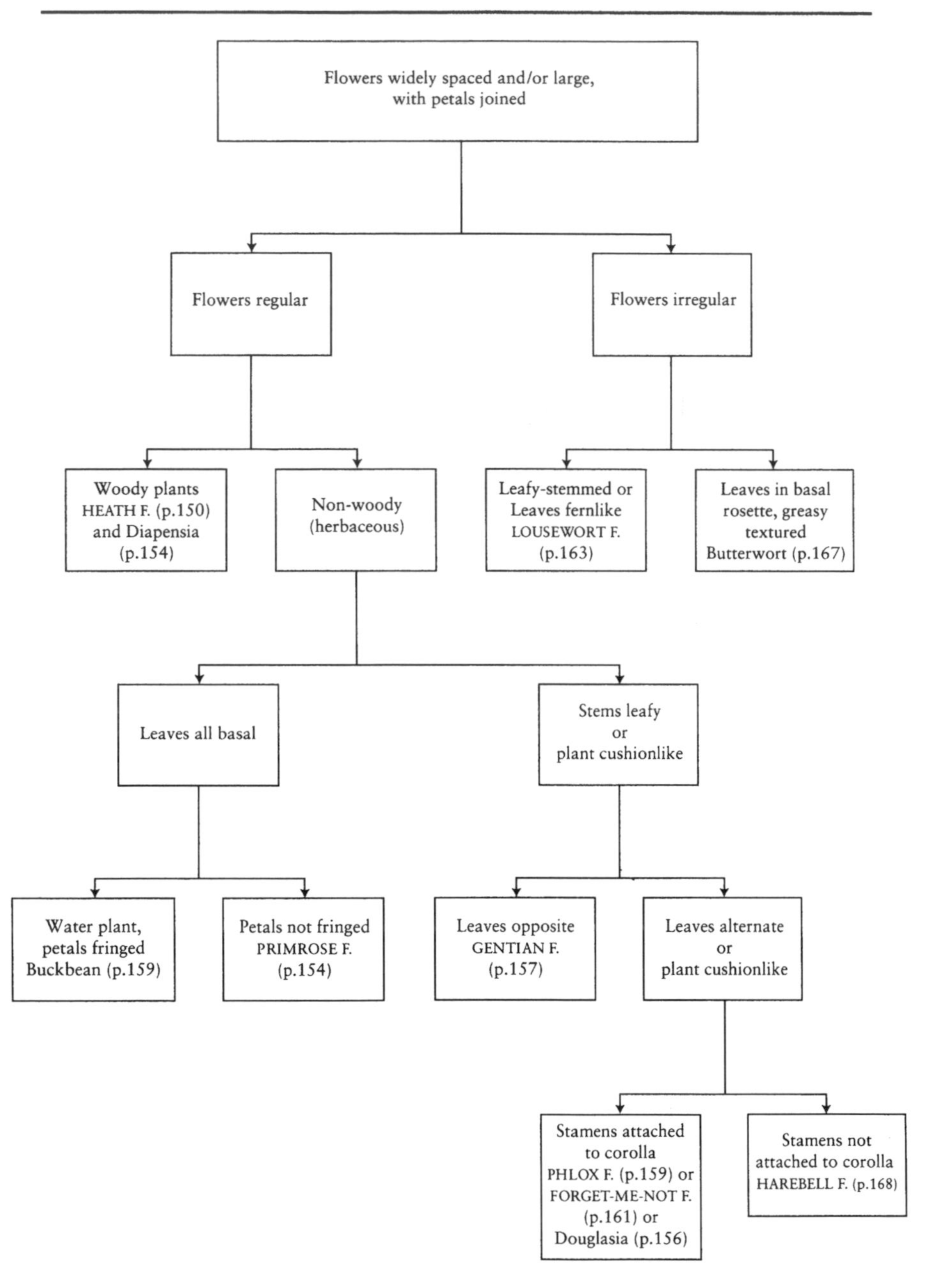
Flowers widely spaced and/or large, with petals joined
Flowers regular
Flowers irregular
Woody plants HEATH F. (p.150) and Diapensia (p.154)
Non-woody (herbaceous)
Leafy-stemmed or Leaves fernlike LOUSEWORT F. (p.163)
Leaves in basal rosette, greasy textured Butterwort (p.167)
Leaves all basal
Stems leafy or plant cushionlike
Water plant, petals fringed Buckbean (p.159)
Petals not fringed PRIMROSE F. (p.154)
Leaves opposite GENTIAN F. (p.157)
Leaves alternate or plant cushionlike
Stamens attached to corolla PHLOX F. (p.159) or FORGET-ME-NOT F. (p.161) or Douglasia (p.156)
Stamens not attached to corolla HAREBELL F. (p.168)

Chart 5

To identify plants with soft fruits

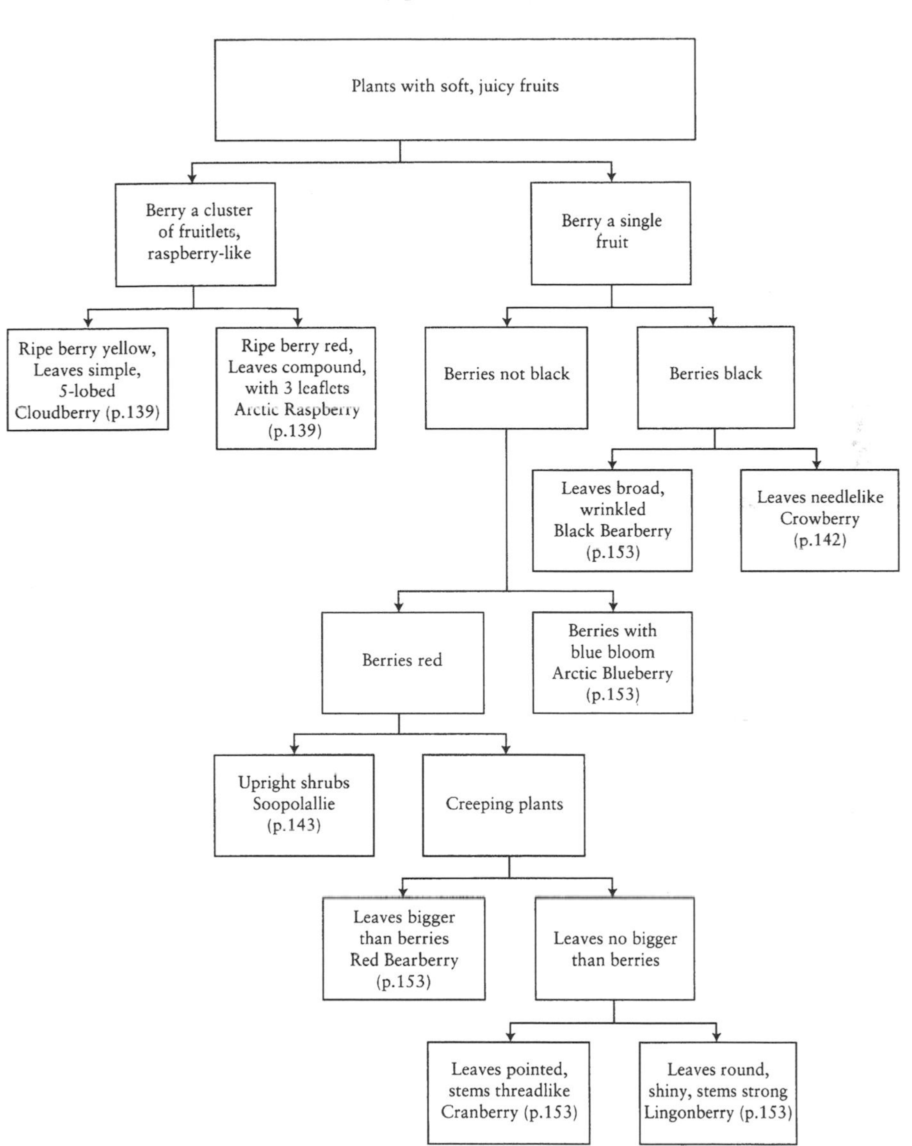

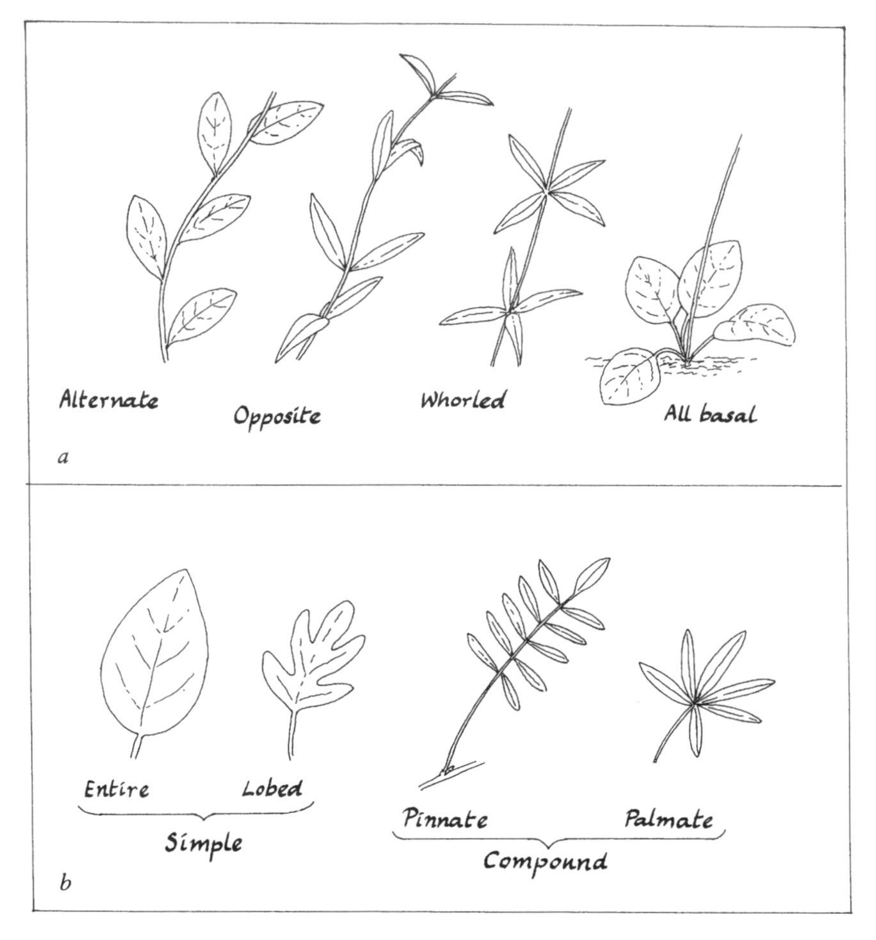

Figure 5.9. (a) Different arrangements of leaves; (b) Different shapes and structures of single leaves.

portant respect from other members of their family, or for plants that are the only arctic representatives of their family. And occasionally, when the differences between two families are too technical to state briefly, the chart leaves you to choose between them by referring to the descriptions and illustrations given later. The page number to turn to is given in parentheses after each name. When consulting the figures, note that the individual plants are labeled in the order in which they are mentioned in the text, but they are arranged on the page to facilitate comparisons, not necessarily in alphabetical order.

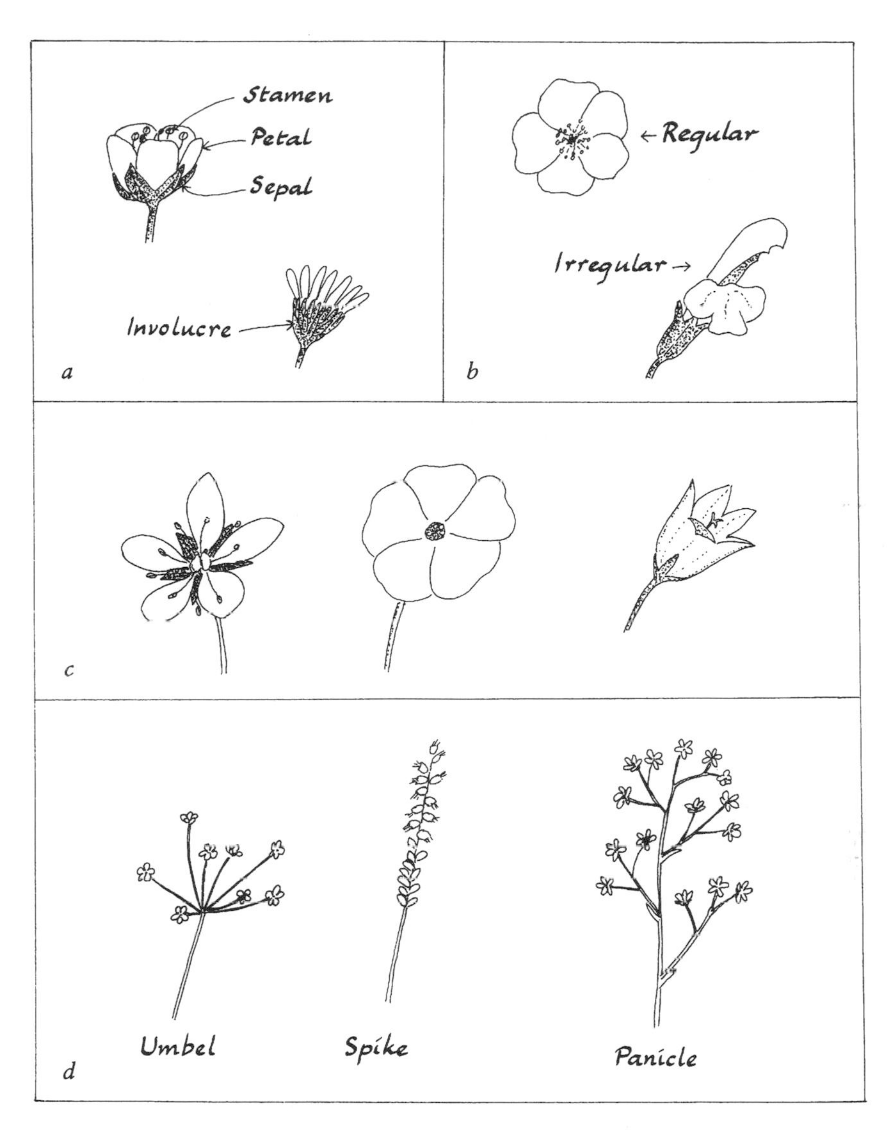

Figure 5.10. Different types of flowers and inflorescences. (a) The stamens, petals, and sepals of a simple flower, and the involucre of a compound flower of the Daisy Family. (b) A regular and an irregular flower. (c) Three types of corolla: in the flower on the left, the petals are obviously separate, and in the flower on the right they are obviously joined. The corolla of the central flower has widely spread lobes that look, at a glance, like separate petals; inspection shows that the lobes are joined at the base. (d) Three types of inflorescence: an umbel (with branches radiating from a single point), a spike, and a loose panicle.

Points to notice in using the charts, and definitions of the few unavoidable technical terms, follow:

1. "Plants" include woody shrubs as well as herbaceous plants.
2. An "inconspicuous" flower (chart 1) is one less eye-catching than other parts of the plant or less likely to be noticed than the plant's fruits. "Small" flowers (chart 2) and "large" flowers (charts 3 and 4) mean flowers less than or more than 5 mm across, respectively.
3. The adjective "colored" includes white.
4. Note the arrangement and structure of the leaves (fig. 5.9).
5. Concerning flower structure: Figure 5.10 shows the parts of a flower, and the details to observe. Note whether the flowers are regular or irregular. A regular flower looks the same from all sides; in an irregular one, the upper and lower halves are clearly different. Next consider the petals: a flower's whole set of petals is called the *corolla,* and the whole set of sepals the *calyx*. The petals may be separate or joined. The fact that they are joined is often obvious (when the corolla forms a tube or a bell, for instance) but occasionally it is not obvious (when the petals spread widely and are joined only at the very base). Judge whether the latter is true by observing a partially withered flower: if the petals are joined, the corolla will fall away as a unit. Note the number of stamens and where they are attached. And if numerous flowers are crowded together into a mass (an *inflorescence*), note its shape.

The Willow Family (Salicaceae)

The Willow Family is made up of trees and shrubs belonging to two genera: the poplars (*Populus*) and the willows (*Salix*). In both genera the flowers are in catkins: a catkin is a short column of several dozen tiny flowers, without petals, densely crowded together. The flowers are unisexual—either female, with seeds, or male, with pollen-producing stamens. All the flowers in one catkin are of the same sex, and likewise all the catkins on one plant (tree or shrub) are of the same sex, so each plant is either male or female. In several species the male catkins are furry enough to earn the name pussy willows.

Two poplars—**Balsam Poplar** [Al], *Populus balsamifera* (fig. 5.2), and **Trembling Aspen** [al], *P. tremuloides* (not illus.)—grow at the southern limits of the arctic region, in treeline woodlands. They are described in section 5.2.

At least twenty different species of willows live in the Arctic; they are identifiable (sometimes) only by experts. They range from tall, upright shrubs forming dense thickets in the wet tundra of the Low Arctic, to dwarf, prostrate sub-shrubs creeping over dry, stony ground in the High Arctic polar desert. Three can be identified easily.

Net-veined Willow [AmLB], *Salix reticulata* (fig. 5.11), is a ground-creeper with easily recognizable leaves; the veins lie in a network of deep grooves. No other arctic willow has leaves like it, but the plant could be confused with bearberry (in the Heath Family) whose leaves are similarly wrinkled. The difference is that in net-veined willow leaves, the leaf blade and leaf stalk are clearly distinct; in bearberry leaves the blade lacks a clear-cut base and merely tapers down into the stalk (see fig. 5.36).

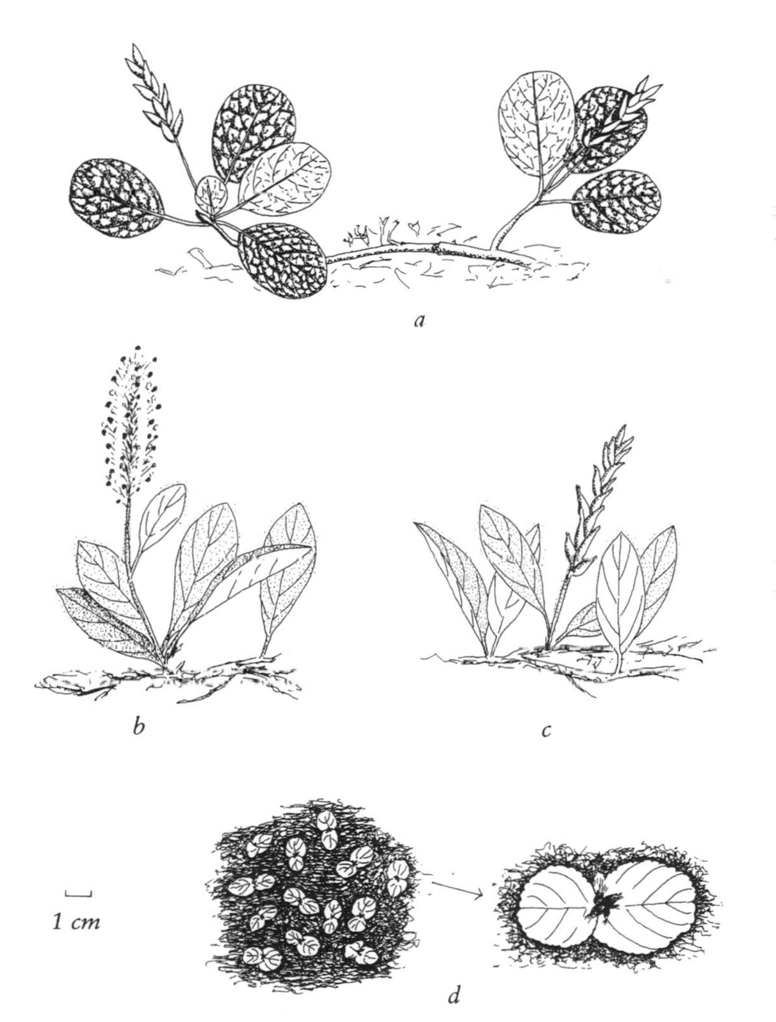

Figure 5.11. (a) Net-veined willow (female). (b) Arctic willow (male). (c) Arctic willow (female). (d) Least willow (female).

Arctic Willow [AḤMLB], *S. arctica* (fig. 5.11), is instantly recognizable in the High Arctic (north of Parry Channel) as the only willow to be found. It is also by far the commonest willow in the Arctic Islands south of Parry Channel. On the mainland it could be confused with other creeping willows. Although it lacks colorful petals, its catkins (both sexes) are attractive to pollinating insects (including bumblebees) because they secrete plenty of nectar. As mentioned in section 5.8, the catkins warm up in the sun, possibly because their "fur" of clear transparent hairs functions like greenhouse windows. In sunshine, female catkins warm to 4° or 5° C above ambient temperature; male catkins, in spite of being furrier, are about 1° cooler. Studies in Devon Island show that male and female plants do best in different habitats. In moist, fertile, sheltered sites females greatly outnumber males; in dry, exposed habitats, males outnumber females, though only slightly. It is conjectured that this *may* result from adaptations that enable female plants to crowd out the males competing with them for space at the better sites; the females need good soil, as they have to nourish ripening seeds.

Least Willow [LB], *S. herbacea* (fig. 5.11), is found growing on beds of damp moss in the eastern Arctic. From trailing stems, buried in the moss, grow short side branches each tipped with a pair of tiny leaves; the leaves are oval, often with a small notch at the tip, and with shallow, rounded teeth on the margin. The leaves of each pair spread out opposite each other and are unequal in size, with the larger seldom longer than 1 cm. Scattered pairs of small, mismatched leaves lying on a bed of moss are certainly this species. In season, a tiny catkin (male or female) can be seen between the leaves; the catkins are miniature, few-flowered versions of those of larger willows.

The Birch Family (Betulaceae)

Two shrubs of the Low Arctic shrub tundra—dwarf birch and green alder—belong to this family. Like willows and poplars, they have separate male and female catkins. But unlike willows and poplars, catkins of both sexes grow on the same plant. Male catkins fall soon after they have shed their pollen, however, after which only female catkins are to be found; the female catkins are short and plump, looking rather like miniature spruce or pine cones.

Dwarf Birch [AmLb], *Betula glandulosa* (fig. 5.12), has small, leathery, round to slightly egg-shaped leaves, with rounded teeth on the margins. The twigs are rough with "warts." The height of the shrub depends on the environment: in moist, sheltered sites it can grow shoulder-high but in dry, exposed sites it may be no more than ankle-high. The female catkins are soft-textured; they wither and fall after the seeds are shed.

Green alder [Al], *Alnus crispa* (fig. 5.12), is a larger shrub than dwarf birch and has larger, thinner leaves, with sharp teeth on the margin. After ripening, the female catkins become hard and woody; they remain on the shrub through winter.

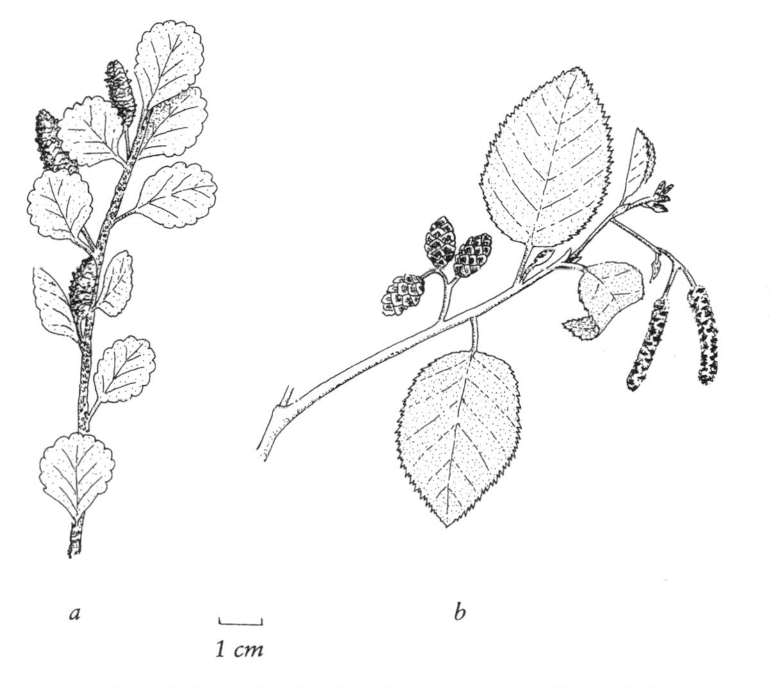

Figure 5.12. (a) Dwarf birch (with female catkins). (b) Green alder (with female and male catkins).

The Dock Family (Polygonaceae)

In temperate latitudes, the best-known members of this family are the docks—large, coarse weeds—but there are many other, more attractive, plants in the family. Four species grow in the Arctic and are colorful even though the flowers are small and devoid of true petals. Their absence is obvious in the first two species described below (arctic dock and mountain sorrel); the flowers have only sepals (calyx parts), which, though colored, lack the texture of ordinary petals. In the second two species (bistort and viviparous knotweed), the sepals look like petals even though, in the technical, botanical sense, they are not.

Figure 5.13. (a) Arctic dock. (b) Mountain sorrel. (c) Viviparous knotweed. (d) Bistort.

Arctic Dock [Aml], *Rumex arcticus* (fig. 5.13), is a western plant, seldom found east of 100° W. It resembles the familiar weedy docks in form, but the thick spikes of massed flowers are bright red. The color remains when the flowers go to seed; the seed (strictly speaking, an *achene,* a seed with a thin outer covering) is plump and has three sharp edges. The leaves are long and narrow.

Mountain Sorrel [AHMLB], *Oxyria digyna* (fig. 5.13), is found throughout the Arctic, to the extreme northernmost limit of land and in a wide range of habitats. It can grow as tall as 30 cm in moist, sheltered sites. But on exposed rocks, it is dwarfed; in such inhospitable places, the plants, though evidently thriving, may grow no taller than 2 cm, with correspondingly tiny leaves. The flower spikes resemble those of docks in general appearance and are dark red to reddish green. Close inspection shows that the seeds (achenes) are flat and thin and are surrounded by a wide, translucent wing. The kidney-shaped leaves are the most characteristic part of a mountain sorrel plant; they are dark green when young, becoming bright red, or even plum-colored, with age; when fresh, they are delicious to eat. Their tart, acid flavor is like that of sheep sorrel, a common garden weed familiar to naturalists and gardeners in temperate latitudes, and they are a rich source of vitamin C.

Viviparous Knotweed [AHMLB], *Polygonum viviparum* (fig. 5.13), is common throughout the Arctic. Its single, tall, slender spike has little white flowers with protruding stamens at the top and green bulblets, in place of flowers, lower down. These bulblets become detached and scattered by the wind, and those that come to rest in a suitable environment grow into new plants (see sec. 5.10). The bulblets are also food for lemmings and ptarmigans. The plant is sometimes called alpine bistort or, confusingly, bistort, because of its close relationship to true bistort.

Bistort [A], *Polygonum bistorta* (fig. 5.13), grows only in the western Arctic. Its flower spike is much thicker than that of viviparous knotweed and is a solid mass of flowers from top to bottom, with no bulblets. The flowers are bright pink, making the whole plant look like an upright, pink bottle brush.

The Pink Family (Caryophyllaceae)

Members of the Pink Family belong to two distinct groups, the chickweeds and the campions. The chickweed group is sometimes treated as a separate family, the Chickweed Family (*Alsinaceae*), which we will consider first.

CHICKWEEDS

Five genera of this group are widespread in the Arctic; all five grow as low cushion plants, some straggly and some compact; all have little white starlike flowers; and all have simple leaves (i.e., without teeth or lobes) growing in pairs opposite each other along the stems. Any plant having these 3 characteristics belongs to the chickweed group.

The first 4 genera are the starworts, sometimes called mouse-ear chickweed (*Stellaria*), the true chickweeds (*Cerastium*), the sandworts (*Arenaria*), and the pearlworts (*Sagina*). They are easily distinguished by examining the flowers (fig. 5.14, upper panel).

Look at the petals first. In starworts and chickweeds (a and b in the figure) the 5 petals are deeply cleft into 2 parts, so deeply that the flowers of some species seem at first glance to have 10 petals. In sandworts and pearlworts (c and d) the ends of the petals are convex or shallowly notched, but never deeply cleft; most species have 5 petals, but one of the pearlworts has only 4.

Next look at the styles (the slender stalks growing up from a flower's center and tipped with the sticky stigmas to which pollen adheres). The starworts and sandworts (a and c) all have flowers with 3 styles. The chickweeds and pearlworts (b and d) all have more than 3; they have as many styles as petals and thus the number of styles is usually 5 but occasionally 4.

Distinguishing the several species within each genus can be troublesome. Some distinctive species of chickweeds and starworts are worth noting.

Alpine Chickweed [HmlB], *Cerastium alpinum* (not illus.), is clammy to the touch, being covered with moist hairs.

Polar Chickweed [aHMb], *C. regelii* (not illus.), has leaves green only at the shoot-tips; the rest of the plant is covered with dead leaves, bleached white. Flowers are seldom found.

Long-stalked Starwort [ahmlb], *Stellaria longipes* (not illus.), has smooth, glaucous (blue-green) leaves forming big cushions, covered with starry flowers.

Seashore Starwort [AHMLB], *S. humifusa* (not illus.), is a small starwort found only on seashores, where it is common.

The fifth genus of the chickweed group has only one species.

Sea-purslane [AMLB], *Honckenya peploides* (fig. 5.14), is an unmistakable plant confined to sandy seashores, where it grows as thick, spreading cushions or carpets of fleshy, apple-green leaves. Its flowers are like those of the sandworts, but are inconspicuous among the mass of greenery because their narrow petals are greenish white.

CAMPIONS

Two genera of campions occur in the Arctic. The first, *Silene,* has only one truly arctic species, moss-campion. The second, *Melandrium,* is represented by several species, collectively called bladder-campions, of which two are widespread and common.

Moss-campion [AHMLB], *Silene acaulis* (fig. 5.15), consists of cushions of tiny, densely packed, dark green leaves, covered at flowering time by brilliant pink flowers which lie flat on the surface of the cushion.

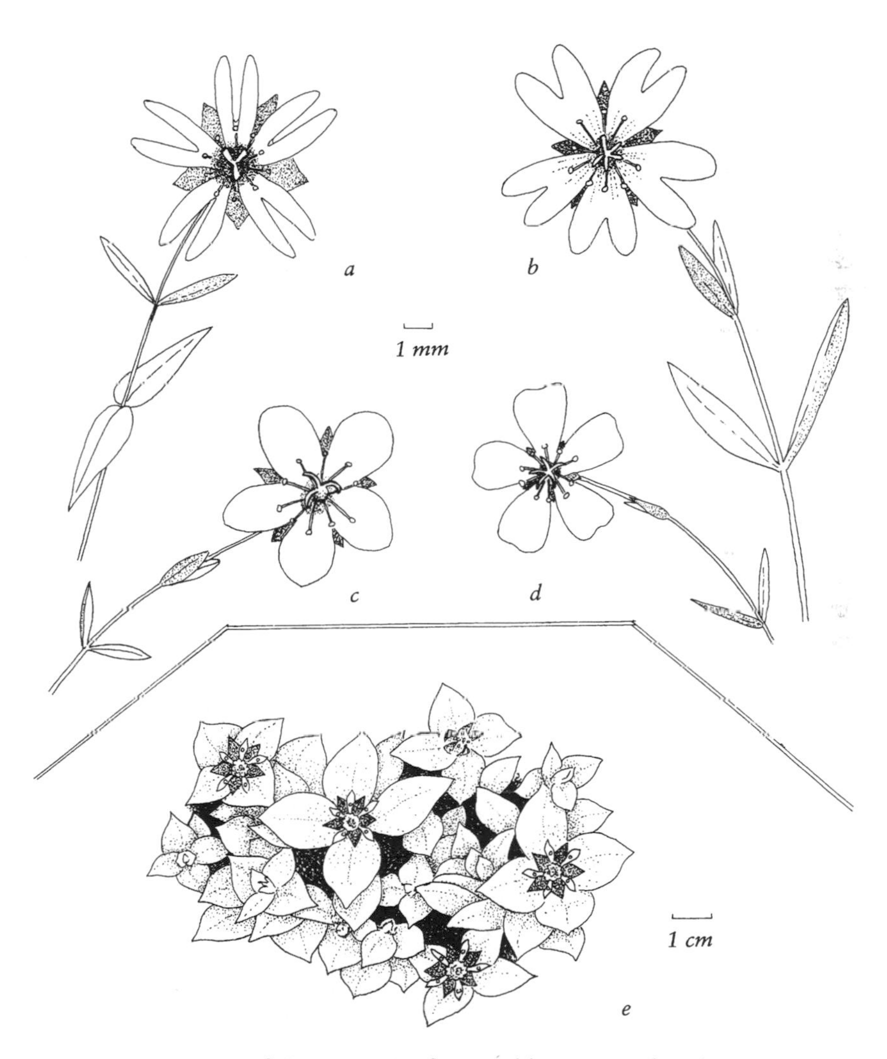

Figure 5.14. *Upper panel:* Representative flowers of four genera; their distinguishing characteristics are described in the text. (a) Starwort. (b) Chickweed. (c) Sandwort. (d) Pearlwort. *Lower panel:* (e) Sea-purslane.

Early in the season a moss-campion cushion bears a few well-spaced flowers; later, when it is in full bloom, the flowers may conceal the leaves completely, so that the plant appears as a pink hemisphere. It is unmistakable.

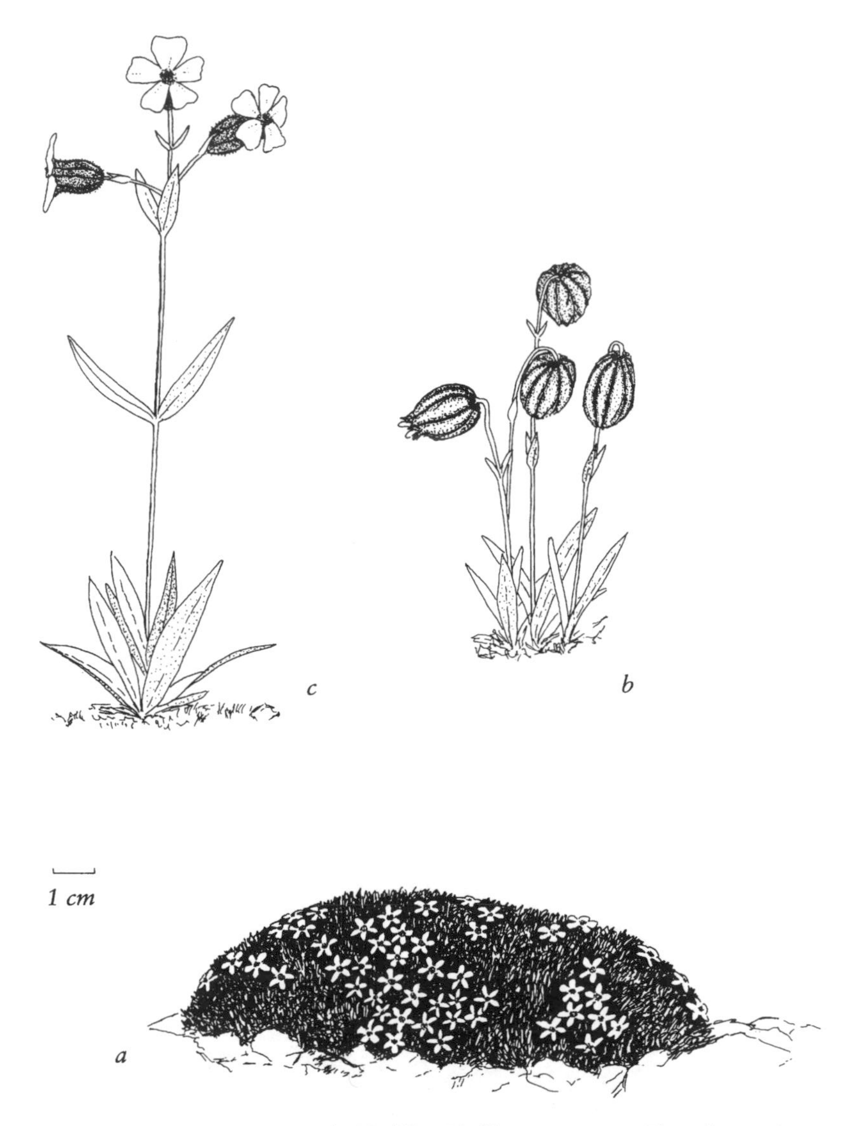

Figure 5.15. (a) Moss-campion. (b) Nodding Bladder-campion. (c) Three-flowered Bladder-campion.

Bladder-campions are so called because of the distinctive striped "bladder," which is the calyx of the flower and is often inflated.

The species with the most conspicuous bladders is **Nodding Bladder-campion** [AHMLB], *Melandrium apetalum* (fig. 5.15); the purplish bladder, which hangs down ("nods") like a tiny Japanese lantern, appears to form the whole flower, because the little purple petals are tucked inside it and are almost too short to be seen. The result is a strikingly unflowerlike flower, impossible to mistake.

In the other bladder-campions the flowers are upright and have conspicuous white or pink notched petals emerging from the bladder. The most widespread species is **Three-flowered Bladder-campion** [AHMLB], *M. affine* (fig. 5.15), in which the stalk is topped by a trio (usually) of flowers, as the name implies. Its petals are white and contrast strongly with the dark bladders, which are furred with dark, sticky hairs.

The Buttercup Family (Ranunculaceae)

The Buttercup Family is a very large one, whose 6 arctic genera show an extraordinary diversity of color and form.

The largest genus is *Ranunculus,* the buttercups and crowfoots. All those with yellow petals (the great majority) are easy to recognize as members of the genus; regardless of the size of the flowers (and some are tiny), the combination of glossy petals and abundant stamens marks them unmistakably as buttercups. Many of the numerous species are easily confused with one another but the following stand out:

The **Sulphur Buttercup** [aHMlB], *Ranunculus sulphureus* (fig. 5.16), and the **Snow Buttercup** [AHMlB], *R. nivalis* (not illus.), are the most strikingly handsome arctic buttercups. Both are low, tufted plants with big (up to 2.5 cm diam.), showy, brilliant yellow flowers, noteworthy for the dense "fur" of dark brown hairs on the calyx, which sets them apart from the common buttercups of temperate latitudes. They can only be distinguished from each other by dismantling a flower. Sulphur buttercup has brown hairs growing on the receptacle (the internal "floor") of its flower, but snow buttercup has none.

Dwarf B [AhMLB], *R. pygmaeus* (fig. 5.16), has tiny flowers whose petals do not extend beyond the calyx.

Seaside B [aml], *R. cymbalaria,* is a small, rather fleshy buttercup, growing on brackish shores and spreading by long runners. Its flowers are no bigger than those of dwarf buttercup, but its leaves are shaped differently (fig. 5.16). (The same species grows in temperate latitudes as a larger plant of inland saline meadows.)

The following species grow in ponds or wet marshes, sometimes in the open and sometimes concealed among taller sedges; some have white

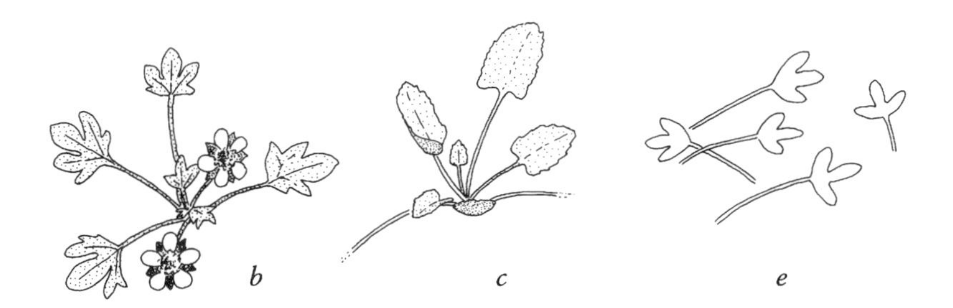

Figure 5.16. (a) Sulphur Buttercup. (b) Dwarf Buttercup. (c) Seaside Buttercup. (d) White Water Crowfoot. (e) Floating Buttercup. (f) Marsh-marigold.

flowers. Most can grow in brackish water as well as fresh. The two crowfoots, **Yellow Water-crowfoot** [Ahml], *R. gmelinii* (not illus.), and **White Water-crowfoot** [AmLB], *R. aquatilis* (fig. 5.16), have finely divided underwater leaves; hence the name "crowfoot." The flowers of the latter are often submerged too. **Floating Buttercup** [AHMLB], *R. hyperboreus,* with tiny yellow flowers, is distinguished from the crowfoots by its floating 3-lobed leaves (fig. 5.16), which make it easy to recognize even when it is not flowering. **Pallas's Buttercup** [Alb], *R. pallasii* (not illus.), a species of coasts and estuaries, has big (up to 2.5 cm diam.), perfumed, white flowers. The flowers are also unusual in having 6 to 10 petals (most buttercups have 5, except floating buttercup, which sometimes has as few as 3). The seeds form part of the food supply of many ducks and shorebirds.

Marsh-marigold [AML], *Caltha palustris* (fig. 5.16), a marsh plant with glossy yellow flowers and rounded leaves, could be (and often is) mistaken for a buttercup. But a look at the underside of the flower shows it to be differently constructed. It seems to lack a calyx and to have petals growing directly from the flower stalk. In fact, the apparent "petals" are really sepals (calyx parts) with the same color and texture as buttercup petals; they are called petaloid sepals. There are no true petals at all. The marsh-marigold of the tundra is a smaller and less luxuriant variety of the familiar temperate zone species; the higher the latitude, the smaller and less showy it becomes.

The anemones (*Anemone*) have 6 common arctic species, though none reaches the High Arctic. The genus is easy to recognize. All species have petaloid sepals, no petals, and a trio of lobed or dissected leaves in a ring around the flower stem; all the other leaves grow from ground level, on longer stalks. The species can be distinguished as follows:

Yellow Anemone [AL], *A. richardsoni* (not illus.), is the only one with yellow flowers. Its petaloid sepals distinguish it from buttercups and its divided leaves from marsh-marigolds.

Both **Northern A** [AmL], *A. parviflora,* and **Drummond's A** [A], *A. drummondi* (both in fig. 5.17), have white flowers, often tinged with pale blue on the outside. But their leaves differ in shape; also, those of northern anemone are glossy and hairless, those of Drummond's covered with soft, silky hairs.

Cut-leaf A [al], *A. multifida* (fig. 5.17), is tall (up to 50 cm), and its narrow leaf segments are distinctive. The flowers are usually white but occasionally pink, bluish-purple, or even red.

Narcissus-flowered A [A], *A. narcissiflora* (not illus.), varies in appearance depending on conditions. In sheltered sites it is conspicuous: tall

Figure 5.17. (a) Northern Anemone. (b) Drummond's Anemone. (c) Cut-leaf Anemone. (d) Pasque Flower.

(up to 60 cm) and with creamy-white flowers growing in threes. In exposed sites it is smaller and the flowers grow singly. In any site, it can always be recognized from the seed head; its "seeds" (technically, achenes) are dark and smooth, whereas in the other white-flowered arctic anemones they are covered with a dense, pale "wool."

Pasque Flower [Aml], *A. patens* (fig. 5.17), is bright purple, and densely clothed, on the leaves and the outside of the flower, with long, silky white hairs. The seed heads look like feather dusters, with long, silky plumes. The plant is often treated as belonging to a separate genus, and called *Pulsatilla ludoviciana*. It is a variety of the familiar prairie flower often called "prairie crocus" (inaptly, as it is entirely unrelated to crocuses).

Meadow rue (*Thalictrum*) is another genus with petaloid sepals and no true petals. But it is utterly unlike an anemone or a marsh-marigold. Instead of a few showy flowers, meadow rues have numerous tiny ones,

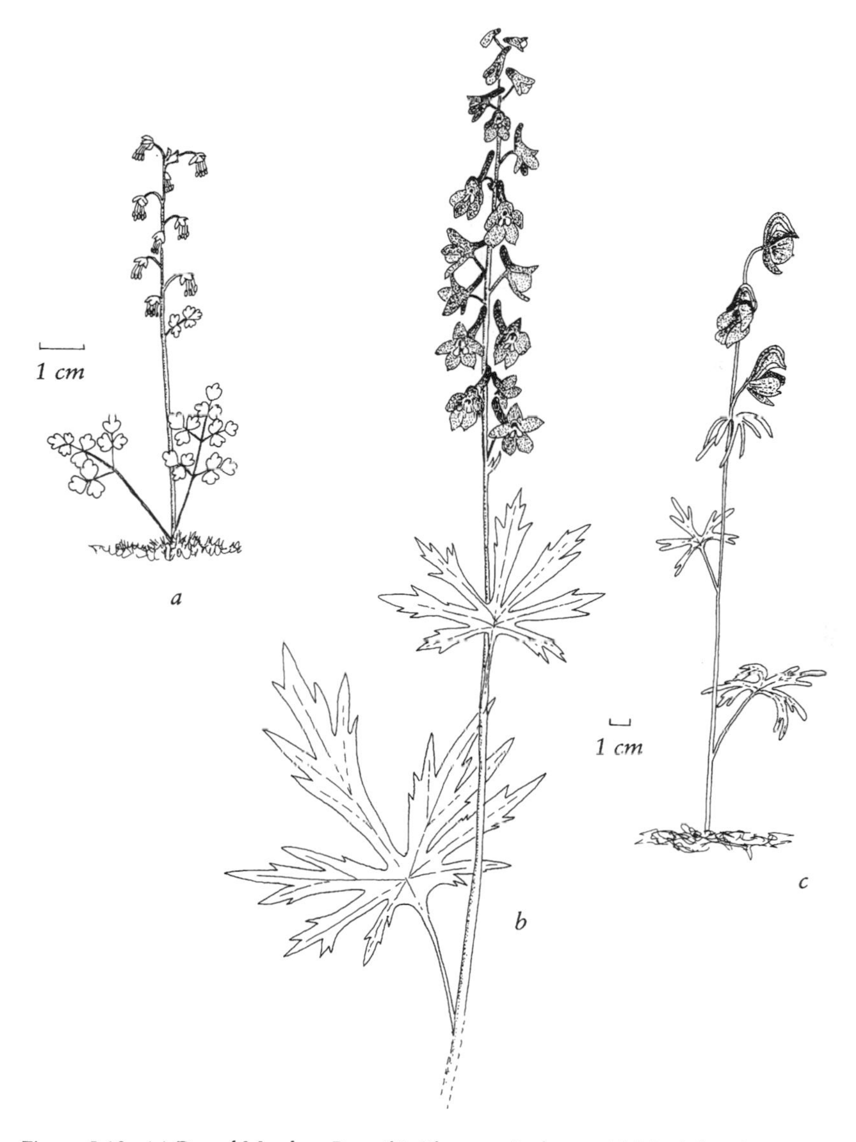

Figure 5.18. (a) Dwarf Meadow Rue. (b) Glaucous Larkspur. (c) Monkshood.

with small, dull-colored sepals. It is the stamens that make the plants eye-catching. **Dwarf Meadow Rue** [A], *T. alpinum* (fig. 5.18), is the single arctic species, and it is found only in the mountains. Its inconspicuous sepals fall off as soon as the flower opens, leaving the bright yellow anthers (the pollen-containing tips of the stamens) dangling and fluttering in the breeze. Winds, not insects, carry pollen from flower to flower.

The two remaining arctic genera of the buttercup family are larkspurs *(Delphinium)* with two species and monkshoods *(Aconitum)* with one. Their handsome blue flowers closely resemble those of the related garden plants, so gardeners recognize them instantly.

The larkspurs are **Glaucous Larkspur** [A], *Delphinium glaucum* (fig. 5.18), and **Alpine Larkspur** [a], *D. brachycentrum* (not illus.). Glaucous larkspur is the commoner and much the larger (up to 200 cm tall), with a tall spike of crowded flowers. Alpine larkspur is a smaller plant (50 cm or less) of the mountains; its flowers are fewer but larger.

Monkshood [A], *Aconitum delphinifolium* (fig. 5.18), has royal blue flowers, each with a "monk's hood" arching over the rest of the flower. Seen from the front, the hood is quite narrow, with its left and right sides joined along a sharp crease.

The Poppy Family (Papaveraceae)

Poppies (genus *Papaver*) are the only members of the family in the Arctic. Their big, bright yellow flowers are familiar to all arctic travelers and their similarity to the Iceland poppies grown in gardens makes them unmistakable. Like garden poppies, each flower has a barrel-shaped seed capsule with radiating "spokes" (rays) on top.

The common wide-ranging species is **Arctic Poppy** [HMLB], *Papaver radicatum,* also called *P. lapponicum* (fig. 5.19), replaced west of the Mackenzie River by Keele's Poppy (*P. keelei*). Botanists have described several other species on the basis of subtle differences such as the shape of the capsule, the number of rays, the shape and hairiness of the leaf, and the color (white or yellow) of the latex that flows from a cut stalk. But as the specialists disagree on nearly all points, and as most of the species have only small, localized ranges, they are not considered here.

Although yellow is the standard color of arctic poppies, white ones are quite common, in some places commoner than yellows. Whites are most numerous in regions where frequent cloudiness and fog obscure the sun. This probably has to do with the way poppy flowers constantly turn to face the sun, whenever it is shining, so that their interiors become appreciably warmer than the surrounding air (see sec. 5.8). Yellow flowers are more efficient than white in absorbing sunshine, so that yellowness is a valuable adaptation in a region with plenty of sunlight, but much less valuable where it is cloudy most of the time. Moreover, producing yellow flowers has its costs; the poppy plant must use up some of its chemical energy (gained by photosynthesis) to create the yellow pigment. Therefore, in cloudy districts, where yellowness is seldom useful, plants with white flowers have an advantage because they can use this energy more economically by devoting it to growth.

Besides existing in yellow and white forms, poppy flowers are often streaked and blotched with green; occasionally an all-green poppy flower will be found. These discolorations are actually "bruises" on the petals, which, because they are thin, are very susceptible to bruising. Petals are easily bruised in the bud, where they are crumpled and crushed together inside the tight 2-sepal calyx.

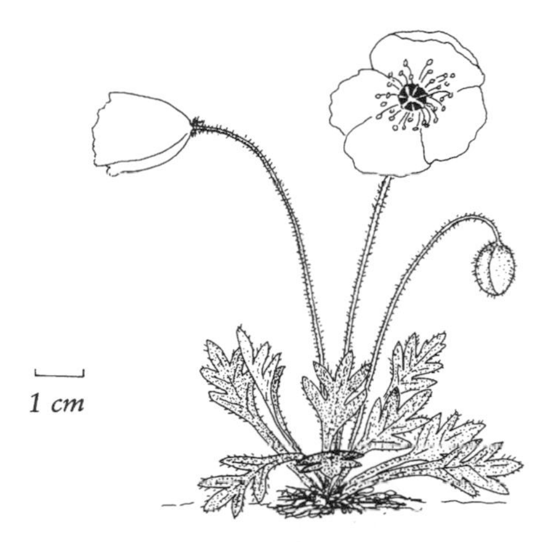

Figure 5.19. Arctic Poppy.

The Mustard Family (Brassicaceae or Cruciferae)

More than forty species of "crucifers," as members of the Mustard Family are called, grow in the Arctic. The flowers of all species have 4 petals, 4 sepals, and 6 stamens, of which 4 are long and 2 short (fig. 5.20). No other plants have flowers of this design. Consequently, members of the family can always be recognized with certainty, provided they are in flower.

In spite of being colorful and beautiful, the separate species in the family are hard to identify. Flower color is no help: the colors range through white, yellow, lilac, purple, and pink. Except for a few noteworthy species described below, identifications cannot be made from the

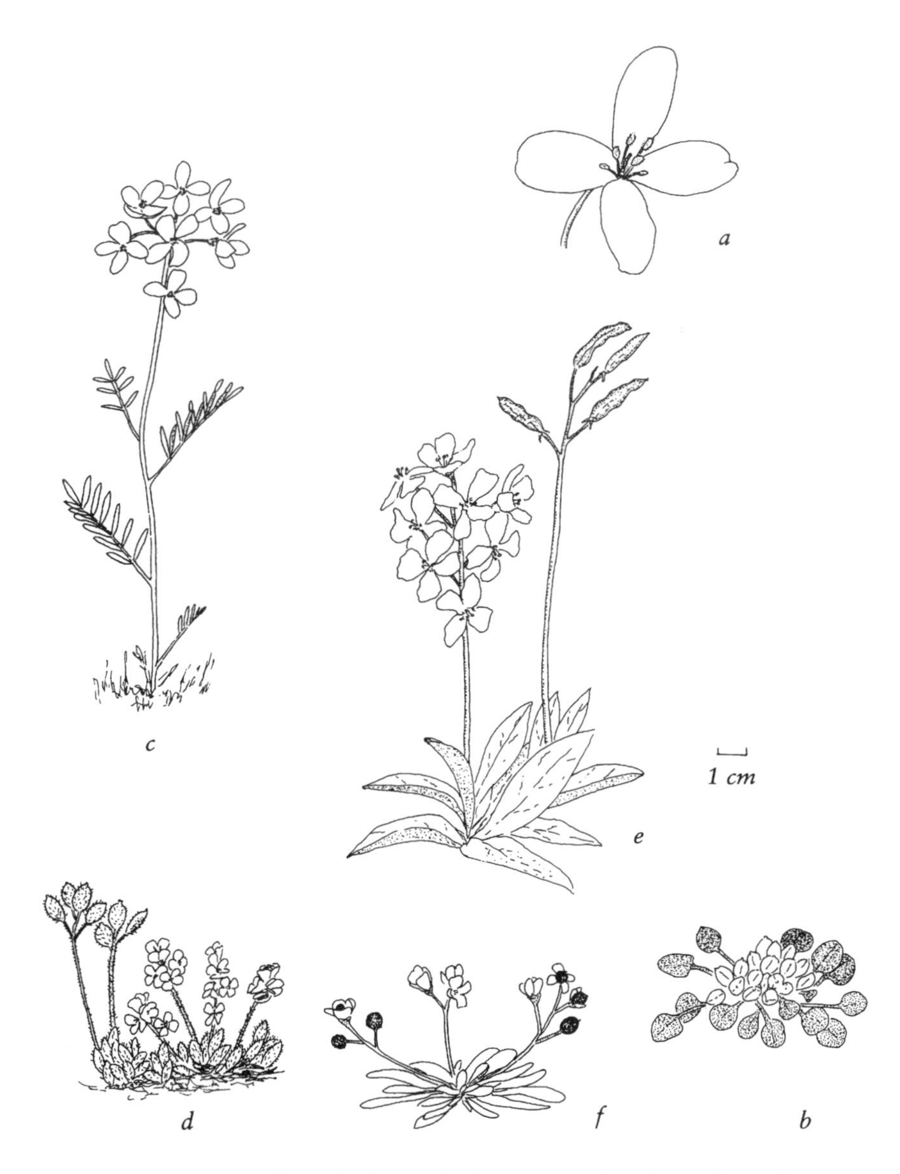

Figure 5.20. (a) A typical crucifer flower. (b) Scurvy-grass, showing pods. (c) Cuckoo Flower. (d) Whitlow-grass (*Draba corymbosa*). (e) Parrya. (f) Arctic Bladderpod.

flowers alone. To put a name to the majority of species is a task for specialists, requiring examination of the seed pods (which may be short or long; flat or rounded); also, with a strong lens, the hairs, if any, on the leaves and stems (which are often forked or branched in various ways). Then a complete technical Flora (see preliminary notes) must be consulted.

Seven members of the family are distinctive enough to be spotted by nonspecialists.

Scurvy-grass [AHMLB], *Cochlearia officinalis* (fig. 5.20), is a succulent seashore plant growing flat on the ground, with clusters of small white flowers followed by plump oval pods looking (when fresh) like clusters of green beads. The species is most easily identified by these distinctive pods. The plant is edible, and is a good source of vitamin C.

Cuckoo Flower [AHMLB], *Cardamine pratensis* (fig. 5.20), is one of several species of bittercress, as members of the genus *Cardamine* are collectively known. Note the pinnate leaves. All the bittercresses (except for one dwarf species) have pinnate leaves, and no other arctic crucifer has them. Thus any pinnate-leaved crucifer is certainly a bittercress. This species, with its large (1 cm or more), flowers is the handsomest; the flowers are usually white, occasionally lilac. The plant grows in damp places.

Whitlow-grasses [AHMLB], *Draba* species (fig. 5.20), form a genus with numerous hard-to-differentiate species; they are widespread and common, and some are likely to be encountered almost anywhere. Most are cushion plants with hairy leaves, yellow or white flowers, and flattish oval pods pointed at both ends; sometimes the pods are hairy.

Parrya [Al], *Parrya nudicaulis* (fig. 5.20), is an outstandingly showy plant with very large (2 to 3 cm), fragrant purple or white flowers. It is confined to the western Arctic but the genus is represented farther east by a species with smaller (1 cm) flowers, **Arctic Parrya** [hMl], *P. arctica* (see fig. 5.7). In this species too, some specimens have purple flowers, some white. Arctic parrya could be confused with Pallas's wallflower (mentioned below) unless seed pods are examined. Those of arctic parrya are short (3 cm or less) and plump, and bulge over the seeds within; those of Pallas's wallflower are long (5 cm or more) and thin, and often curved.

Pallas's Wallflower [AHMlb], *Erysimum pallasii* (see fig. 5.4), when in flower, resembles arctic parrya except that, while they are young, the flowers are close to the ground. The contrast between the seed pods is obvious (see preceding paragraph).

Yellow Wallflower [Al], *E. inconspicuum* (not illus.), is an attractive and conspicuous yellow wallflower; its Latin name is inexplicable.

Arctic Bladderpod [AhMLB], *Lesquerella arctica* (fig. 5.20), a conspicuous little plant with a rosette of silvery leaves, has bright yellow flowers and globular seed pods.

The Stonecrop Family (Crassulaceae)

The single arctic member of this family is **Roseroot** [A], *Rhodiola integrifolia* (fig. 5.21). It has fleshy upright stems, up to 30 cm tall, densely covered with succulent bluish-green leaves; sometimes (not always) the leaf margins have a few widely spaced small teeth. The flowers are of an unusual color, a dark maroon red. They are crowded into a flat or slightly domed cluster atop each stem. Most of the flowers are unisexual, having only the stamens functional (male flowers), or only the ovaries functional (female flowers). The latter develop into plump seed capsules (technically, *follicles*) shaped like tiny wineskins and matching the petals in color.

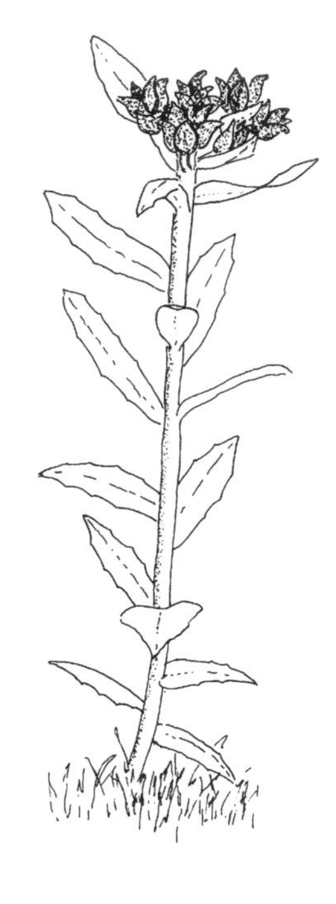

Figure 5.21. Roseroot.

The Saxifrage Family (Saxifragaceae)

By far the most important of the 4 arctic genera of this family is *Saxifraga,* the saxifrages. It has a score of arctic species, many of them easily recognizable and strikingly beautiful. All have unmistakable flowers, with 5 petals and 10 stamens surrounding a two-part seed capsule formed of divergent "horns"; these are easily seen with a lens, even in small-flowered species, and are diagnostic. The petals are often (not always) spotted. Depending on the species, the flowers are differently shaped: some are saucerlike, some cuplike, and some funnellike. Figure 5.22 gives an enlarged view of a typical saxifrage flower.

To sort out the widespread, easily recognizable species, we classify them by color.

One species has deep purple flowers. It is **Purple Saxifrage** [AHMLB], *S. oppositifolia* (fig. 5.22); its brilliant flower-covered cushions are common throughout the Arctic. They are the earliest arctic flowers of the season.

Five species have yellow flowers:

Yellow Mountain S [hMlB], *S. aizoides* (fig. 5.22), has narrow petals with orange dots; its stems are covered with fleshy leaves.

Thyme-leaved S [Ah], *S. serpyllifolia* (not illus.), has single flowers on thin stems, with 1 or 2 small leaves, arising from tufts of small fleshy leaves.

Bog S [AHMLB], *S. hirculus* (fig. 5.22), has cup-shaped flowers with rounded petals; the stem below the flower is fuzzy with rust-colored hairs. It could be mistaken for a buttercup until you look at the typical saxifrage flower interior.

Spiderplant [AHM], *S. flagellaris* (fig. 5.22), spreads like a strawberry plant, by means of long, arching, bright red runners (stolons).

Cushion S [Am], *S. eschscholtzii* (not illus.), is like a compact pin cushion; its flowers are tiny, narrow-petaled, on very short stems.

Five species have white or creamy-white flowers:

Tufted S [AHMLB], *S. caespitosa* (fig. 5.22), has white flowers with yellowish centers; its leaves have 3 blunt lobes and are clammy to touch.

Nodding S [AHMLB], *S. cernua* (fig. 5.23), has pure white flowers one at the top of each stem, not nodding when fully open; clusters of little beadlike red-purple bulblets appear in axils of stem leaves; the leaves are distinctively shaped.

Grained S [AHMLB], *S. foliolosa* (not illus.), is rather like nodding saxifrage, but its flowers are smaller and it has no leaves on the stems; the bulblets are green, at tips of short side branches.

Prickly S [AHMLB], *S. tricuspidata* (fig 5.23), grows as cushions of stiff, prickly, 3-toothed leaves, usually copper colored; it has creamy-white flowers on tall stems; its petals have yellow-orange dots.

Spotted S [A], *S. bronchialis* (not illus.), has flowers rather like those of Prickly Saxifrage; its leaves are soft, gray-green, edged with fine white hairs.

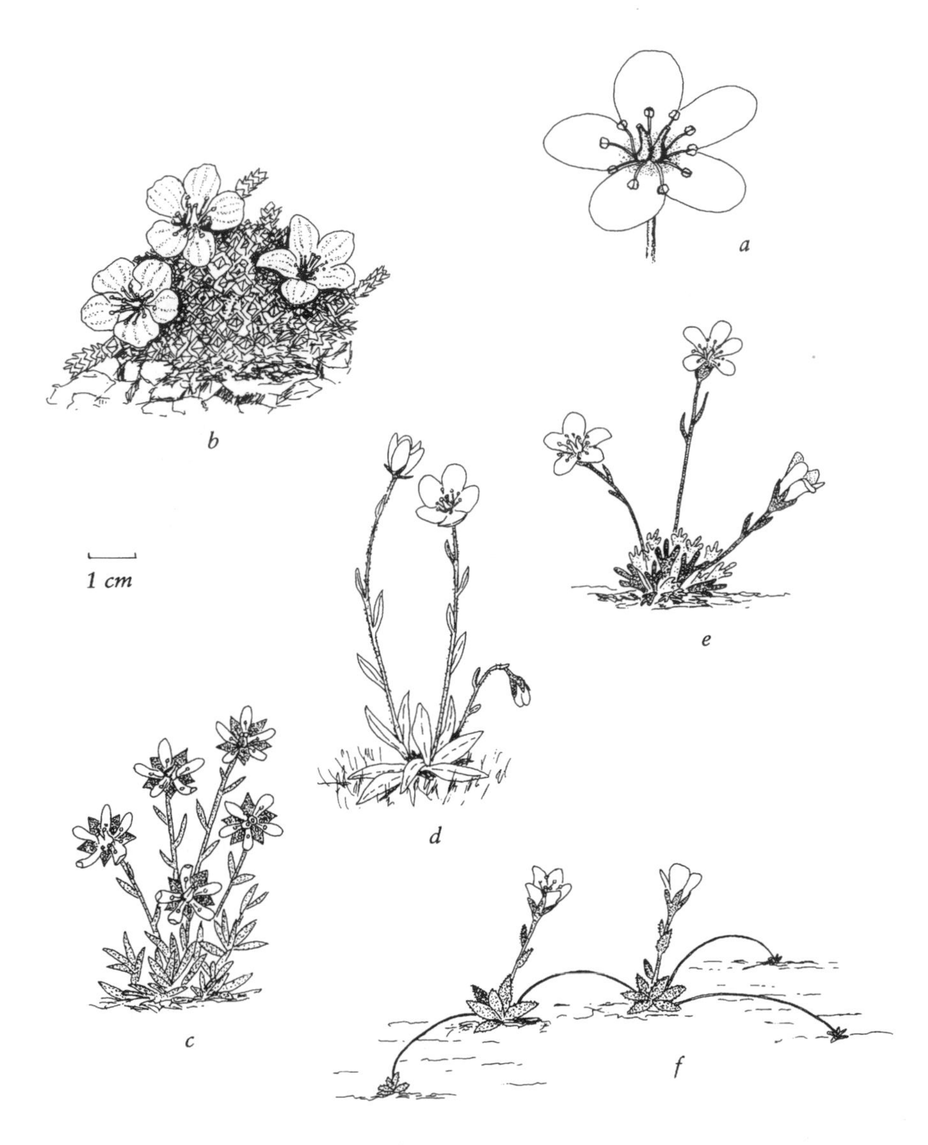

Figure 5.22. (a) A typical saxifrage flower. (b) Purple saxifrage. (c) Yellow Mountain S. (d) Bog S. (e) Tufted S. (f) Spider Plant.

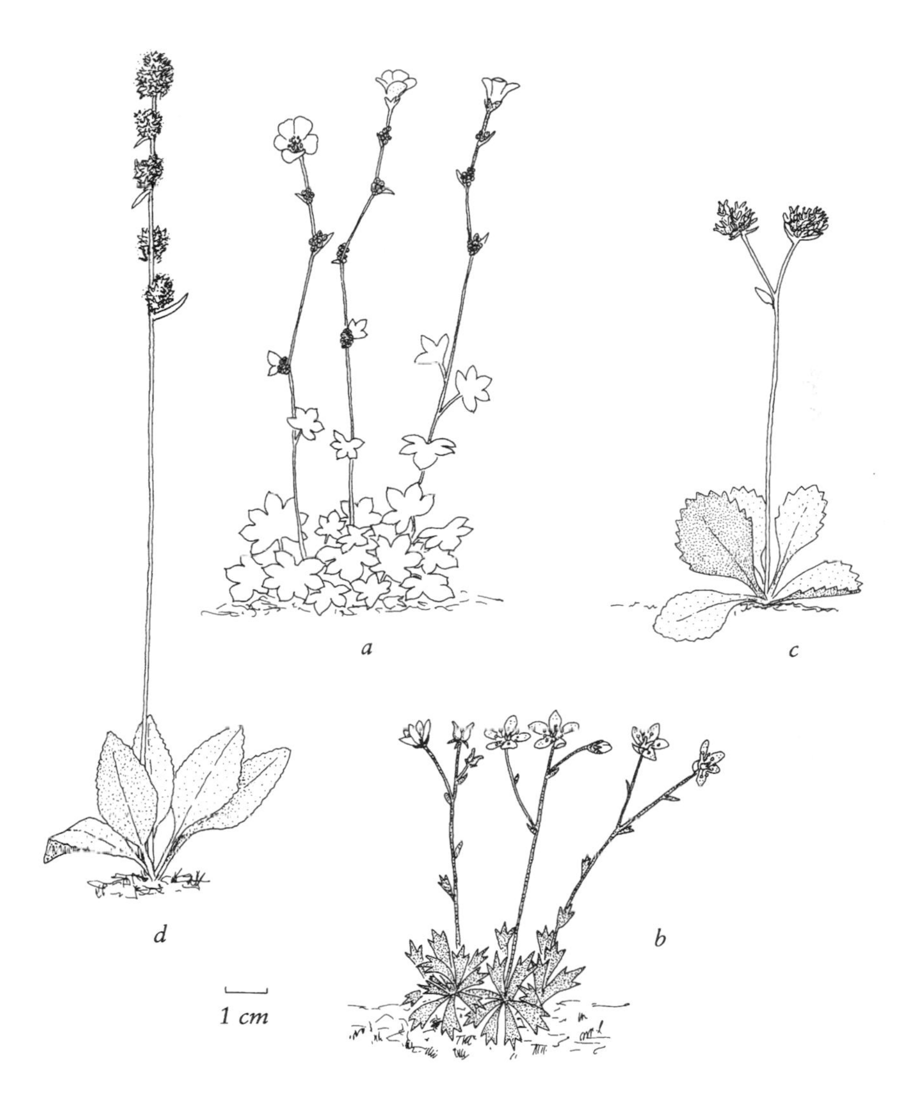

Figure 5.23. (a) Nodding Saxifrage. (b) Prickly S. (c) Alpine S. (d) Stiff-stemmed S.

One species sometimes (not always) has pale pink flowers. It is **Brook S** [AHMLB], *S. rivularis* (not illus.), which grows in wet habitats. Only specimens in which the small flowers are pink rather than white are easy to identify.

Two species have inconspicuous flowers.

Alpine S [AHMLB], *S. nivalis* (fig 5.23), has flowers with white (sometimes red) petals that are too short to be showy; flowers grow in a cluster at the tip of a leafless flower stem growing from a rosette of toothed, leathery leaves.

Stiff-stemmed S [Ahmlb], *S. hieracifolia* (fig. 5.23), has flowers, with small greenish petals, in a spike at the top of a leafless stem growing from a rosette of leaves.

In addition to the saxifrages described here there are several others, especially in the western mountains. Most have white (or pale mauve) flowers densely or loosely clustered, or in spikes; and fan-shaped or wedge-shaped, toothed leaves. The differences among them are slight, and members of the same species often vary from one locality to another. A detailed Flora is needed to sort them out.

Four other species (in three genera) of the Saxifrage Family are found in the Arctic. Two of the species belong to the grass-of-Parnassus genus, *Parnassia:* **Northern Grass-of-Parnassus** [AL], *P. palustris,* and **Kotzebue's Grass-of-Parnassus** [AmL], *P. kotzebuei* (both in fig. 5.24). Both have solitary white flowers atop tall, slender flower stalks with, at most, one small leaf growing from the stalk itself; the other leaves grow as a tuft, directly from the ground. All the leaves have smooth, hairless surfaces and smoothly rounded outlines. The flowers of the 2 species differ markedly. That of northern grass-of-Parnassus has big (6 mm or more long) white petals marked with greenish veins; there are 5 stamens, and they alternate with 5 *staminodes,* organs like miniature yellowish petals fringed with tiny "fingers," each tipped with a glistening, sticky gland; the staminodes are worth examining with a hand lens. The flower of Kotzebue's grass-of-Parnassus is not nearly so beautiful. The white petals are shorter than the sepals, and the staminodes are less conspicuous; the egg-shaped seed capsule is large, and grows to form the most prominent part of the flower, more noticeable than the small petals.

Northern Water Carpet [AhMLB], *Chrysosplenium tetrandrum* (fig. 5.24), although it is often called "golden saxifrage," is most unlike a saxifrage. The name Water Carpet describes the moist habitat it favors. Small and delicate, these inconspicuous plants often escape notice because their yellowish-green flowers don't contrast with the leaves. The lobed, kidney-shaped leaves are distinctive. And the flowers, though not colorful, deserve a close look; they have no petals, only a cup-shaped

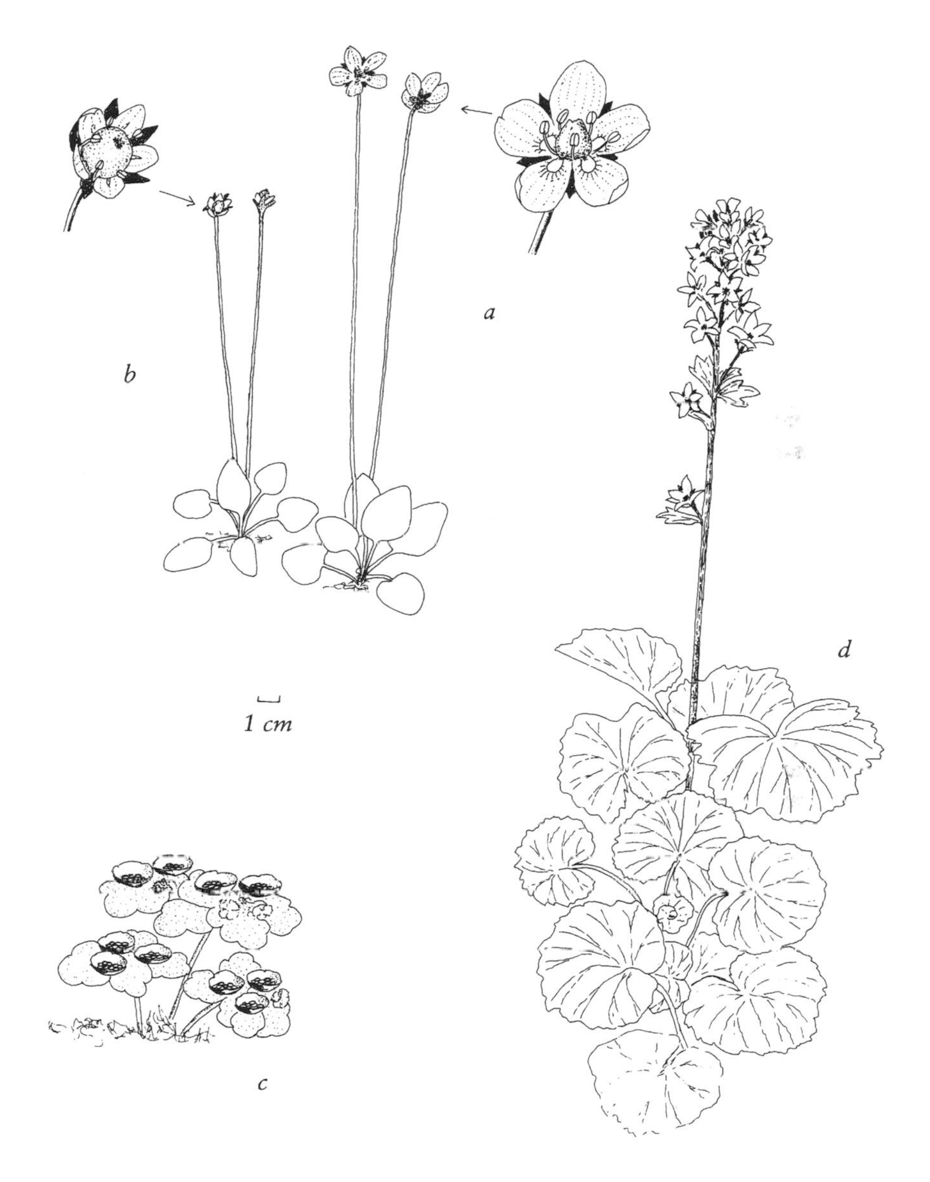

Figure 5.24. (a) Northern Grass-of-Parnassus. (b) Kotzebue's G-of-P. (c) Northern Water Carpet. (d) Alaska Boykinia.

calyx, in which the numerous, shiny seeds lie in an exposed heap when they are ripe. This cup acts as a "splash-cup"; when a falling raindrop lands in it, the seeds are splashed out in all directions, and in this way the plant spreads.

Alaska Boykinia [A], *Boykinia richardsonii* (fig. 5.24), is a tall (up to 80 cm) plant with a showy spire of white (or pale pink), dark-centered flowers and masses of big, almost circular leaves much enjoyed by grizzlies. The flower resembles a saxifrage flower in having a seed capsule with two long, divergent "horns," but there are 5 instead of 10 stamens. Alaska boykinia is notable for being endemic to its region (Alaska and adjacent Yukon)—i.e., it grows nowhere else. It survived the last ice age (and probably for a few million years before that) in the place where it grows now and, surprisingly, has not managed to spread.

The Rose Family (Rosaceae)

Numerous members of the Rose Family are present in the Arctic, and many are strikingly beautiful.

The best known is **Arctic Dryad** [AHMLB], *Dryas integrifolia* (fig. 5.25); unfortunately, it is often called "mountain avens," which causes needless confusion with plants properly called avens, the members of the genus *Geum*. Arctic dryad is the territorial flower of the North West Territories and huge areas of the tundra are dotted with it, covered, in early summer, with creamy-white, roselike flowers. (A flower is called "roselike" if its flowers resemble those of wild roses, which are single, unlike the compound flowers of most garden roses.) The flowers are followed by distinctive seed heads (fig. 5.25) in which the plumes topping the "seeds" (actually, achenes) are at first twisted in a tight spiral; as they dry, they spread wide like a feather duster. The plant is unmistakable in all stages. The leaves are like tiny arrowheads, with margins tightly curled under; the upper surfaces are a shiny dark green and the lower covered with dense white "wool."

Botanists recognize several closely related *Dryas* species in the western Arctic; they differ from the common arctic dryad in the shape and hairiness of the leaf, and in the prominence of the leaf veins. They are poorly differentiated species that often hybridize, giving a continuous range of intermediate forms.

Many species of cinquefoil (*Potentilla*) grow in the Arctic, perhaps as many as 20. Their little roselike flowers, yellow in all species but one, make the genus as a whole easy to recognize even though, to the uninitiated, they look like buttercups; this error is avoided by noting the texture of the petals, which are always glossy in buttercups but never in cinquefoils. Except for a shrubby species (mentioned below), the yellow-flowered cinquefoils are hard to identify as to species; to do so, you must note flower size, leaf shape, and whether and where the plant is hairy, and then consult a technical Flora. As to leaf shape, all species have compound leaves (with several leaflets). The leaves of some, e.g., **Arctic Cinquefoil** [AHMLB], *P. hyparctica* (fig. 5.25), have 3 leaflets and resemble strawberry leaves; some have 5 or 7 leaflets, arranged fanwise, e.g., **Red-stemmed Cinquefoil** [aHMlb], *P. rubricaulis* (fig 5.25); and some have leaflets in pinnate formation, in 2 rows, e.g., **Silverweed** [al],

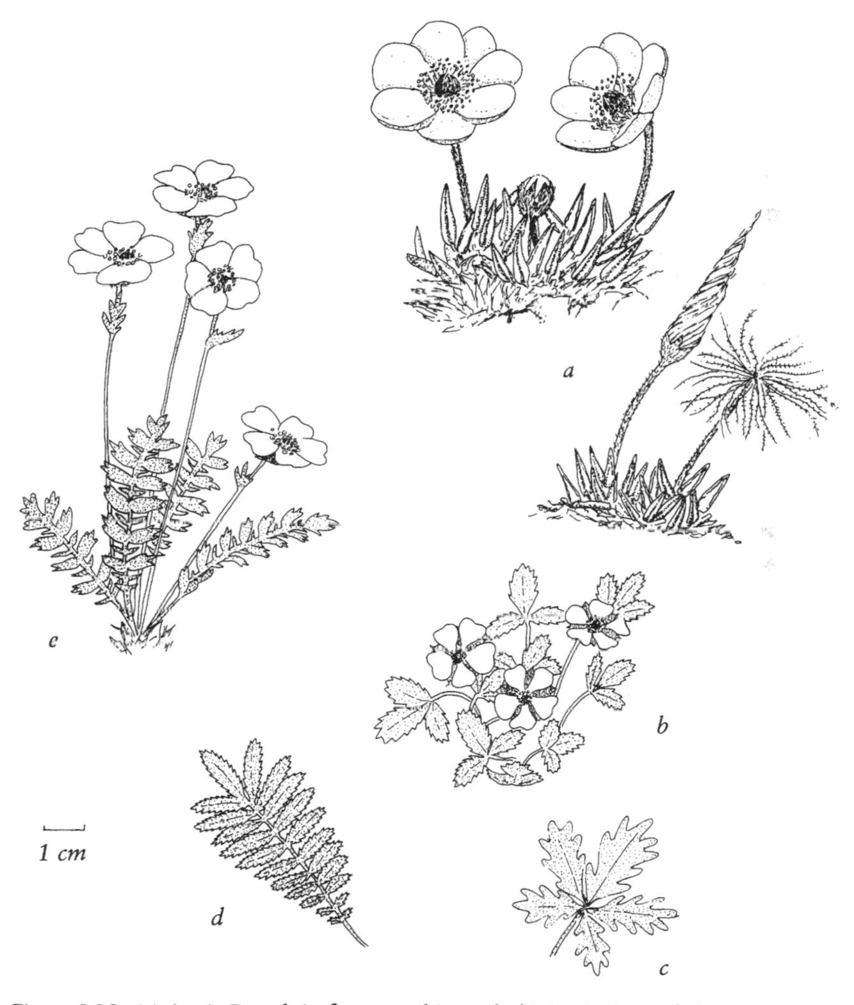

Figure 5.25. (a) Arctic Dryad, in flower and in seed. (b) Arctic Cinquefoil. (c) Leaf of Red-stemmed Cinquefoil. (d) Leaf of Silverweed. (e) Alpine Avens.

P. anserina (fig. 5.25), and **Seaside Cinquefoil** [amLb], *P. egedii* (not illus.). **Vahl's Cinquefoil** [HMLB], *P. vahliana* (fig. 5.7 in sec. 5.12), is one of the more showy species. It has orange-centered flowers; and the flower stalks and 3-part leaves (mostly the undersides) are covered with long, silky, yellowish hairs, which form tufts at the tips of the leaf teeth.

The two instantly recognizable species are: **Shrubby Cinquefoil** [AL], *P. fruticosa* (not illus.), the only cinquefoil that grows as a shrub; its flowers are yellow and it is a familiar garden ornamental in temperate

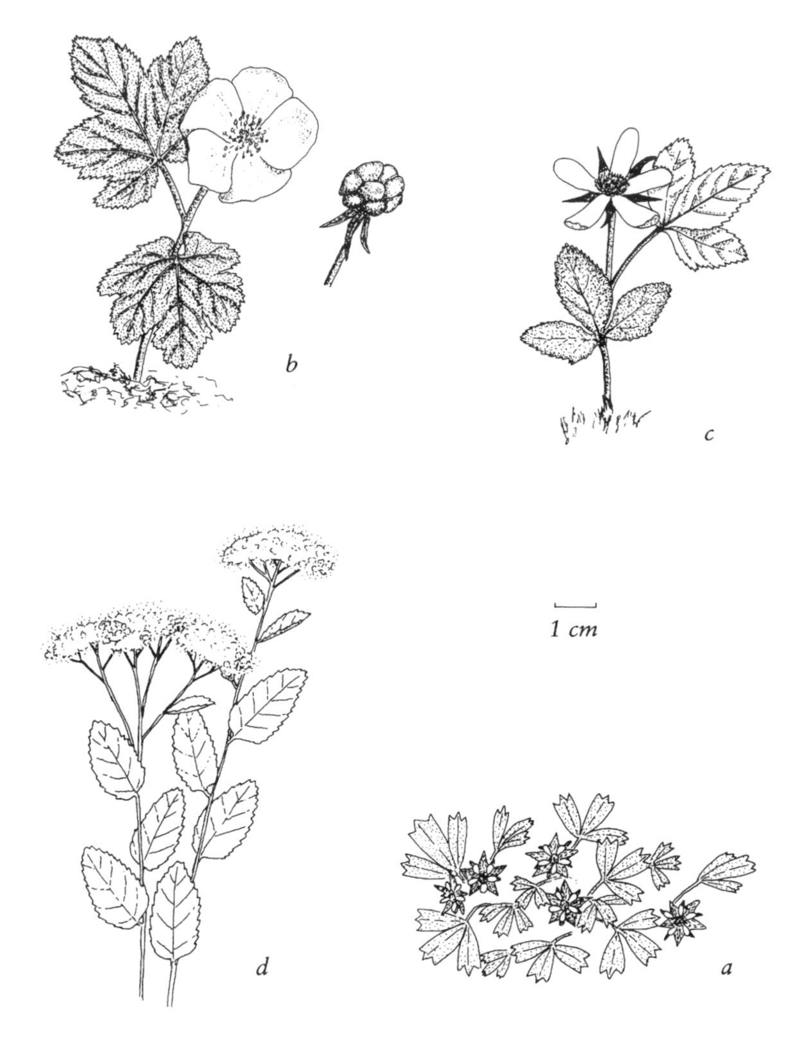

Figure 5.26. (a) Sibbaldia. (b) Cloudberry. (c) Arctic Raspberry. (d) Alaska Spiraea.

latitudes; and **Marsh Five-finger** [AL], *P. palustris* (not illus.), a marsh plant with flowers of an unusual maroon-purple color.

Two species of avens (*Geum*) occur in parts of the Arctic and their showy yellow flowers are almost identical with those of the cinquefoils except for being larger. The larger-flowered of the two is **Glacier Avens** [A], *G. glaciale* (not illus.), with flowers up to 4 cm in diameter; its pinnate leaves are shaggy with long, silky hairs. **Alpine Avens** [AH], *G. rossii* (fig. 5.25), has flowers up to 3 cm in diameter and is a hairless plant with distinctively shaped pinnate leaves.

A creeping plant with small greenish-yellow flowers and trifoliate leaves like those of some of the cinquefoils is **Sibbaldia** [aLb], *Sibbaldia procumbens* (fig. 5.26); its wedge-shaped, 3-toothed leaflets are its identifying character. The tiny flowers, with petals shorter than the sepals, are inconspicuous.

The genus *Rubus* (raspberries, blackberries, and their kin) has two Low Arctic species:

Cloudberry or **Baked-apple Berry** [ALb], *Rubus chamaemorus* (fig. 5.26), is restricted to acid, peaty soils, and is famous for the beauty of its large white flowers and the excellence of its juicy yellow berries. The plants are unisexual. Female plants bear fruit but their stamens are atrophied; male plants have functioning, pollen-producing stamens but yield no fruit. The male flowers are larger and richer in nectar than the female flowers; therefore, insects, which feed on both pollen and nectar, are more strongly attracted to them.

Arctic Raspberry [Al], *R. acaulis* (fig. 5.26), has narrow-petaled flowers and trifoliate leaves closely resembling (in shape) those of garden raspberries. But the flowers are an eye-catching pink; and the ankle-high plants, which yield small, tasty wild raspberries, are devoid of prickles.

The final member of the Rose Family to look for is **Alaska Spiraea** [Al], *Spiraea beauverdiana* (fig. 5.26), a low shrub of boggy shrub-tundra, which any gardener will recognize as a spiraea from both the fragrance and the appearance of the attractive white flowers. Though individually tiny, they grow in wide, flat-topped panicles crowned with a mist of protruding stamens.

The Pea Family (Fabaceae or Leguminosae)

Any arctic plant with flowers shaped like those of a pea plant is a member of the Pea Family, which makes the family instantly recognizable. Five genera concern us: lupine (*Lupinus*), with one arctic species; liquorice-root (*Hedysarum*) with two; milk-vetch (*Astragalus*) and oxytrope (*Oxytropis*), each with several species; and beach pea (*Lathyrus*), with one.

Arctic Lupine [Aml], *Lupinus arcticus* (fig. 5.27), closely resembles a garden lupine. It has tall spires of bright blue and white flowers, and its leaves are palmately compound, i.e., each consists of several leaflets radiating from a common center like the spokes of a wheel.

The two rather similar species collectively called liquorice-roots have spikes of purple-pink flowers and pinnately compound leaves; i.e., each leaf consists of pairs of leaflets arranged in a row (as in fig. 5.27). These characters are not diagnostic of the genus, however, since the milk-vetches and oxytropes also have pinnate leaves, and some of them have

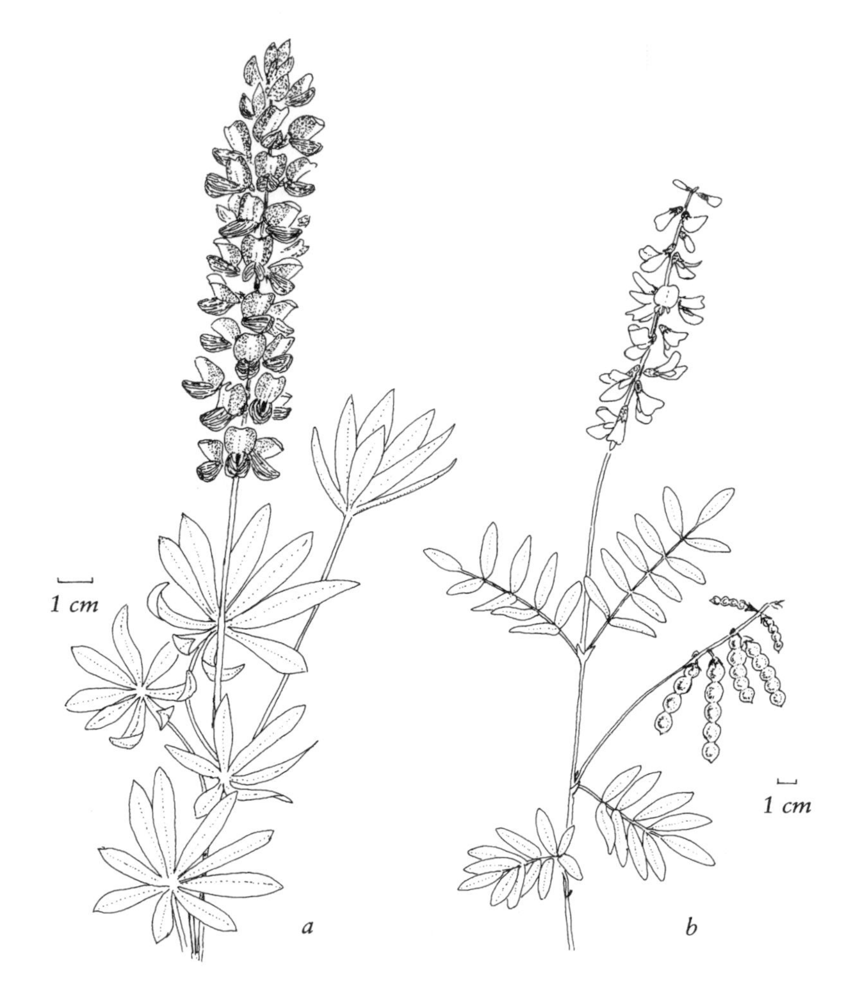

Figure 5.27. (a) Arctic Lupine. (b) Wild Sweet Pea.

purple flowers. The unique characteristic of the liquorice-roots is the shape of the seed pod, in which each seed occupies its own disc-shaped section of the pod, separated from the next section by a narrow constriction. This means that late in the season when the pods are large, liquorice-roots can be recognized with no trouble; earlier in the season a careful search may be needed to find an immature pod concealed among the flowers when, no matter how tiny the pod, its characteristic "beaded" shape can be seen.

The two species of liquorice-roots are **Wild Sweet Pea** [AML], *Hedysarum mackenzii* (fig. 5.27), and **Bear Root** [AL], *H. alpinum* (not illus.). To distinguish them, look at the backs of the leaflets; the veins of the leaf show as conspicuous dark lines in bear root but are not visible in wild sweet pea. Wild sweet pea is a bushy plant, the flowers are big and bright, and the roots are poisonous. Bear root is taller and more straggly, the flowers are smaller and paler, and the roots are edible. They are a favorite food of grizzly bears—hence the plant's name. Where bear root grows, patches of torn-up, roughly dug soil are a sure sign that grizzlies have been feeding.

A plant of the Pea Family that is neither a lupine nor one of the liquorice-roots must (unless it is growing on a sea beach) be either an oxytrope or a milk-vetch. The two genera look much alike but can be distinguished by looking at the flower stalks: if the flower stalk is leafless, with all leaves growing from ground level or very near it, the plant is an oxytrope; if the flower stalk is leafy, the plant is a milk-vetch. Close examination of a flower shows a second distinguishing character; the keel of the flower is tipped with a sharp "tooth" in oxytropes but not in milk-vetches (fig. 5.28).

To sort out the many species of these two genera requires attention to numerous minor details and consultation of a technical Flora. Flower color may be purple, lilac, white, cream, or yellow. And the species also differ in the sizes, shapes, colors, hairiness, and attitude (upright or dangling) of the pods; in the number and arrangement of the leaflets of each leaf; in the shape and hairiness of the calyx, and likewise of the stipules (the pair of small leaflike structures at the bottom of the leaf stalk); and in the size and shape of the whole plant—it may be a compact cushion or a spreading mat, or it may straggle. All these characteristics should be noted by a traveler planning to consult a Flora after returning from the Arctic. One quite easily recognizable oxytrope deserves attention because of its interesting geographic range (see sec. 5.12 and fig. 5.7). It is **Silvery oxytrope** [hMlb], *Oxytropis arctobia,* with purple flowers followed by dark, hairy pods, atop a cushion of densely hairy, silvery-gray leaves.

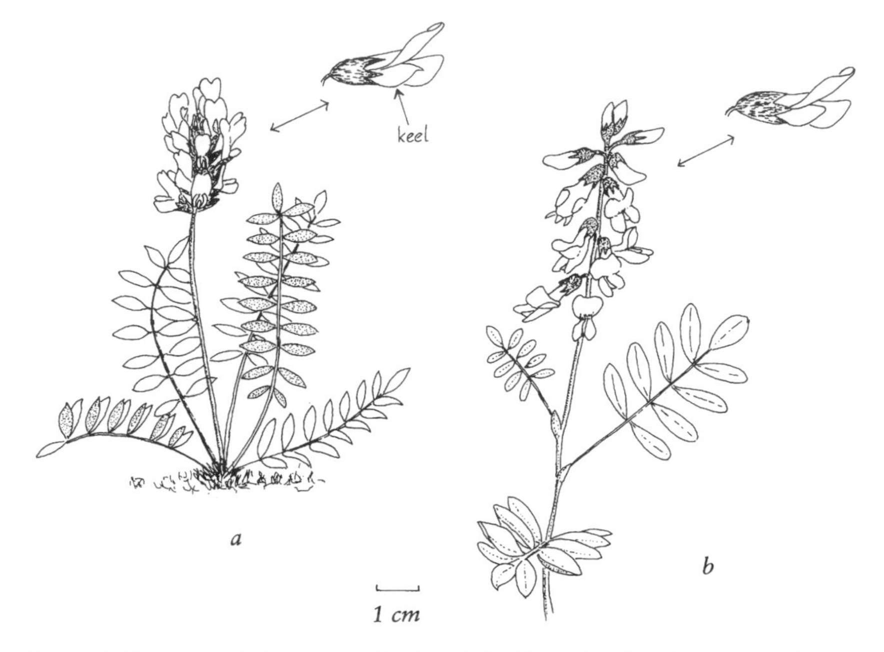

Figure 5.28. (a) A typical oxytrope. (b) A typical milk-vetch. The enlargements of individual flowers each show a flower with one side petal removed to expose the "keel." The keel is tipped with a tooth in oxytropes, but not in milk-vetches.

Beach pea [l], *Lathyrus japponicus* (not illus.), has a restricted range in the Arctic, being found only at a few beaches on the mainland shore of Canada. When you find it, sprawled across a sandy beach, there can be no doubt about its identity. It closely resembles an ordinary garden sweet pea with blue-purple flowers. As with sweet peas, its leaves are tipped with coiled tendrils.

The Flax Family (Linaceae)

The only member of this family to grow in the Arctic is **Wild Flax** [Aml], *Linum perenne* (fig. 5.29), a tall plant (up to 60 cm) with big, sky-blue flowers, and short, narrow, grayish-green leaves all up the stem. The plant grows in dry soil, as it does throughout the mid-latitude prairies of North America.

The Crowberry Family (Empetraceae)

Crowberry [AhmLb], *Empetrum nigrum* (fig. 5.30), is the only arctic member of its family. It is a low, matted evergreen shrub, whose tiny glossy leaves make it look like heather. The dark purple flowers are inconspicuous. The plants become more noticeable when the shiny, jet-black berries ripen; they often grow in great profusion. The berries are juicy and edible, though rather tasteless. They are eaten by numerous species of birds (e.g., wheatear, gray-cheeked thrush, jaegers, gulls, geese) and mammals (e.g., voles, lemmings, sik-siks, grizzly bears, and even polar bears).

The Oleaster Family (Elaeagnaceae)

This is a family of shrubs and small trees with one representative in the Arctic, the shrub **Soopolallie** (also called **Soapberry** and **Buffalo Berry**) [al], *Shepherdia canadensis* (fig. 5.30), whose range extends only a short distance into the tundra. South of treeline it is a common tall (over 3 m) shrub, but it seldom grows more than 1 m tall in the north.

Its conspicuous features are its leaves and, in season, its berries; the flowers are small and undistinguished. The leaves grow in pairs, facing each other, up the twigs; they are elliptical in shape, and leathery and stiff in texture. What makes the plant easily recognizable is the "scurf" of glistening rust-colored scales on the backs of the leaves. The scurf is most noticeable on the young, small leaves, still upright near the ends

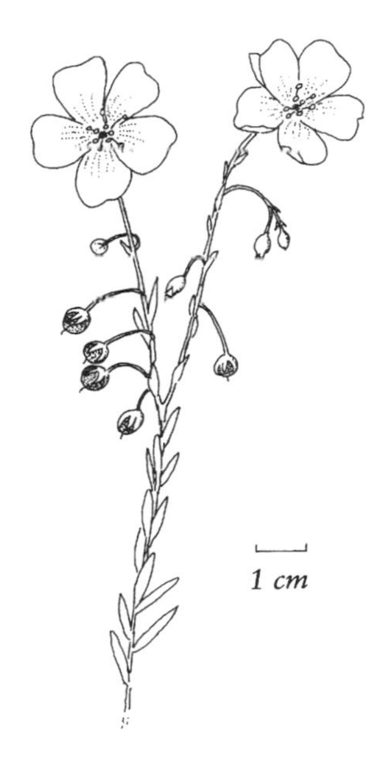

Figure 5.29. Wild Blue Flax.

of the twigs; when full-grown, the leaves spread horizontally, with the scurfy side underneath. The youngest pair of leaves, at the very tip of the twig, are pressed together like hands folded in prayer.

The berries are a bright orange-red, shiny and translucent. Also, they are edible (an acquired taste!). Berries will only be found on some, not

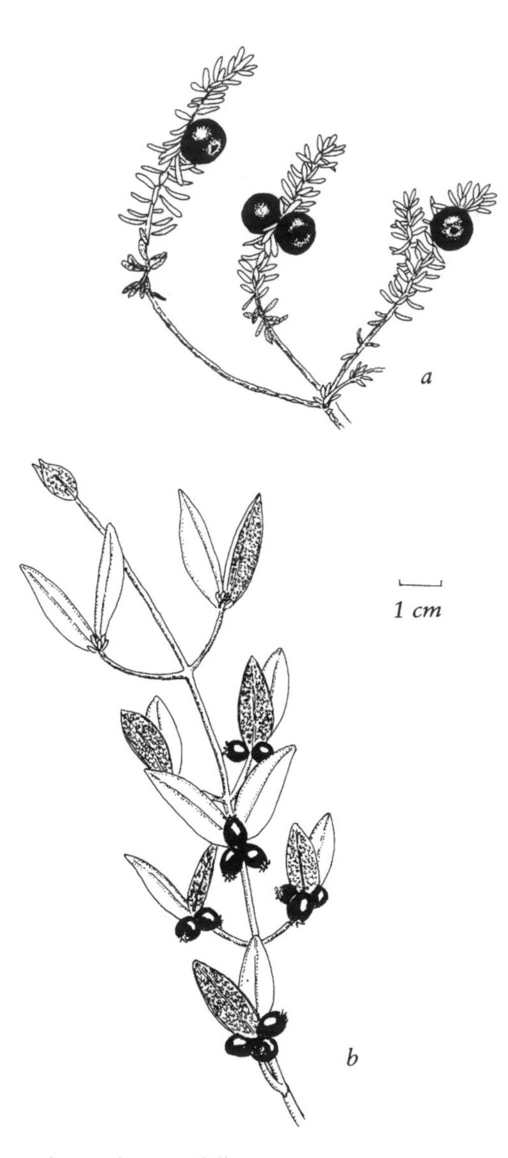

Figure 5.30. (a) Crowberry. (b) Soopolallie.

all, soopolallie shrubs, however, as the shrubs are unisexual: some have only female flowers, from which berries develop; the rest have only male, pollen-producing flowers.

The Willowherb Family (Onagraceae)

Two of the most beautiful and best-known arctic flowers belong to this family. Both are in the genus *Epilobium*. They are **Fireweed** [ALb], *Epilobium angustifolium* (fig. 5.31), and **River Beauty** [AhmLB], *E. latifolium* (fig. 5.31). Both have a number of different English names. Fireweed is often called **Willowherb;** River Beauty is also called **Dwarf Fireweed** and **Broad-leaved Willowherb.**

Fireweed, with its tall spikes of vivid purple-pink flowers, is a common "weed" over much of the northern hemisphere; it quickly colonizes disturbed land, especially old burns—hence its name. It is the territorial flower of Yukon Territory; huge, brilliant patches of it can be seen from great height as you fly into Whitehorse during its flowering season. It is not found in the Arctic Islands, except for southernmost Baffin Island.

River beauty is obviously related to fireweed: the form of their flowers is identical, with four big, bright petals and, below the petals and alternating with them, four narrow, dark red sepals. However, river beauty's flowers are larger and somewhat paler than those of fireweed, and do not grow in tall spikes; the plant is shorter, and its leaves are pale bluish green (glaucous), waxy smooth, and slightly fleshy. River beauty, like fireweed, is a plant pioneer, sometimes on ground that has existed for too short a time to have acquired a cover of vegetation (e.g., newly formed gravel bars in rivers), and sometimes on disturbed ground (e.g., the sandy, gravelly, well-manured ground where colonies of arctic ground squirrels have dug their burrows).

Two more species of *Epilobium* to watch for are *E. palustre* [AL] (fig 5.31) and *E. arcticum* [ahmlb] (not illus.). They are small-flowered and inconspicuous—hence the lack of English names. They are very similar to each other; both have upright, leafy stems and small flowers which may be pink or white. A close look at a flower (preferably with a lens) shows at once that they are miniature versions of the flowers of fireweed and river beauty and share with them the characteristic feature of all *Epilobium* species: what looks like the flower stalk is actually a long, narrow, 4-sided seed capsule (pod) with the 4 sepals and 4 petals rising from the top. When they are ripe, the capsules split open and release quantities of tiny seeds, each bearing a tuft of long, silky white hairs. These two small species can be distinguished by noting that *E. palustre* (but not *E. arcticum*) has runners (stolons) growing from the base of the stem.

Fireweed offers an interesting demonstration of one of the ways by which plants avoid self-fertilization with their own pollen. Its stamens always ripen and shed their pollen before the pistil (the female part of the flower) is ready to be pollinated, which prevents a flower from pollinating itself. While the pistil is unready, its 4-lobed stigma (the organ to

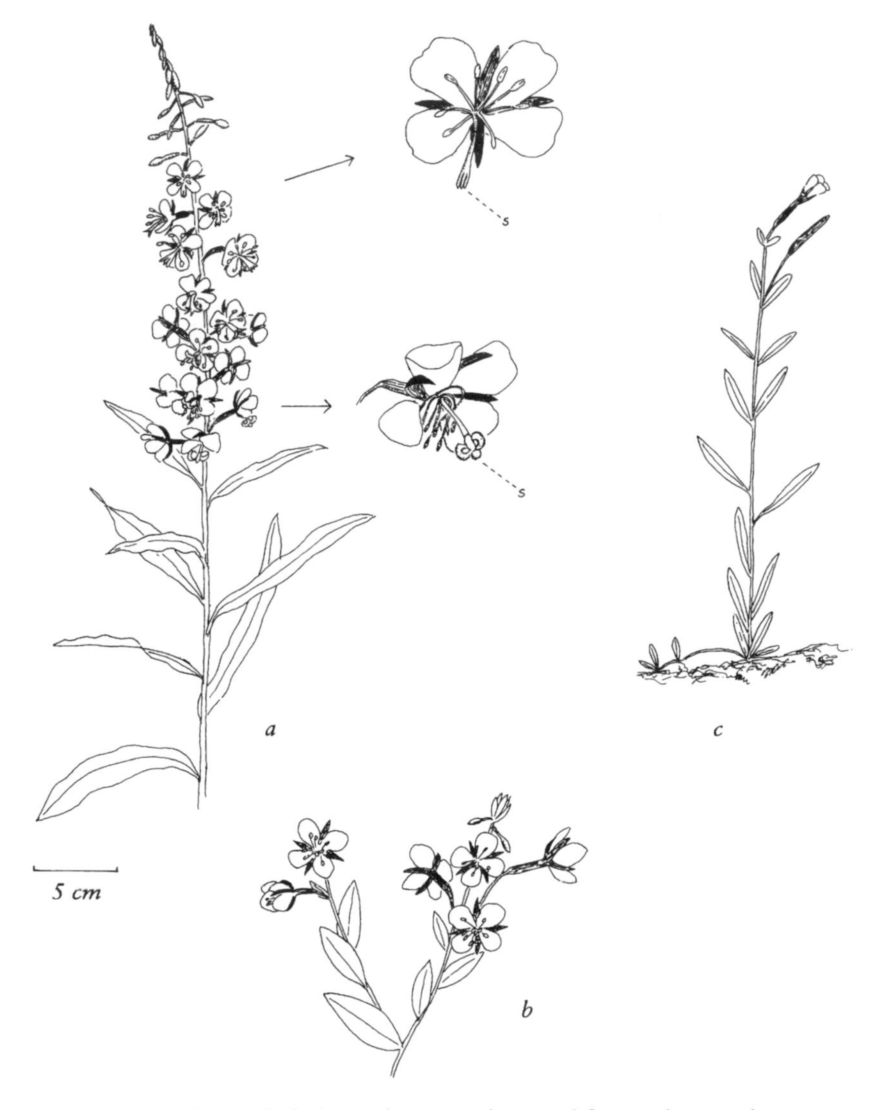

Figure 5.31. (a) Fireweed; the insets show a newly opened flower, above, and a mature flower, below. The stigma is labeled s. (b) River Beauty. (c) *Epilobium palustre.*

which pollen adheres) is closed and tucked in below the stamens (see flower details in fig. 5.31). Then, as the flower ages and the spent stamens, having shed their pollen, droop and wither, the stigma straightens up and its lobes curl back like rams' horns to receive pollen; at this stage, also, it secretes copious nectar. The cross-pollinating is done by bumblebees, which visit the flowers to feed on pollen and nectar, and in so doing carry pollen from flower to flower on their hairy bodies. This would not necessarily ensure cross-fertilization, of course, unless some mechanism prevented pollen from being carried from one flower to another on the same plant.

There is such a mechanism and it works like this. When bumblebees visit fireweed flowers they habitually start at the bottom of a flower spike and work their way upward. The plant is adapted to benefit from this behavior. The flowers at the bottom of a spike always open first and development proceeds upward, so that when the flowers at the top of the spike are at the pollen-producing stage, those at the bottom have reached a later stage, with stigmas curled back to receive pollen, and nectar available to attract the bees. Now consider a bumblebee leaving a spike it has just "finished" and flying to another. It comes from the top of the first spike, where it has just picked up a load of pollen, and alights at the bottom of the next spike, on a flower ready to receive pollen. The bee therefore carries pollen from the flowers of one plant to the flowers of another and cross-pollination is usually achieved (not invariably; occasionally a plant has two or more flower spikes in bloom simultaneously).

River beauty, oddly enough, lacks this adaptation. Its flowers are not in long spikes, and their stamens and pistils ripen at roughly the same time.

The pollen of both fireweed and river beauty is of a beautiful turquoise color; it is well worth examining with a lens.

The Parsley Family (Apiaceae or Umbelliferae)

This is the family whose flowers grow in umbrella-shaped *umbels,* with numerous tiny flowers at the ends of radiating "spokes." Two familiar examples, found throughout mid-latitude North America, are Queen Anne's lace and cow parsley. Worldwide, it is a huge family, and sorting out the species can be troublesome. Only two members of the family grow in the Arctic, however, and they are found only in the western Arctic, a short way beyond treeline. The difference between them is obvious, making recognition easy:

Thoroughwax [Al], *Bupleurum americanum* (fig. 5.32), has compact umbels of bright yellow flowers; the leaves are simple and entire (i.e.,

Figure 5.32. (a) Thoroughwax. (b) Hemlock-parsley.

smooth-edged, with no teeth or lobes). The plants are seldom more than 40 cm tall.

Hemlock-parsley [Al], *Conioselenium cnidiifolium* (fig. 5.32), has flowers that are purple when young, but soon turn pale yellow to creamy white. The leaves are extremely finely divided and lacy. The plants grow up to 60 cm tall.

Because of the strongly contrasting leaves of the two species, they are easily identified even when they have gone to seed.

The Wintergreen Family (Pyrolaceae)

Two members of this family are quite common in the Low Arctic and are easy to recognize:

Large-flowered Wintergreen [AhmLB], *Pyrola grandiflora* (fig. 5.33), has a spike of fragrant, cream-white flowers at the top of a leafless stalk growing up from a clump of wintergreen leaves (see sec. 5.7). The leaves are dark green, shiny, and leathery, with rounded tips. The distinctive characteristic of the plant, making identification certain, is the long, thick, downward-curving style at the center of each flower (see the inset figure).

Figure 5.33. (a) Large-flowered Wintergreen; the enlarged single flower shows the curved style, s. (b) One-sided Wintergreen.

A similar, related plant found in moist places in the southern tundra is **One-sided Wintergreen** [AmL], *Orthilia secunda* (fig. 5.33). The plant looks like a stunted specimen of large-flowered wintergreen; its inconspicuous greenish flowers are only one-half the size of those of its showier relative, and they are ranged in a row along one side of the flower spike.

The Heath Family (Ericaceae)

Many of the Arctic's most spectacular plants belong to this family, which is mostly confined to the mainland Arctic and islands close to the mainland. Only two species are to be found in the High Arctic.

Arctic members of the family can conveniently be divided into three groups: Group I has 3 species with showy, upward-facing flowers. Group II has 4 species with hanging, bell-shaped flowers. Group III has 5 species with juicy berries (in the first two groups the seed capsules are dry when ripe).

Group I consists of **Lapland Rosebay** [AmLB], *Rhododendron lapponicum;* **Alpine Azalea** [ALb], *Loiseleuria procumbens;* and **Labrador Tea** [AmLB], *Ledum decumbens* (all in fig. 5.34). They are all dwarf shrubs with strikingly beautiful flowers.

Lapland rosebay's flowers are big (up to 2 cm), fragrant, purple funnels; their resemblance to garden rhododendrons makes them easy to recognize.

Alpine azalea's flowers, though somewhat similar, are paler—pink rather than purple—and smaller than those of Lapland Rosebay. These two flowers can be distinguished with certainty by noting that the azalea has 5 stamens, the rosebay 10.

Labrador tea is unmistakable; its 5 white petals are separate from each other; its flowers are in neat, round clusters; its narrow leaves have their edges rolled under, and their lower surfaces densely covered with rust-colored "wool." This plant is a dwarf version of the common Labrador tea (*Ledum groenlandicum*) found in, but not beyond, the northern forests, and is instantly recognizable by anybody familiar with the forest species.

Group II contains two common and widespread members: **Arctic White Heather** [AHMLB], *Cassiope tetragona,* and **Bog Rosemary** [AL], *Andromeda polifolia* (both in fig. 5.35). Two much less common plants in the group are also mentioned below.

Arctic white heather is one of the two members of the Heath Family that grows in the High Arctic, where it is confined to sites protected by deep snowbanks in winter (see sec. 5.7). It is a low sub-shrub with pure white flowers and stiff, quadrangular, dark green branches clothed in 4 rows of scalelike leaves that overlap each other like tiles on a roof. In the same genus, but found only in the eastern Arctic, is **Moss Heather** [lb], *Cassiope hypnoides* (not illus.). Its flowers closely resemble those of the common species, but the twigs trail instead of standing stiffly upright, and the leaves are thinner and softer.

Bog rosemary is a low, spreading shrub. Its evergreen leaves are long, narrow, and curled under along the edges; their lower surfaces are a

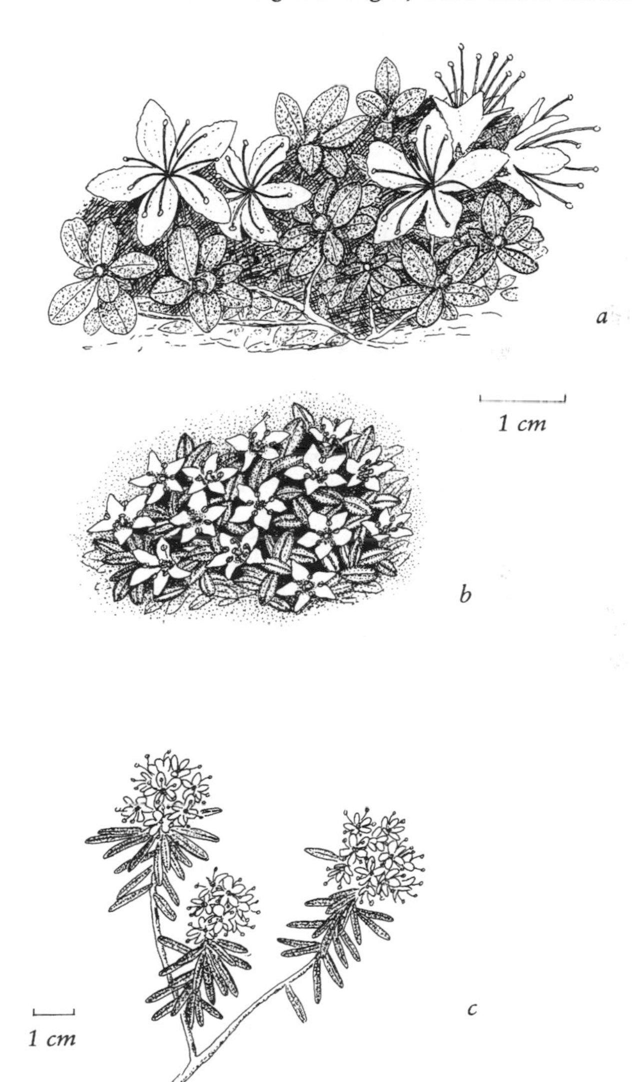

Figure 5.34. (a) Lapland Rosebay. (b) Alpine Azalea. (c) Labrador Tea.

light turquoise color which combines with the vivid pink of the flowers to make a notably beautiful color combination.

A rare heather to be looked for in southern Baffin Island and near treeline on the mainland is **Mountain Heather** [mLB], *Phyllodoce coerulea* (not illus.). Its flowers are purple-pink and its leaves crowded, evergreen, needlelike, and shiny.

Group III contains three genera: *Arctostaphylos, Vaccinium,* and *Oxycoccus.*

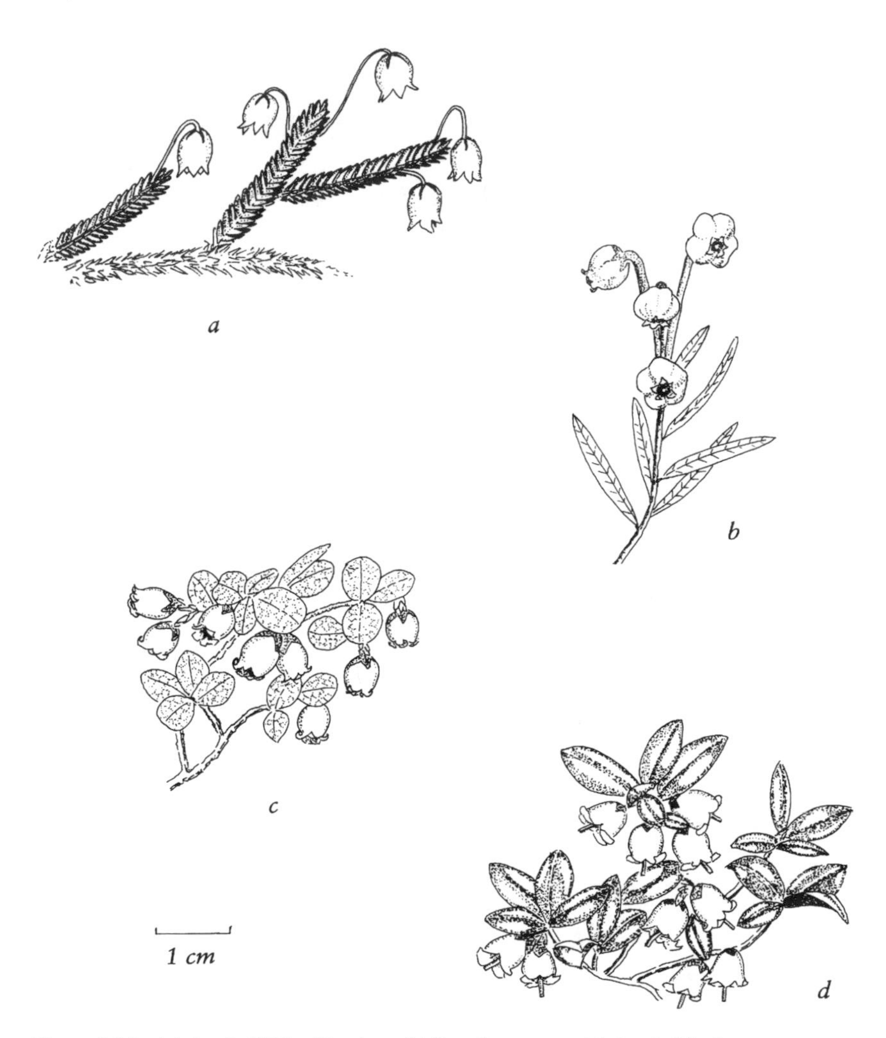

Figure 5.35. (a) Arctic White Heather. (b) Bog Rosemary. (c) Arctic Blueberry. (d) Lingonberry.

Vaccinium is the genus to which bilberries, blueberries, huckleberries, and cranberries belong (the English names are inconsistent and confusing: everybody has a private version). The two arctic species are **Arctic Blueberry** [AHmLB], *V. uliginosum,* and **Lingonberry** or **Mountain Cranberry** [AmLb], *V. vitis-idaea* (both in fig. 5.35). They are low, ground-hugging sub-shrubs that often carpet the ground. When in fruit they are unmistakable. Both have delicious juicy berries. Arctic blueberry has blue-black berries with a blue bloom and nonshiny deciduous leaves that turn plum-colored in the fall; apart from arctic white heather (described above) it is the only member of the heath family in the High Arctic. Lingonberry has shiny, scarlet berries and shiny, evergreen leaves. Both species have downward-looking bell-shaped flowers (like those of the plants in Group II), white to pale pink in color.

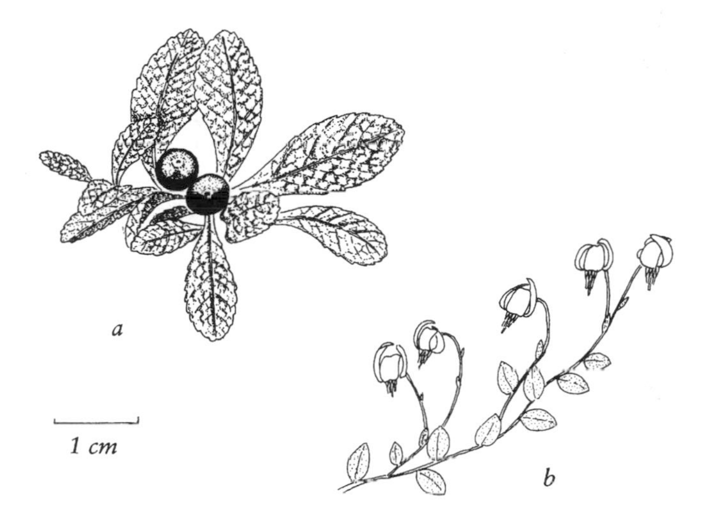

Figure 5.36. (a) Black Bearberry. (b) Cranberry.

Arctostaphylos is the bearberry genus. Its 2 arctic species are **Black Bearberry** [AmLB], *A. alpina* (fig. 5.36), and **Red Bearberry** [Amlb], *A. rubra* (not illus.), named for the colors of their berries. They are easily identifiable when the berries are ripe. Both are trailing shrubs with white, bell-shaped flowers and wrinkled leaves that turn brilliant red in the fall; the leaves are thicker and more deeply wrinkled in black bearberry and stay on the plant for several years; those of red bearberry fall off before winter. Bearberries without fruit could be mistaken for net-veined willows, as they have similarly wrinkled leaves. To distinguish them, look at the junction of the leaf blade and leaf stalk: the junction is abrupt in net-veined willows (see fig. 5.11), whereas in bearberries the base of the blade tapers down into the stalk (fig. 5.36). (Note: bearberries belong to the same genus as kinnikinick, *A. uva ursi,* a prostrate, evergreen, red-berried shrub familiar to all visitors to the evergreen forests of Canada and northern United States.)

Oxycoccus has one common species, **Cranberry** [al], *O. microcarpus* (fig. 5.36). Its pink flowers are unlike those of its berry-bearing relatives (described above) in having the petals curled back. The plant as a whole is tiny, with trailing threadlike stems, often hidden in moss, and tiny oval leaves, widely spaced along the stem, and only half as big as the cranberries. It is found only in the southern tundra, not far north of treeline.

The Diapensia Family (Diapensiaceae)

The two arctic members of this family are both diapensias, ground-hugging dwarf shrubs, covered (in season) with white, upward-facing flowers, each a 5-lobed bell with 5 stamens. The flowers are much larger than the crowded dark green leaves. To avoid confusion between diapensias and members of the Heath Family, note that most of the latter have 10 stamens, rather than 5; the exceptions are alpine azalea, whose flowers are pink, and Labrador tea, which can have 5 or 10 stamens, but whose leaves are long and narrow, with rust-colored "fur" on the undersides.

The two diapensias are rather similar, but at no risk of being confused with each other because of their well-separated geographic ranges. Some botanists treat the western species as a subspecies of the eastern.

The eastern species is **Lapland Diapensia** [LB], *Diapensia lapponica* (fig. 5.37), which grows no farther west than the western boundary of the Canadian (Precambrian) Shield.

The western species is **Alaska Diapensia** [A], *D. obovata* (not illus.), which is not found east of the Mackenzie River.

The Primrose Family (Primulaceae)

Four genera belonging to this family grow in the Arctic, but only one, the primrose genus, *Primula,* consists of obvious "primroses," recognizable as such by any gardener.

Consider the 3 arctic species of *Primula* first. All have similar flowers, with 5 deeply notched, pinkish-lilac (or, occasionally, white) petals and a yellow "eye" at the center. In only one western species are the flowers large enough (1.5 cm across) to be showy. This species is **Northern Primrose** [Al], *P. borealis* (fig. 5.37), a plant that flourishes in the brack-

Figure 5.37. (a) Lapland Diapensia. (b) Northern Primrose. (c) Rock-jasmine. (d) Fairy Candelabra. (e) Ochotsk Douglasia. (f) Northern Shooting Star.

ish water of coastal meadows. The other 2 species have flowers of the same shape and color, but seldom more than half as large. They have no English names. They are *P. egalikensis* [AL] (not illus.) and *P. stricta* [amL] (not illus.). Both are plants of wet meadows. They can be distinguished by noting that *P. egalikensis* has, and *P. stricta* lacks, tiny glandular hairs on the calyx. (A glandular hair is a hair ending in a minute, sticky globe, hard to see without a lens.)

The remaining genera of the family are not obvious primroses. But all have 5 petals joined at the base; and 5 stamens, each one aligned with a petal.

The genus *Androsace* has 2 arctic species, **Rock-jasmine** [Aml], *A. chamaejasme,* and **Fairy Candelabra** [AhmLb], *A. septentrionalis* (both in fig. 5.37). Rock-jasmine is strikingly beautiful; it is a western species, not found east of 100° W longitude. Its flowers form a cluster atop a leafless, hairy flower stalk that grows from the center of a rosette of small, hairy leaves. The flowers are creamy white to begin with, and turn pink before withering; each has a bright yellow "eye." Although the color pattern sometimes resembles that of northern primrose (see above), rock-jasmine's petals have the tips rounded instead of cleft. Fairy candelabra has flowers of the same form as those of rock-jasmine, but they are smaller and are not crowded together. Instead, they are arranged in "candelabra" (technically, *umbels*): the stalks of the individual flowers radiate from a single point at the top of a larger stalk, which itself grows up, with others like it, from a rosette of leaves.

The genus *Douglasia,* found only in the west, also has 2 arctic species. Both are dense cushion plants, sometimes completely covered with bright purple-pink, five-petaled flowers. The only plants with which either could be confused are moss-campion (see the Pink Family) and alpine azalea (see the Heath Family). To disinguish them, note that moss-campion has notched petals that are completely separate from each other, whereas in the *Douglasia* species the rounded petals are joined at the base; in alpine azalea, the petals, though joined, are pointed. The species are **Ochotsk Douglasia** [A], *D. ochotensis* (fig. 5.37), and **Arctic D** [a], *D. arctica* (not illus.). The latter is apparently limited (in the Arctic) to the Yukon north slope, the only area where both species occur. To distinguish them, note that Ochotsk douglasia, but not arctic douglasia, has hairs on the upper surfaces of the leaves. Both are spectacularly beautiful plants.

Northern Shooting Star [A], *Dodecatheon frigidum* (fig. 5.37), is so unlike a primrose that it comes as a surprise to nonbotanists to learn that it belongs in the primrose family. Its brilliant pink-purple petals, instead of cupping the stamens, point backward. The stamens, which are joined to form a tapering tube, point forward like a rocket's nose cone. Several other shooting stars (other species of *Dodecatheon*) grow in temperate western North America, but this is the only one found north of treeline.

The Leadwort Family (Plumbaginaceae)

Thrift [AhMLB], *Armeria maritima* (fig. 5.38), is the only arctic member of this family. It is commonest near the seashore (though seldom on the beach), but can sometimes be found on gravelly lakeshores far inland on the tundra. The plant consists of a cushion or mat of narrow, slightly fleshy leaves growing in dense tufts. The bright pink flowers are individually small, but are packed into showy hemispherical heads, each atop a stiff, leafless stalk about 15 or 20 cm long. The plant will be familiar to gardeners as it is often grown as an ornamental.

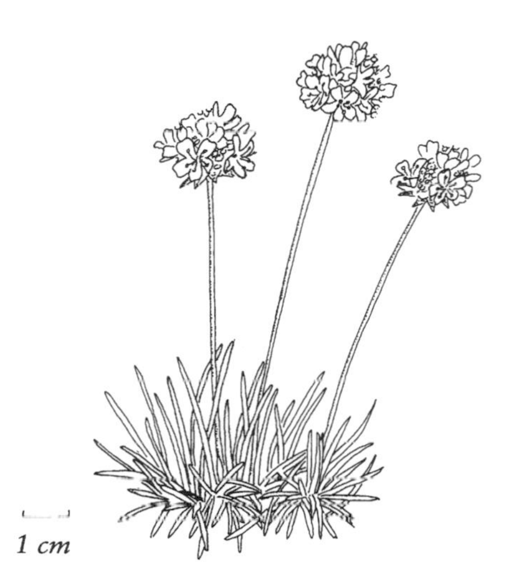

Figure 5.38. Thrift.

The Gentian Family (Gentianaceae)

Two genera belonging to this family are found in the Arctic.

The true gentians (*Gentiana*) have 4 arctic species. None has flowers as large or as showy as those of cultivated gentians, or the big, royal blue, wild gentians of the western mountains in more southerly latitudes. On the contrary, the arctic species are rather inconspicuous. They are most easily recognized as gentians by examining the flowers when they are closed, i.e., when they are still in bud or have closed because the sun has stopped shining. In most gentians, the flowers open only in sunshine, and in some species not even then. The flower has 4 or 5 petals (either number is possible, even within a single species) which are united into a tube at the base and spread apart at the tips when the flower is open; when it is closed the petal tips fold in, turning the flower into a narrow box with a pyramidal top. The leaves, some with side branches growing from their axils, are arranged in a *decussate* pattern, i.e., they are in opposite pairs, with each pair at right angles to the pair above and below it. The 4 species are quite distinct:

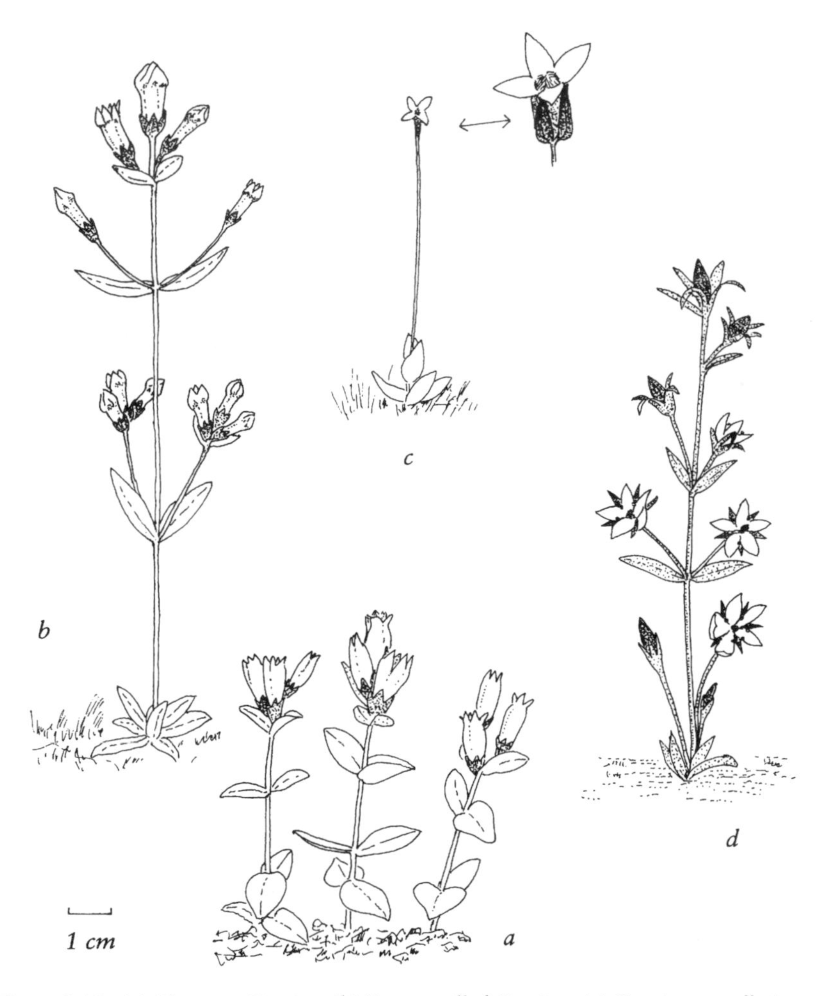

Figure 5.39. (a) Glaucous Gentian. (b) Four-petalled Gentian. (c) *Gentiana tenella* (note fringe in throat). (d) Marsh Felwort.

Glaucous Gentian [A], *Gentiana glauca* (fig. 5.39), seems to be unique among plants in having flowers of a greenish-blue, almost aquamarine color. The remarkable color makes the plant unmistakable.

Four-petaled G [AL], *G. propinqua* (fig. 5.39), has deep blue or purplish-blue flowers at the top of upright stems and (usually) on one or more pairs of side branches. The number of petals is occasionally 5, in spite of the plant's name. This gentian is seldom found far beyond treeline.

Moss G [A], *G. prostrata* (not illus.), is a small plant with weak stems that lie along the ground for most of their length. The flowers are pale blue, and open only in sunshine.

Gentiana tenella [al] (fig. 5.39), because it is uncommon, has no English name. It is small and inconspicuous and found only near the seashore. Its noteworthy feature is the "fringe" at the base of the 4 petals, closing the top of the flower tube.

One other member of the Gentian Family grows in the Arctic, **Marsh Felwort** [AL], *Lomatogonium rotatum* (fig. 5.39), also called **Star Gentian.** The latter name is descriptive: the plant is gentianlike in most respects, but its flowers open wide to form a 5-pointed star; the petals are joined only at the base, rather than being united into a tube for most of their length as are the petals of the true gentians. Marsh felwort is strictly a seaside species. Being small and dark-colored, it often goes unnoticed, but it is worth watching for because of its unusual color: the whole plant is purplish blue. The flowers match the rest of the plant when closed, but in sunshine they open into sky-blue "stars."

The Buckbean Family (Menyanthaceae)

Buckbean [AL], *Menyanthes trifoliata* (fig. 5.40), is the only member of its family to grow in the Arctic. It is a strikingly beautiful plant found growing in bogs and ponds throughout the world's north temperate zone, and for a short distance north of treeline into low arctic tundra. Sometimes it grows in wet soil, sometimes in shallow ponds with the base of the plant submerged and only the leaves and flowers emerging above water. Its handsome white flowers are grouped into a spike at the top of a leafless stalk and are unusual in having "bearded" petals; the "beards" consist of numerous thin, white filaments growing up like curly hairs from the petals' surfaces. The plants have distinctive leaves, too. Each has 3 oval leaflets.

The Phlox Family (Polemoniaceae)

Several members of this family, all with showy, beautiful flowers, grow in the Arctic, but only in the west. They belong to two genera, phlox (*Phlox*) and Jacob's ladder (*Polemonium*).

Two species of phlox occur, one common (within its restricted range) and one rare. The common species is **Siberian Phlox** [A], *Phlox alaskensis* (fig. 5.41), found in arctic Yukon and Alaska. It grows as a compact mat, covered, in season, with large (2 cm diameter) circular flowers ranging in color from deep pink to pale pinkish white; the center of the flower is often outlined with a ring of deeper color. The rare species is **Richardson's Phlox** [aml], *P. richardsonii* (not illus.), known only from a few sites on the coast of the western arctic mainland and Banks Island; its flowers are only half the size of those of Siberian phlox and are fragrant.

Jacob's ladder, so called because of its supposedly ladderlike leaves (they are pinnate; see the figure) has three species in the western Arctic: **Northern Jacob's Ladder** [Ahml], *Polemonium boreale* (fig. 5.41), **Tall J L** [Al], *P. acutiflorum* (not illus.), and **Showy J L** [Al], *P. pulcherrimum*

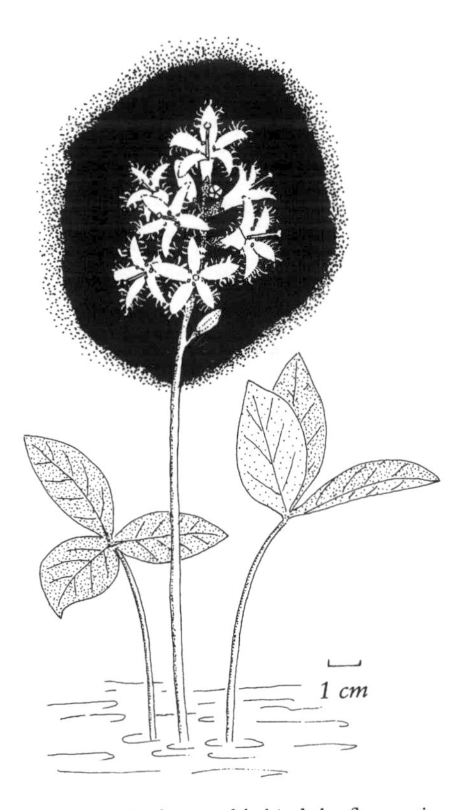

Figure 5.40. Buckbean (the dark background behind the flowers is to emphasize their fringed petals).

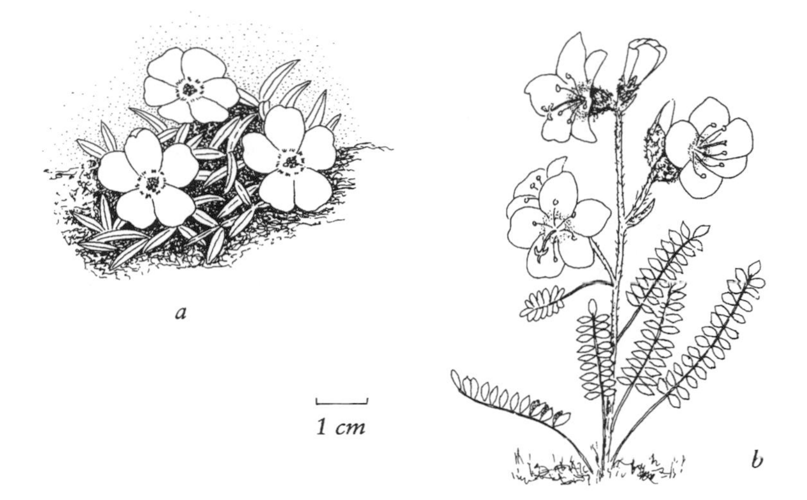

Figure 5.41. (a) Siberian Phlox. (b) Northern Jacob's Ladder.

(not illus.). All have cup-shaped blue flowers with prominent stamens and the distinctive ladderlike leaves. To tell the species apart note that tall Jacob's ladder is the only tall (up to 40 cm) species, and it has hairless leaves. The other two species have yellow-centered flowers and leaves with sticky hairs. Where they differ is that in northern Jacob's ladder the calyx is densely covered with soft hairs, and the leaves usually have fewer than 10 pairs of leaflets; in showy Jacob's ladder the calyx is only slightly hairy, if at all, and the leaves have 10 to 15 pairs of leaflets.

The Forget-me-not Family (Boraginaceae)

Members of this family are easy enough to recognize individually, though the feature they have in common, which unites them in one family, is not at all conspicuous. It is that all of them have flowers producing exactly four seeds, each separately packaged as a "nutlet"; a quartet of these nutlets, arranged like the quarters of a miniature hot-cross bun, can be seen (with a lens) in the bottom of each flower. The only other family bearing such quartets of nutlets is the Mint Family, which has no arctic representatives. (The distinction between the families is that the leaves grow alternately up the stem in the Forget-me-not Family and are arranged in opposite pairs in the Mint Family.)

Four members of the Forget-me-not Family are quite common in the Arctic. All have blue flowers.

First come the forget-me-nots (fig. 5.42), recognizable by anyone familiar with their garden relatives. The two arctic species (found only in the western Arctic) are especially bright and handsome. Surprisingly, they belong to different genera even though their flowers—small, brilliantly blue, and with yellow "eyes"—look almost identical. The distinction is in their leaves.

Arctic Forget-me-not [A], *Eritrichium aretioides*, has a compact cushion of tiny leaves that look pale gray because they are densely covered with hairs.

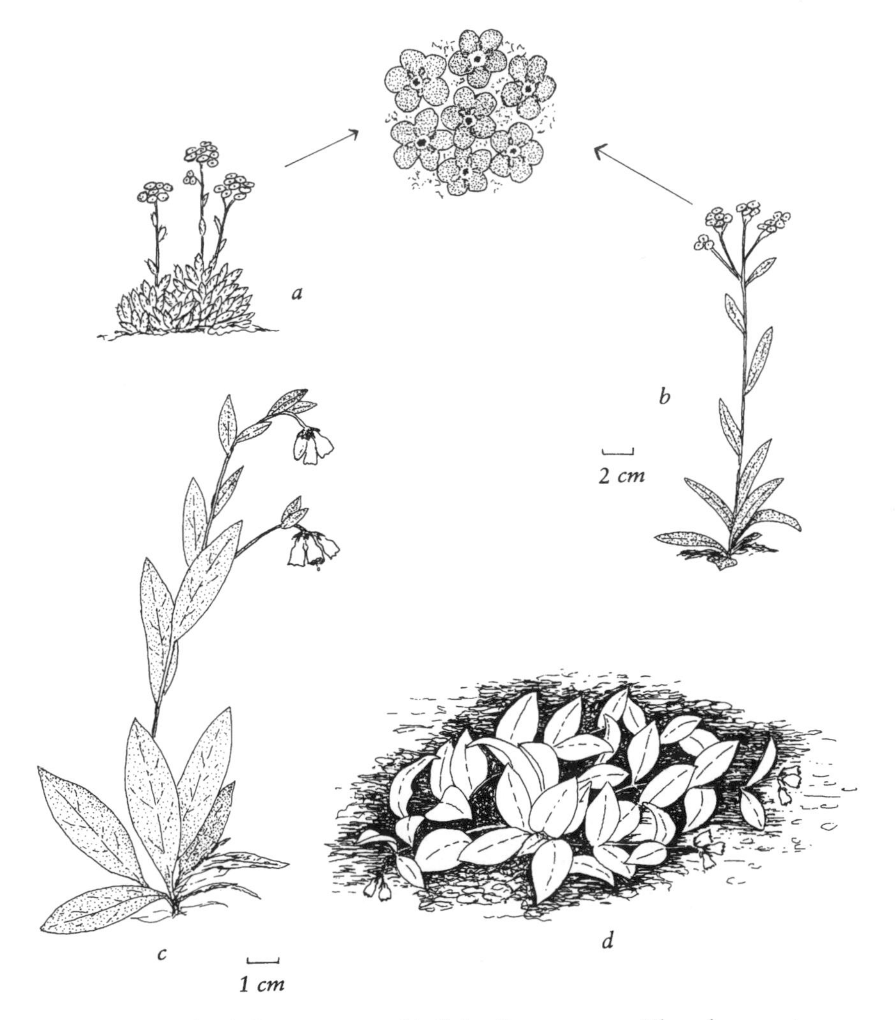

Figure 5.42. (a) Arctic Forget-me-not. (b) Alpine Forget-me-not. (The enlargement shows the flowers of both Forget-me-not species.) (c) Lungwort. (d) Sea Lungwort.

Alpine Forget-me-not [A], *Myosotis alpestris,* is taller, with long, uncrowded leaves which, though fairly hairy, are green in color. This is the state flower of Alaska.

The two other members of this family to watch for both belong to the genus *Mertensia,* in which the unopened flower buds are often pink or purple, gradually turning blue as they open.

Lungwort [al], *Mertensia paniculata* (fig. 5.42), doesn't grow far beyond treeline. Its flowers are hanging blue bells, and its leaves are rough to the touch because covered with bristly hairs.

Sea Lungwort or **Oysterleaf** [AmLB], *M. maritima* (fig. 5.42), is confined to seashores of the Low Arctic. It is a sprawling plant with smooth, fleshy, blue-green leaves and vivid blue, bell-shaped flowers.

The Lousewort Family (Scrophulariaceae)

Five genera of this big family grow in the arctic: louseworts (*Pedicularis*), Indian paintbrushes (*Castilleja*), eyebright (*Euphrasia*), speedwell (*Veronica*) and lagotis (*Lagotis*). Only the louseworts reach the High Arctic.

Louseworts as a group can be recognized by their spikes of colorful flowers and by their finely cut leaves, easily mistaken for tiny fern fronds. The individual flowers are *irregular* (see preliminary notes); the color depends on the species, ranging from cream or yellow to deep red or purple. Nine species grow in the Arctic.

Capitate Lousewort [AHMLb], *Pedicularis capitata* (fig. 5.43), has big (up to 4 cm) cream-colored flowers sometimes tinged with pink.

Labrador L [ALb], *P. labradorica* (fig. 5.43), is the only species having (when full-grown) a branching stem with a flower spike at the tip of each branch; all other arctic louseworts have a single flower spike at the top of an unbranched stem. Its flowers are yellow, tinged with brown on top of the "helmet."

Sudetan L [AhMLB], *P. sudetica* (fig. 5.43), has bicolored flowers. The helmet of the flower is deep crimson, the lip pale pink spotted with red. The flower helmets are all twisted to one side, so that seen from the top, the spike shows a spiral pattern (see the figure).

Whorled L [A], *P. verticillata* (not illus.), is the only lousewort with "verticillate" leaves, i.e., with whorls of 3 or more leaves radiating from the stem like spokes. The flowers are bright reddish purple.

Lapland L [ALb], *P. lapponica* (not illus.), has fragrant, pale yellow flowers and bronze stems and leaves.

Flame L [LB], *P. flammea* (not illus.), a small, vividly colored species, has brilliant yellow flowers except for the tip of the helmet, which is bright purple-red. It is an eastern species, rare west of 100° W longitude.

The three remaining species, arctic, hairy, and woolly louseworts, are all rather alike. To distinguish them, note that **Arctic Lousewort**

[AHMlb], *P. arctica* (not illus.), is the only one of the three lacking a dense fuzz of hairs. **Hairy L** [HmlB], *P. hirsuta* (not illus.), and **Woolly L**, *P. lanata* (fig. 5.43), are hard to separate unless they can be compared side by side. Both are coated with white woolly hairs when young, especially woolly lousewort, which is much the woollier of the two and also has darker flowers. Note too that hairy lousewort is an eastern species, unlikely to be found west of 100° W longitude.

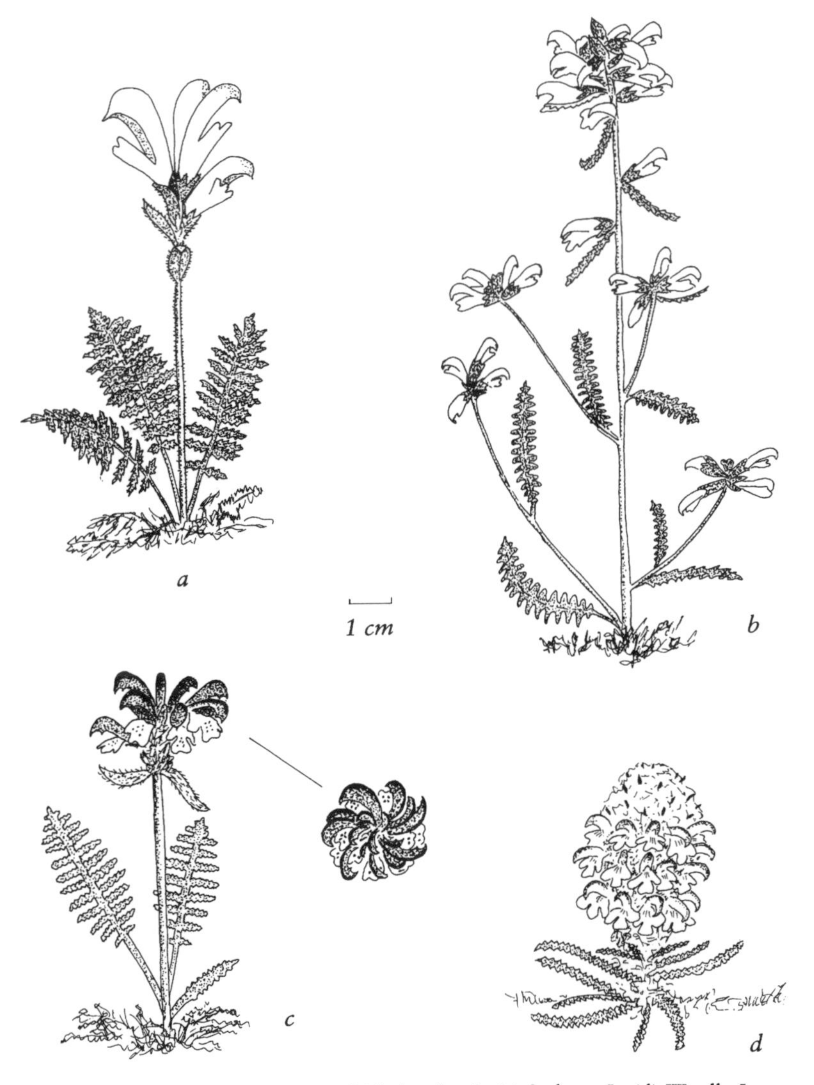

Figure 5.43. (a) Capitate Lousewort. (b) Labrador L. (c) Sudetan L. (d) Woolly L.

Indian paintbrush is another genus with several arctic species; they are hard to differentiate without a technical Flora and not easy even with one. All the paintbrushes seem, at first sight, to have spikes of brightly colored flowers; close inspection shows that it is the bracts (special leaves that shield the flowers) that are colorful; the true flowers,

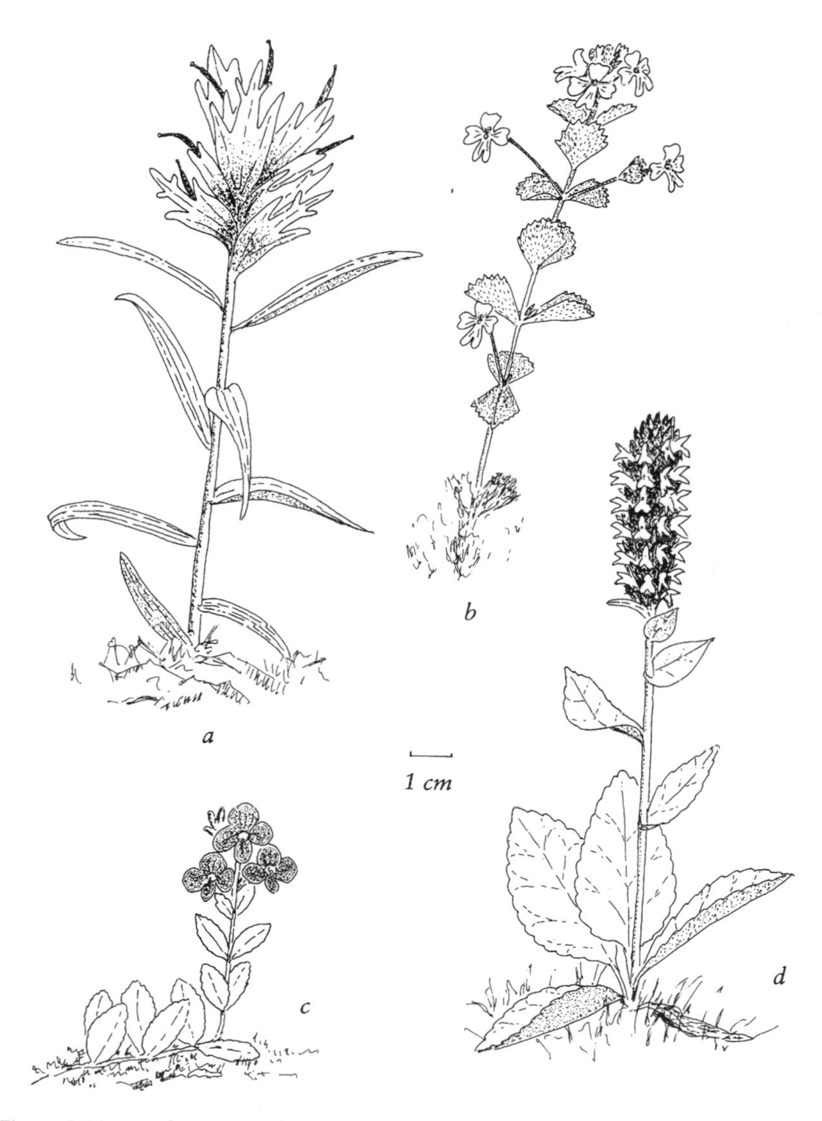

Figure 5.44. (a) Elegant Paintbrush. (b) Arctic Eyebright. (c) Alpine Speedwell. (d) Lagotis.

hidden among the bracts, are inconspicuous and resemble tiny lousewort flowers in structure. The most wide-ranging species is **Elegant Paintbrush** [AML], *Castilleja elegans* (fig. 5.44), a very beautiful plant whose bracts shade from deep pink at the base to pure white at the tip.

Three other members of the lousewort family (all in fig. 5.44) are worth watching for by naturalists who visit their ranges. All are easy to recognize. They are **Arctic Eyebright** [lb], *Euphrasia arctica;* **Alpine Speedwell** [lb], *Veronica wormskjoldii;* and **Lagotis** [A], *Lagotis stelleri.* All three are widespread in forested latitudes and have localized ranges north of treeline.

Arctic eyebright is an eastern species found in southernmost Baffin Island and on the shores of Hudson Bay. Its flowers are white with blue veins; its leaves, arranged in pairs up the stem, are wedge-shaped and deeply toothed.

Alpine speedwell occupies much the same territory as arctic eyebright north of treeline, but south of treeline it ranges across the continent. Its brilliant blue flowers have 4 petals and only 2 stamens; the petals are slightly unequal in size, making the flower weakly irregular. The leaves are softly hairy.

Lagotis is fairly common beyond treeline in Yukon and Alaska. It has irregular blue flowers, arranged in a fat, tightly packed spike. Its large basal leaves (growing directly from the ground) are slightly fleshy.

The Broomrape Family (Orobanchaceae)

This family consists wholly of nongreen, parasitic plants; lacking chlorophyll, they cannot make their own food, but must suck the food-containing sap from other, self-reliant plants. This they do through tubes that grow out from the parasitic plant's base and attach themselves like leeches to the host plant's roots. Only one member of the family grows in the Arctic.

Poque [Al], *Boschniaka rossica* (fig. 5.45), is found only a short way north of treeline, parasitizing alder shrubs. (South of treeline, it also parasitizes spruce trees.) Its other name—**Ground Cone**—describes its appearance. The whole plant looks like a pine cone, up to 20 cm long, standing upright on the ground; the thick stem is densely covered with dark flowers, each with a dark-brown, scalelike leaf below it. The flower color is variable, usually dark reddish brown with a paler, sometimes yellowish, interior. Look for it wherever alders grow; dead and living specimens of poque will often be found. At first glance, living specimens look dead, but they turn out to be fresh and resilient when you touch them.

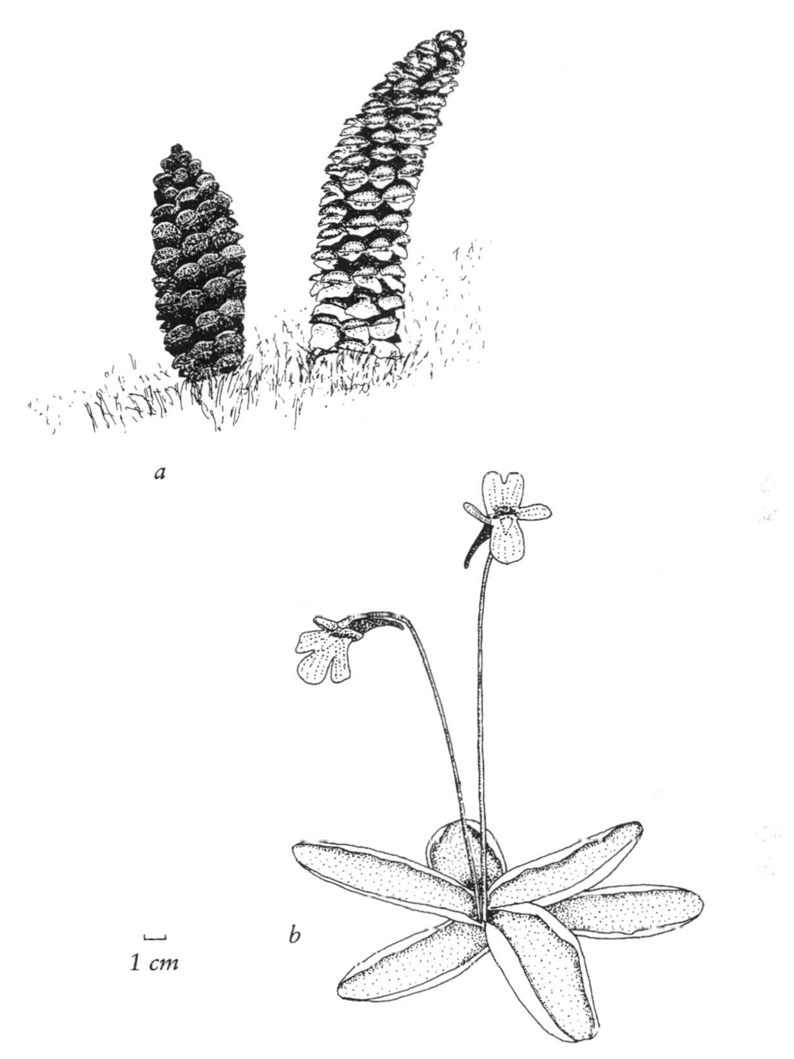

Figure 5.45. (a) Poque. (b) Common Butterwort.

The Butterwort Family (Lentibulariaceae)

The only conspicuous member of this family is **Common Butterwort** [AL], *Pinguicula vulgaris* (fig. 5.45), which grows in wet ground and has a range extending only a short distance north of treeline. Its attractive flowers are easy to mistake for violets because of their bright purple color and their shape (they are spurred).

A look at the leaves makes it obvious that these are not violet plants, however. The leaves are spread in a rosette and are flat to the ground except for their curled-up edges; they are pale yellowish green in color and have shiny, greasy-looking upper surfaces, which (on examination with a 10× lens) seem to have a minutely pebbled texture.

The plants are carnivorous. The leaf surface is covered with mucilage droplets—hence the greasy, pebbled appearance—which prevents small insects that settle on it from flying away; leaves are usually speckled with the bodies of captured insects, in the process of being digested and absorbed as nourishment by the plant.

Another butterwort to watch for is **Hairy Butterwort** [al], *P. villosa* (not illus.). It resembles a small, inconspicuous version of common butterwort but is less abundant, being restricted to acid bogs. Its flowers are less than 1 cm long (compared with 1.5 cm for the common species), and the lower part of the flower stalk is slightly hairy (hairless in the common species).

The Bedstraw Family (Rubiaceae)

The only arctic member of this family is **Northern Bedstraw** [Al], *Galium boreale* (fig. 5.46). Its whorls of 4 narrow leaves, radiating from the stem like spokes, resemble those of the low latitude bedstraws (often called cleavers or goose grass) which are familiar weeds in populous regions. But whereas the weedy bedstraws have tiny, widely spaced flowers and long stems that trail weakly over other vegetation, northern bedstraw is a conspicuously attractive plant. It has upright stems, and the little white flowers, like 4-pointed stars, are crowded together in showy clusters.

The range of the plant extends only a short way beyond treeline in the eastern Arctic; it reaches the coast in the west.

The Valerian Family (Valerianaceae)

The only arctic member of the family is **Mountain Heliotrope** [A], *Valeriana capitata* (fig. 5.46); it is confined to the western Arctic.

Its small, fragrant flowers are pale pink to lilac when the buds first open, but become white with age. They are packed into showy hemispherical heads at the top of tall (occasionally up to 1 m) stems which bear one or two pairs of opposite leaves. These stem leaves are unstalked and bear tapering marginal teeth. The leaves from the base of the plant, in contrast, have stalks and are smoothly elliptical in shape.

The Harebell Family (Campanulaceae)

Three members of this family, all belonging to the genus *Campanula,* grow in the Arctic. All have sky-blue flowers. To distinguish them, consider their geographic ranges first.

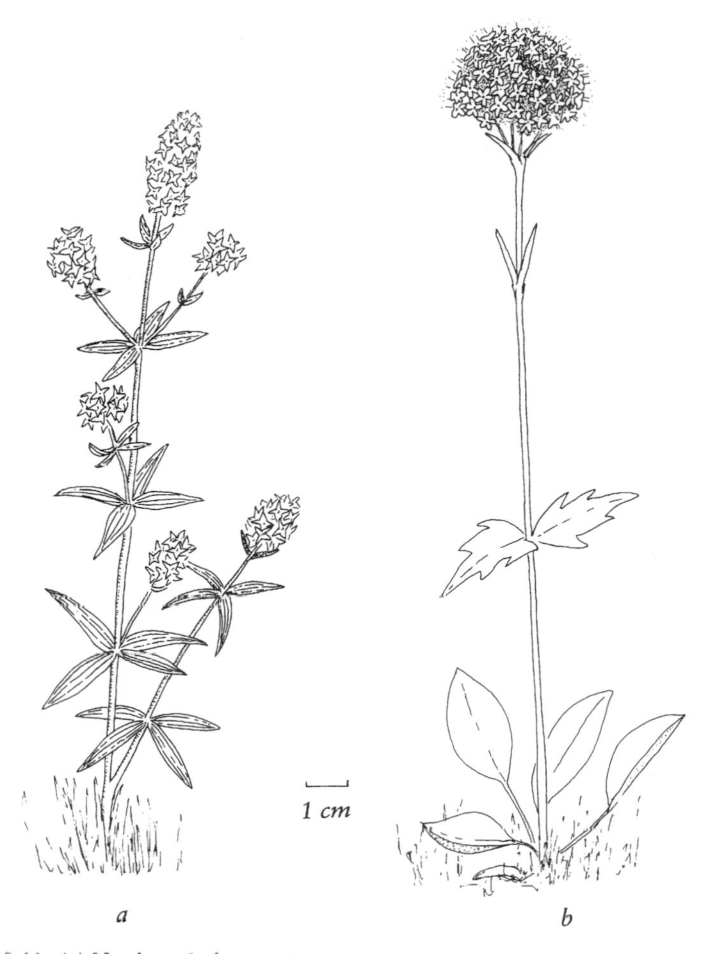

Figure 5.46. (a) Northern Bedstraw. (b) Mountain Heliotrope.

Common Harebell [b], *Campanula rotundifolia* (fig. 5.47), is not uncommon on grassy hillsides throughout the north temperate zone; it is found north of treeline only in southeastern Baffin Island. **Western H** [A], *C. lasiocarpa* (fig. 5.47), ranges throughout the western mountains, and occurs north of treeline only on the arctic slope of Alaska overlooking the Beaufort Sea. **Arctic H** [ahmLB], *C. uniflora* (fig. 5.47), is the one truly arctic species, found throughout the barren lands, the Arctic Islands, and the Alaskan arctic slope. Thus there is no risk of confusing common and western harebells as they never grow together: they are separated by the width of the continent. Arctic harebell is the only species of the three that might be found in company with one of the others. It is easily recognized by the shape of its corolla (the tube of joined petals), which is like a narrow funnel; in the other two species the corolla is bell-shaped.

Other points to notice are these: common harebell has a smooth (non-hairy) calyx and often (not always) has several flowers on one stem; in the other two species, the calyx bears a fuzz of hairs, and there is never more than one flower per stem. The Latin name *rotundifolia* for common harebell is misleading; only the lowermost leaves are "rotund," and they wilt and disappear when spring is over; thereafter, all the leaves are long and narrow. Arctic and western harebells, besides having corollas of different shapes, can also be told apart by their leaves: the leaf edges are smooth in arctic harebell, toothed in western harebell.

Figure 5.47. (a) Common Harebell. (b) Western Harebell. (c) Arctic Harebell.

The Daisy Family (Asteraceae or Compositae)

The Daisy Family is one of the biggest in the whole plant kingdom (about 25,000 species, worldwide) and has plenty of arctic representatives. In all plants of the family each apparent flower (known as a *head*) is really a mass of minute flowers (*florets*), all held in a "cup" (the *involucre;* see fig. 5.10) formed of a number of narrow bracts. Only those with colorful flowers will be described here.

In most species the heads are daisies; that is, they have a central *disk*, which consists of numerous tiny, tubular florets densely packed together, and an outer, radiating circle of *rays,* each ray being a floret with a strap-shaped petal. In other species, the heads are not daisylike; instead of having a central disk surrounded by rays, each head consists only of strap-shaped florets (dandelions and hawksbeard), or only of tubular florets (saussurea, and some abnormal groundsels). The margins of tubular florets are toothed or lobed, and in some species the lobes are long and narrow; but a tubular floret is always symmetrical, and a strap-shaped floret always lopsided.

The contrasted types of head form a convenient basis for subdividing the members of the family into those with both kinds of florets ("daisies"), those with strap-shaped florets, and those with tubular florets. Daisies are so numerous that it helps to subdivide them further, into those with yellow rays and those with rays of some other color. This gives four groups, as follows:

Group I (daisies with yellow rays). This group has three genera: the arnicas (*Arnica*), the groundsels (*Senecio*), and the goldenrods (*Solidago*).

Arnicas are big, showy, yellow daisies, whose three arctic species are clearly differentiated. **Alpine Arnica** [AhMLB], *Arnica alpina* (fig. 5.48), can always be told by the fact that, besides numerous leaves growing up from ground level, its stem bears a pair (sometimes two pairs) of leaves growing opposite each other at the same level. In the western Arctic are two additional arnicas, both recognizable from the fact that their heads "nod" (face downward). In one of them, **Lessing's A** [A], *A. lessingii* (fig. 5.48), the disk is dark purple; in the other, **Lake Louise A** [Al], *A. louiseana* (not illus.), the disk is yellow like the rays.

Groundsels form a large and confusing genus that, with one exception, cannot easily be distinguished without a technical Flora. All that can be said is that yellow daisies that are not arnicas (described above) and that grow from rosettes of leaves on the ground with only tiny leaves (if any) on their stems are groundsels of one species or another. In some species the daisy heads often lack rays; they are merely disks that look as if the rays had been pulled off (see Group III). The exceptional groundsel, which has big leaves all up the stems, is **Mastodon Flower** [AhMLB], *Senecio congestus* (fig. 5.48). It grows in damp places, including seashores, often reaching a height of more than 1 m. The leafy stems are topped with big globular clusters of yellow daisies, embedded in a dense fuzz of white hairs. Indeed, the whole plant is shaggy with white hair.

The third genus of yellow daisies is the only truly arctic goldenrod, **Northern Goldenrod** [AmLb], *Solidago multiradiata* (fig. 5.48). It has

Figure 5.48. (a) Alpine Arnica. (b) Lessing's Arnica. (c) Mastodon Flower. (d) Northern Goldenrod. (e) Entire-leaved Chrysanthemum.

leafy stems, topped with tight-packed clusters of little yellow daisies, often so closely bunched that their daisy shape isn't obvious without close inspection. The plant is seldom more than 30 cm tall and is not hairy, so it cannot be confused with mastodon flower.

Group II (daisies with white, pink, or purple rays). There are five genera: chrysanthemums (*Chrysanthemum*), chamomile (*Matricaria*), coltsfoots (*Petasites*), asters (*Aster*), and fleabanes (*Erigeron*).

Chrysanthemum "flowers" are big daisies with yellow disks and wide white rays, easy to recognize because they are so like the common ox-eye daisies which are weeds in temperate latitudes. The two species that grow in the Arctic are best told apart by their leaves: **Entire-leaved Chrysanthemum** [ahMLB], *C. integrifolium* (fig. 5.48), has rosettes of smooth-edged leaves; and **Arctic Daisy** [Yukon coast and Mackenzie Delta; Hudson Bay shores], *A. arcticum* (not illus.), a seashore plant, has larger (up to 4 cm) daisies and rosettes of fleshy, lobed leaves.

Seashore Chamomile [ALB], *Matricaria ambigua* (fig. 5.49), is another white-rayed daisy confined to seashores, but it differs from the chrysanthemums in having finely divided, feathery leaves.

In the coltsfoots, the flower heads are white to pale pink, with no color contrast between rays and disks. Indeed, the daisy form of coltsfoot "flowers" is sometimes not at all obvious, especially when they are just beginning to open and are tightly clustered; then the whole cluster of heads looks like an upturned brush. The three arctic species are rather similar. They are best distinguished by the contrasted shapes and textures of their basal leaves (growing from ground level), but these are small and undeveloped (or even invisible) at flowering time. The two relatively common species are **Arctic Coltsfoot** [Ahml], *Petasites frigidus* (fig. 5.49), and **Arrowleaf C** [aL], *P. sagittatus* (fig. 5.49). Both have basal leaves felted with white hairs underneath; the characteristic leaf shape for each species is shown in the figure, but because the species hybridize, intermediate shapes are often found. A third, very local species, *P. arcticus* (not illus.), grows only in the Mackenzie Delta and is immediately recognizable because its leaves are hairless, and green on both sides.

The remaining daisies, the asters and fleabanes, also have narrow rays but are obvious daisies nonetheless. In color, the rays range from white through palest pink to lilac and purple. In the treeline zone and in the Low Arctic for a short distance north of treeline, the two genera are hard to tell apart and each contains numerous, bewilderingly similar species. But farther north the numbers quickly dwindle, and from the arctic coastline onward only two species of asters and three of fleabanes are to be found.

The asters are **Siberian Aster** [Al], *Aster sibiricus* (fig. 5.49), and the

Figure 5.49. (a) Seashore Chamomile. (b) Coltsfoot, before the basal leaves have expanded. (c) Arctic Coltsfoot leaf. (d) Arrowleaf Coltsfoot leaf. (e) Siberian Aster.

rare **Pygmy A** [ml], *A. pygmaeus* (not illus.). Both have purple rays; the leaves of Siberian aster are toothed, whereas those of pygmy aster are not.

The fleabanes all have a fringe of innumerable tiny rays, white to pale lilac in color, circling the relatively large disk. They are **Cutleaf Fleabane** [AHmlb], *Erigeron compositus* (fig. 5.50), the only species with lobed leaves; **Mountain F** [AmLb], *E. humilis* (fig 5.50), recognizable from the purple-black "wool" of curly hairs concealing the involucre; and **Arctic F** [AHMlB], *E. eriocephalus* (not illus.), with a spreading involucre embedded in a "wool" of curly gray hairs.

Group III (flower heads of strap-shaped florets only). This group includes the dandelions (*Taraxacum* species), of which at least a dozen, hard-to-differentiate species grow in the Arctic. All of them are obviously

Figure 5.50. (a) Cut-leaf Fleabane. (b) Mountain Fleabane. (c) Dwarf Hawksbeard. (d) Saussurea. (e) Arctic Groundsel.

dandelions, and some are interestingly colored (cream, or even pinkish cream) instead of the standard bright yellow. The only other member of the group is **Dwarf Hawksbeard** [Ahmlb], *Crepis nana* (fig. 5.50), a most attractive plant. It is a neat, circular cushion of dark green, spoon-shaped leaves, among which grow numerous yellow flowers like miniature dandelions.

Group IV (flower heads of tubular florets only). In temperate latitudes this group is typified by thistles, but in the Arctic, where thistles don't grow, the group's representative is **Saussurea** [AL], *Saussurea angustifolia* (fig. 5.50). Its purple stems are topped by several flower heads, each like a slender thistle; protruding from every bright purple floret is a tiny "rod," with a black base and a white, forked tip. The black part is the anthers, the pollen-bearing organs at the tips of the stamens; the white part is the style, the pollen-receiving organ. The plant is spectacular enough to deserve the name sometimes applied to it **Fireworks Flower.**

The only other plants in this group are abnormal—though still common—groundsels, in which some (not all) of the flowerheads lack ray florets. The commonest of the groundsels to produce these rayless heads is a species of the western and central Arctic, **Arctic Groundsel** [Aml], *Senecio atropurpureus* (fig. 5.50). The plant attracts attention because of its oddity; the solitary flower head is merely a green "cup" (the involucre) holding a sheaf of tubular florets tipped with minute, curly lobes.

The Lily Family (Liliaceae)

This family has 5 arctic species, all easy to recognize. Their membership in the Lily Family can be checked—using a lens for small-flowered species—by noting that all have flowers with 6 "petals" and 6 stamens. (The "petals" are technically *tepals,* undifferentiated petals and sepals that look exactly alike.)

Mountain Death Camas [Al, west of Canadian Shield], *Zygadenus elegans* (fig. 5.51), is a tall (up to 60 cm) plant with conspicuously handsome flowers. Each cream-colored petal is crossed by a green band near the base. The whole plant is poisonous.

Alp Lily [A], *Lloydia serotina* (fig. 5.51), has flowers that are creamy-white cups, with purple veins. Each flower stalk grows from a tuft of slightly fleshy, grasslike leaves and bears a single flower at the top.

Wild Chives [A], *Allium schoenoprasum* (fig. 5.51), is the ancestor of garden chives and looks no different. Each flower stalk is topped by a globe of small, tightly packed, pink-purple flowers; the round, hollow leaves, which gradually taper to a point, smell and taste of onions.

The two remaining species belong to the false asphodel genus, *Tofiel-*

dia. Both are dwarf plants in which the tiny flowers are densely clustered into an oval (sometimes almost spherical) head at the top of a leafless stalk. The leaves, all at or near the base, are folded and enclose each other, like minuscule iris leaves. The species are distinguished by the color of their flowers: In **Western False Asphodel** [AmLB], *Tofieldia pusilla* (fig. 5.51), the flowers are white; in **Northern False A** [AmLB], *T. coccinea* (not illus.), they range from pale pink to deep maroon, often being an unusual, and attractive, old rose color.

Figure 5.51. (a) Mountain Death Camas. (b) Alp Lily. (c) Wild Chives. (d) Western False Asphodel (and close-ups of flower, and enfolding leaves).

The Orchid Family (Orchidaceae)

Two species of orchids grow in the mainland Arctic. Both are small, rather inconspicuous plants, each with a single spike of pale greenish-yellow flowers.

Northern Bog Orchid [AL], *Platanthera obtusata* (fig. 5.52), is a miniature version of the tall bog orchids of temperate latitudes. Each flower has a long, curved spur. The whole plant is seldom more than 15 cm tall and has a single oval, green leaf at the base. Northern bog orchids are often pollinated by mosquitoes; see section 9.4 for details.

Northern Coral-root [AL], *Corallorhiza trifida* (fig. 5.52), is a *saprophyte,* i.e., instead of creating, by photosynthesis, the carbon compounds it needs for growth, it obtains them ready-made from decayed plant remains in the soil. Therefore, the plant has no need of chlorophyll and

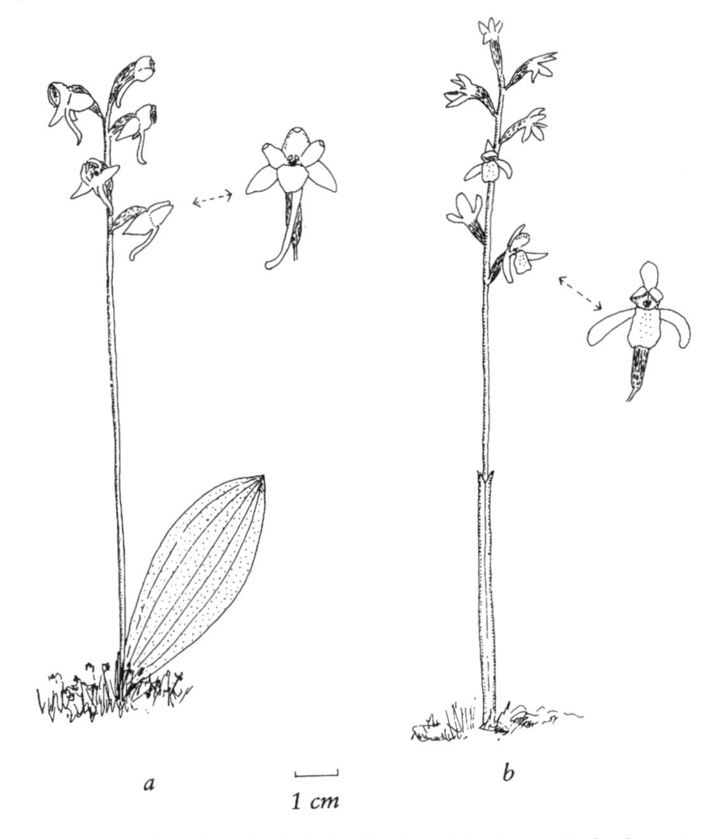

Figure 5.52. (a) Northern Bog Orchid. (b) Northern Coral-root. (In both species, what looks like a fleshy flower stalk below each flower is a developing seed capsule.)

lacks green leaves. In their place, the pinkish-buff stem is sheathed by a pair of long brownish scales. The flowers differ from those of northern bog orchid in being without a spur, and sometimes (not always) the lip of the flower is speckled with tiny purple dots.

Grasses, Sedges, and Rushes

Identification of the great majority of grasses, sedges, and rushes, in the Arctic as everywhere else, is not practicable in the field except by people already familiar with them. And to become familiar takes plenty of time indoors, with specimens in various stages of development, a dissecting microscope and a technical Flora.

The cottongrasses are an exception. As a group, they are instantly recognizable: their fluffy heads, often brilliantly white in the sunshine, contribute greatly to the beauty of the tundra. The "cotton" grows atop a compact spike of flowers whose enormously modified petals and sepals have the form of soft, fine, greatly elongated bristles.

Cottongrasses are not true grasses, but members of the Sedge Family (*Cyperaceae*). The family contains several genera, the biggest by far being sedges themselves, in the genus *Carex*. The cottongrasses belong to the genus *Eriophorum,* with a number of arctic species. Four species are common, but as they have no English names, they are listed by their scientific names.

Eriophorum angustifolium [AmLB] has a cluster of several heads at the top of the stalk. The only other many-headed cottongrass to be found north of treeline, and the only one that grows in the High Arctic, is *E. triste* [aHMlB]), which some botanists regard as merely a subspecies (or geographic race) of *E. angustifolium*. Therefore, all many-headed arctic cottongrasses can be called *E. angustifolium*.

The other cottongrasses have only a single head of "cotton" at the top of each stalk. They can be placed in two groups: those that form tussocks and those that don't. The commonest nontussock species, and the only one in the High Arctic, is *E. scheuchzeri* [AHMLB]. The commonest tussock-forming species is *E. vaginatum* [AmLb]. It forms big, dense tussocks, which often grow in millions, over huge areas (see fig. 5.5). As hikers and backpackers well know, stepping over the tussocks is exhausting, and stepping on them can lead to a sprained ankle, as they are not firm enough to make dependable "stepping stones." A prolonged stint of tussock-hopping is a memorable experience.

5.14 A Field Guide to Arctic Ferns and Their Relatives

The great majority of arctic plants (disregarding mosses and lichens) have flowers and bear seeds. A small minority of plants do not: they are ferns and their relatives, flowerless plants that bear spores instead of seeds. These spore plants are much more primitive, in the evolutionary sense, than the seed plants. Four families are represented in the Arctic: the ferns, the horsetails, the clubmosses and the spikemosses.

Ferns (Aspidiaceae)

Only 3 species of ferns grow in the far north. Although widespread, they are not abundant and not particularly conspicuous. But they are well worth looking for. In all three, the spores are produced on the backs of some of the fronds.

Fragrant Shield Fern [AhmLB], *Dryopteris fragrans* (fig. 5.53), is more noticeable than the others because its dense tuft of green fronds protrudes from a ring of brown, dead fronds left over from previous years; the dead fronds persist for a long time curled over around the base of each fresh, green tuft. The tops of living spore-bearing fronds also curl backward at the tip. The fronds are not at all delicate and are wintergreen (see sec. 5.7). The backs of the fronds and the stem are covered with a dense chaff of rust-colored, papery scales.

The other two ferns are, at first glance, rather like each other and often grow in damp rock crevices, sometimes both on the same cliff. They are **Fragile Fern** [AhmLB], *Cystopteris fragilis,* and **Smooth Cliff Fern** [ahmlb], *Woodsia glabella* (both in fig. 5.53), and are easy to distinguish if they can be examined side by side, as fragile fern is greener and (usually) larger than smooth cliff fern. Fragile fern differs from smooth cliff fern in having wider fronds with longer *stipes* (the stipe is the part of the stalk below the first pair of pinnae or "leaves"). Smooth cliff fern, which is rather rare, is a tiny fern, often no more than 5 cm tall; its yellowish-green fronds are narrow, with pinnae that are fan-shaped rather than elongate.

Horsetails (Equisetaceae)

Horsetails, as a group, though not particularly beautiful, are distinctive and easy to recognize. All the species belong to a single genus, *Equisetum.* All have hollow, jointed stems and appear leafless because the leaves consist of no more than blackish, papery teeth joined in a ring around the nodes (joints) of the stems. Sometimes (not always) the stems bear whorls of green branches at each node, growing out like the spokes of a wheel. The spores are borne in conelike structures at the tips of the stems. Three species are quite common in the Arctic.

Common Horsetail [AhmLB], *Equisetum arvense* (fig. 5.54), is notable for having two kinds of stems on the same plant. The stems are either fertile (cone-bearing) or sterile (without cones). The fertile stems appear early in spring and don't live long; these stems have no branches, and each has a cone at the top. Lacking chlorophyll, they are a pinkish-buff color; they cannot photosynthesize but obtain their nourishment, via the plant's underground parts, from the sterile stems. The sterile stems grow later, after the fertile have begun to wither; they are bright green with chlorophyll, hence capable of photosynthesis. They don't bear cones.

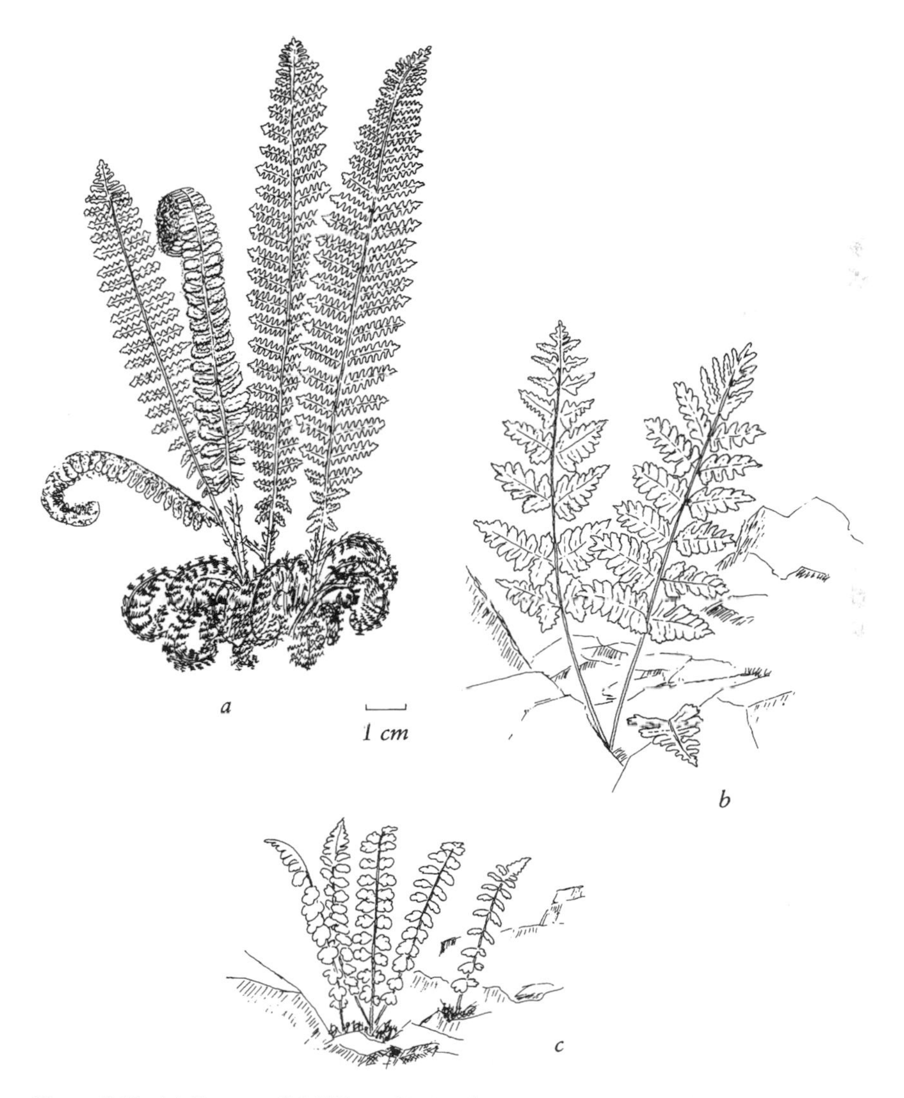

Figure 5.53. (a) Fragrant Shield-fern. (b) Fragile Fern. (c) Smooth Cliff-fern.

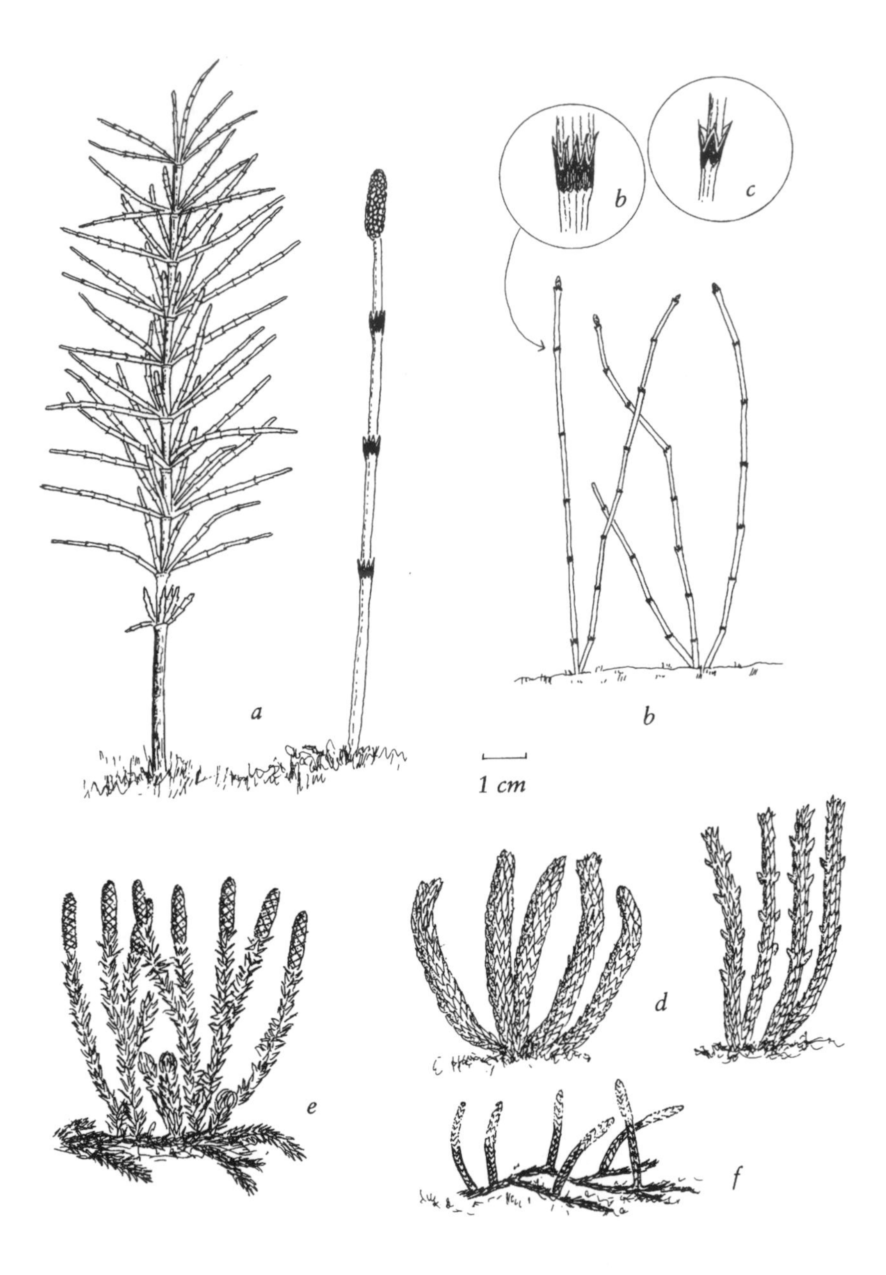

Figure 5.54. (a) Common Horsetail (sterile stem on the left, fertile stem, with cone, on the right). (b) Variegated Horsetail, with (inset) close-up of node. (c) Dwarf Scouring-rush; close-up of node. (d) Mountain Clubmoss (two forms). (e) Bristly Clubmoss. (f) Siberian Spikemoss.

And they have whorls of branches, the only arctic horsetails to have them.

The other two horsetails, **Variegated Horsetail** [AHMLB], *E. variegatum,* and **Dwarf Scouring-rush** [AmLb], *E. scirpoides* (both in fig. 5.54), are rather similar to each other. Both have stems of only one kind: all stems are fertile (capable of bearing cones), green, and unbranched. Dwarf scouring-rush is the smaller of the two; it has thin, flexible stems that curl together in matted tangles, and only 3 or 4 papery teeth at each node. Variegated horsetail, besides being larger, has stronger, stiffer, more upright stems, and 6, 7, or 8 teeth per node. In both species the cones at the tips of the stems (not necessarily on every stem) are small and inconspicuous.

Clubmosses and Spikemosses (Lycopodiaceae and Selaginellaceae)

Members of these two families of plants are outwardly similar, so they are treated together here. The distinction between them, although botanically important—clubmoss spores are all alike, spikemoss spores are of two kinds—is hard to see without a microscope. Plants of both kinds look like large mosses, but their leaves and stems are relatively harsh and stiff, never delicate and filmy. Their spores are borne in spore cases (yellow or tan-colored when ripe) that look like pinheads hidden at the bases of the leaves near the tips of the branches. Sometimes these leaves are like the lower leaves; sometimes they are distinctly different, making the branch tips look like small cones. The two arctic clubmosses belong to different genera: *Huperzia,* without "cones," and *Lycopodium,* with "cones"; the single arctic spikemoss belongs to the genus *Selaginella.* All 3 species are shown in fig. 5.54.

Mountain Clubmoss [AhmLB], *Huperzia selago,* is the only species of the three found in the High Arctic. The plant is a tuft of stiff little upright, yellow-green branches. It occurs in two forms that could easily be mistaken for two different species. The branches of one form are short and thick, of the other longer, thinner, and with regularly spaced buds; both forms are shown in the figure. Unlike the next two species, the stems of this species never creep along the ground. The leaves bearing concealed spore cases look the same as all the other leaves, so the plant has no discernible cones.

Bristly Clubmoss [Alb], *Lycopodium annotinum,* is a ground-creeper with upright branches tipped with distinct, straw-colored cones.

Siberian Spikemoss [A], *Selaginella sibirica,* is also a ground-creeper, forming spreading mats. Its small, scalelike leaves are pressed to the stem; its branches are tipped with cones that differ only slightly from the rest of the branch.

5.15 Mosses, Lichens, and Nostoc

With increasing latitude, mosses and lichens become more and more important as members of arctic plant communities. Their small size makes it difficult to identify the many species without a dissecting microscope, and for the lichens, chemical tests are often required as well. Here, we can consider only the species most likely to draw attention to themselves because of some noteworthy characteristic.

Mosses and lichens are utterly different botanically, but resemble each other, nevertheless, in one important respect: They all lack roots. The roots of higher plants are quite elaborate structures anatomically, being constructed of several different kinds of cells, some specially adapted to convey water upward to the rest of the plant. Mosses and lichens simply absorb water when it is available, and shrivel into a dried-up state of dormancy when it is not; unlike higher plants, they are capable of "coming back to life" when they are moistened again, even after drying up completely. In place of roots, mosses have *rhizoids,* and lichens have *rhizines;* both are merely delicate hairlike strands that anchor the plants to the surface they are growing on.

This is the only point of resemblance between mosses and lichens. The contrasts are much more noteworthy. Mosses are simple, primitive green plants, i.e., plants with chlorophyll in their cells enabling them to feed themselves by photosynthesis. They produce spores, but do so in a way wholly unlike that of the more highly evolved spore-producers, the ferns and their relatives (see sec. 5.14). In most species the spores are contained in small oval capsules, each atop a wiry stalk growing up from the plant.

Lichens are entirely different: they are compound rather than simple. Each is made up of separate organisms of two kinds, one a fungus, the other several cells of a green alga, or alternatively, several bacteria of a kind (the blue-green *cyanobacteria*) capable of photosynthesis; the photosynthesizing cells live embedded in the fungus, which provides a moist home for them and supplies the mineral nutrients they need. In return, the algae or cyanobacteria synthesize food for the fungus, which, lacking chlorophyll, cannot create food for itself. In effect, a lichen is a fungus that grows its "vegetables" inside itself.

A few mosses, and a somewhat larger number of lichens, are both eye-catching and reasonably easy to recognize. Some that are worth looking for are described below.

Some Notable Mosses

A truly beautiful moss is **Red Moss,** *Bryum cryophyllum* (not illus.), which forms great ruby-red cushions of an astonishing brightness. Sometimes it is found among cushions of apple-green mosses, making a dazzling mosaic. And sometimes it grows submerged in water, as do many other arctic mosses. In late summer when arctic flowers are over, vivid cushions of red moss may be the only splashes of color in the High Arctic tundra.

One of the best-known mosses in the north is sphagnum, or **Peat Moss** (several species of the genus *Sphagnum;* fig. 5.55). The genus is easy to recognize, but the several species are hard to differentiate. All

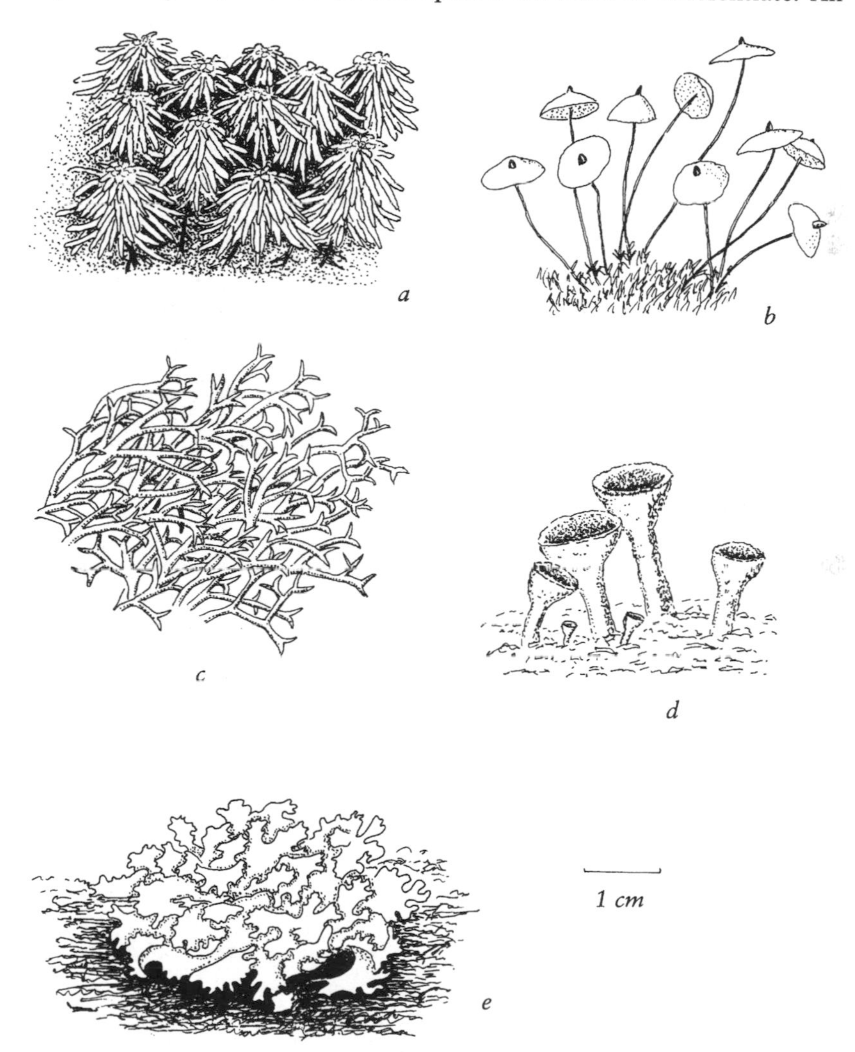

Figure 5.55. (a) Peat Moss (Sphagnum). (b) Mushroom Moss. (c) Reindeer Lichen. (d) Pixie cup Lichen. (e) Snow Lichen.

the species grow as moist, spongy carpets made up of thousands of little mound-shaped rosettes. Sphagnum is not the only moss to form peat; in the Arctic, other kinds of mosses are much more important as peat-formers. Sphagnum is more a moss of the subarctic forest than of the Arctic and becomes scarcer the higher the latitude, being rare in the Arctic Islands. It is fairly common in the Low Arctic.

A most unusual moss is to be found in the western mainland Arctic. It has no popular name, but could well be called mushroom moss; scientifically, it is *Splachnum luteum* (fig. 5.55). Its spore capsules are its distinctive character: the capsule, with its stalk, resembles a tiny mushroom, a miniature version of the kind that form fairy rings on lawns. The spore-containing part of the capsule is a small oval spindle, like that of most mosses, but around the base there is a spreading flange which resembles a mushroom cap so closely that it is often mistaken for one. This pale yellow "mushroom cap" sits atop a tall, wiry red stalk. The moss invariably grows on animal dung, usually that of moose. It is spread by flies that feed on the moss spores and lay their eggs in dung. The moss therefore benefits from flies' visits, because when the flies settle to lay eggs, they are sure to deposit some of the moss spores clinging to their hairy bodies on the precise spot—another pile of dung—where they can germinate and grow. It is believed that the flanges around the capsules evolved as convenient landing platforms for visiting flies.

Another moss to watch for in the western mainland Arctic is **Wool Moss,** *Racomitrium lanuginosum* (not illus.). It is pale gray to almost white, because its tiny leaves taper to long, curly, white, hairlike tips, of the color and texture of sheep's wool. Although small specimens are not particularly striking, the plant can grow into cushions as big as, and remarkably like, the back of a sleeping (or dead) sheep.

Some Notable Lichens

Reindeer Lichen is probably the best known of the arctic lichens,[1] because of its importance in the diet of caribou. Unfortunately, it is often misleadingly called "reindeer moss." The name reindeer lichen is usually taken to mean any species of the genus *Cladina,* although it strictly applies only to the species *C. rangiferina* (fig. 5.55), which is commoner in the subarctic forests (where the caribou overwinter) than in the true Arctic. (Until recently, all reindeer lichens were included in the big genus *Cladonia,* but the latter name is now limited to the pixie cup lichens and their relatives; see below.) All reindeer lichens have bushy thalli that branch repeatedly; depending on the species, their colors range from

1. The "body" of a lichen is called a *thallus,* plural *thalli.*

dirty yellow through gray to greenish gray. In the Low Arctic, and near treeline especially, they often form an unbroken groundcover over a large area, making the whole land surface uncharacteristically light-colored.

The **Pixie Cup** lichens are several species of *Cladonia,* made easily recognizable by their pixie cups (fig. 5.55). Other species of the genus lack cups and aren't so easy to identify.

Mane Lichen, *Alectoria ochroleuca,* (not illus.), looks, at first sight, like a tangled mass of coarse hair lying on the ground, because the branches of its elaborately branched thallus are fine and wiry. The species merits attention for its surprising colors: the branches are yellowish tan at the base and darken upwards to become blue-green and then, at the tips, blue-black. Similar tangles of all-black "hair" are *A. nigricans.*

Snow Lichen, *Cetraria nivalis* (fig. 5.55), and the related **Yellow Lichen,** *C. tilesii* (not illus.), are bushy lichens whose branches, instead of being round, end in flattened lobes. Yellow lichen is notable for its brilliant yellow color, a welcome sight when, as often happens, it is the only bright spot in an otherwise lifeless field of gray rocks. Snow lichen is less bright—more ivory than yellow—and its thallus lobes have crinkly margins.

Rock Tripe, *Umbilicaria* species (fig. 5.56), is the edible lichen that saved an early Franklin expedition (from Great Slave Lake to Coronation Gulf and back, 1819–22) from starvation, though it is not particularly nourishing and certainly not tasty. The lichen grows on rock; the thallus is a dull black "saucer," often with a tattered, uneven margin, and is attached at the center to the underlying rock.

Worm Lichen (fig. 5.56) is the name of two lichens (*Thamnolia subuliformis* and *T. vermicularis*) that cannot be told apart except by chemical tests. The former species is the commoner in the Arctic. The English name describes its appearance. The thallus grows from the ground as a tuft of soft, tapering, hollow, white "worms"; where worm lichen is abundant, the ground is often littered with "worm" fragments, which are a favorite nest-building material of birds such as ptarmigans, plovers, and sandpipers.

Finger Lichen, *Dactylina arctica* (fig. 5.56), is another lichen favored by nest-building birds. It also grows on the ground. Each of its numerous light brown "fingers" looks like the finger of a doll-sized rubber glove, often with a small hole at the tip.

Sunburst Lichen, *Arctoparmelia centrifuga* (fig. 5.56), grows on rock and becomes conspicuous when it has been growing long enough to form a series of concentric circles (sometimes only short arcs of the original circles remain, as in the figure). Such a lichen began as one small thallus whose lobes, branching repeatedly, grew outward to fill an expanding circle; in time, the aging base of each branch died and withered, while the young tips continued to fork and grow outward. The result was a ring-shaped lichen with fresh, light-colored tissue on the outside, and progressively older, darker tissue toward the inside. While this ring continued to expand, a fragment of the original thallus remaining at the center began growth anew, creating a second expanding ring inside the first. Then a third ring started, and so on. As many as 6 or 7 concentric rings are often found. Sunburst lichens are sometimes used to determine the length of time the rocks they are growing on have been exposed (see below).

Map Lichen, *Rhizocarpon geographicum* (fig. 5.56), is one of the commonest arctic lichens. It grows as a crust on rocks. Each specimen consists of a black patch of irregular shape (hence the name "map") speckled with angular dots of pale yellow. When a large, smooth expanse of rock is entirely covered by map lichen, its presence is often unrealized, because it looks more mineral than vegetable. The rate at which the patches of map lichen grow has been studied, leading to the conclusion that some must be 4,500 years old. This result is highly speculative, of course, since it rests on the assumption that the growth rate is correctly known and remains constant over many centuries. Attempts have been made to infer, from the size of lichen crusts growing on rock, when the last ice of the most recent ice age disappeared from a region; the argument is that the largest, and presumably oldest, lichen in the region couldn't have begun growth while the rock was encased in ice; therefore, knowing the lichen's growth rate, it is simple to calculate the age of the oldest lichen, which gives a lower limit to the time that must have passed since the ice disappeared.

Another lichen used for determining age is **Jewel Lichen,** *Xanthoria elegans* (fig. 5.56), a vivid orange lichen, which flourishes most luxuriantly on rocks "fertilized" with bird droppings. Nesting cliffs used by birds that nest in big colonies can be recognized from afar by the orange stains descending from every ledge; and any isolated rock colored orange by jewel lichen can safely be judged to be a perch for carnivorous birds in need of a lookout over the surrounding tundra. Jewel lichen grows on Beechy Island (just off the southwest corner of Devon Island), on the gravestones of three members of Franklin's last expedition who died of illness during the winter of 1846. The lichens are assumed to have started growth when the gravestones were first raised, a likely possibility as the stones, once set upright, immediately became useful bird lookouts. The lichens were found to have grown to 4.4 cm in diameter 144 years later, enabling the growth rate of this particular lichen species, growing in Beechy Island conditions, to be calculated with unusual precision.

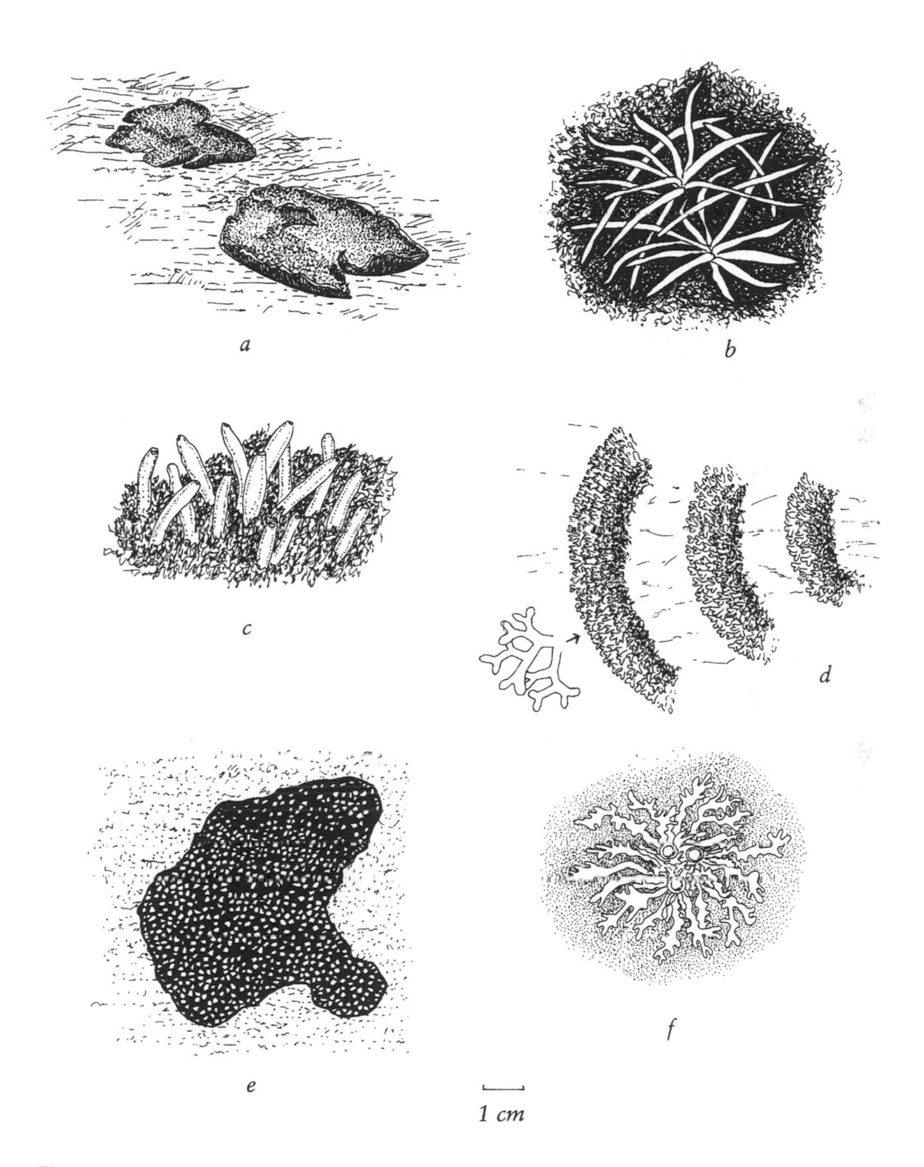

Figure 5.56. (a) Rock Tripe. (b) Worm Lichen. (c) Finger Lichen. (d) Sunburst Lichen. (e) Map Lichen. (f) Jewel Lichen.

Map lichen and jewel lichen are a useful pair of geological indicators. Map lichen grows only on acid rocks such as granite, sandstone, and quartzite, jewel lichen only on rocks rich in carbonate, chiefly limestone. This rule is broken only on "bird rocks" where the abundant supply of nitrogenous fertilizer enables jewel lichen to flourish whatever the chemistry of the rock.

Another fairly easily recognizable lichen found on acid rocks is **Bloodspot Lichen,** *Haematoma lapponicum* (not illus.). It grows as a pale gray crust dotted with circular red spots the color of garnets or, as whoever first named it thought, the color of dried blood.

The final "lower plant" to mention here appears on moist ground or in shallow water as torn sheets of what look like the remains of a green-brown plastic garbage bag or, equally improbably, as fragments of kelp (brown seaweed). In fact, these sheets consist of thousands of blue-green bacteria (cyanobacteria) of the genus *Nostoc.* The cells remain attached to each other after dividing and redividing numerous times (that is their form of reproduction) until quite an extensive sheet of cells is formed. The organism is not only capable of photosynthesis, but also of "nitrogen-fixing"; i.e., it absorbs nitrogen from the air and converts it to a form that other life can use. Thus *Nostoc* is more than a mere curiosity; it is an organism that contributes to the fertility of poor tundra soils. *Nostoc* and other cyanobacteria are believed to be crucial in maintaining soil fertility in the High Arctic "oases."

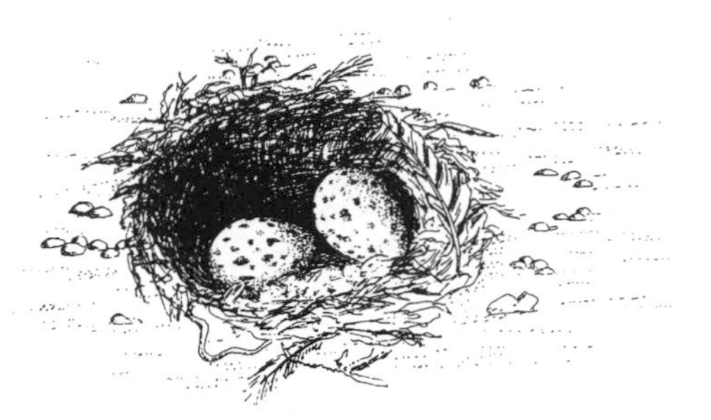

6 *Birds*

6.1 Residents and Migrants: The Quest for Food

Only eleven species of birds are capable of living in the Arctic all the year round: gyrfalcon,[1] raven, rock and willow ptarmigans, snowy owl, redpoll, Ross's and ivory gulls, thick-billed murre, dovekie, and black guillemot. Most of them shift their ranges slightly southward *within* the Arctic during winter, probably to avoid a prolonged period of continuous darkness. And they don't necessarily spend every winter in the Arctic: they move south in years when food is short. Of these eleven species, only gyrfalcons, snowy owls, and the two gulls have breeding ranges wholly (or almost wholly) confined to the Arctic; the other species breed in temperate latitudes as well as in the Arctic.

All other arctic breeders (about 90 species) are migrants: they are temporary rather than permanent arctic residents. They come north (some from enormous distances) only for the short breeding season and are not adapted to feed themselves during an arctic winter, when so much of the food that abounds in summer—growing plants and a myriad of small invertebrates—is dormant and unavailable.

Birds that feed on fresh, green water plants (e.g., tundra swans, geese, many ducks), on aquatic invertebrates such as insect larvae, mollusks, and crustaceans (e.g., phalaropes, king eider, and oldsquaw ducks), or on freshwater fish (e.g., loons, red-breasted mergansers) obviously cannot survive when tundra lakes and ponds are frozen. Many birds that, often or habitually, feed at sea (e.g., fulmar, common eider, jaegers, several gulls, arctic tern) are cut off from their food when the sea ice forms

1. Pronounced "jer-falcon."

(though some seabirds—the last five in the list of permanent arctic residents given above—find the open water they need in polynyas; see sec. 4.4). The shorebirds (plovers, sandpipers, and their relatives) find much of their food (worms, mollusks, crustaceans, insect larvae) in mud and wet sand; in winter the mud and sand are frozen, and the animals dormant. The perching birds (e.g., snow buntings and Lapland longspurs) feed on insects and seeds; in winter, the insects are dormant and concealed, and it takes only a thin layer of snow to hide most of the seeds. The remaining migrants (peregrine falcons, golden eagles, rough-legged hawks) are meat-eaters, and it is not obvious why they don't live full-time on their arctic breeding grounds as gyrfalcons, snowy owls, and ravens do.

The six land birds among the permanent residents consist of three meat-eaters and three vegetarians. The meat-eaters are gyrfalcons, snowy owls, and ravens, which in winter often have to eke out their preferred live prey (ptarmigans, lemmings, and voles, respectively) with carrion left by polar bears, grizzly bears, and wolves. The vegetarians are the two ptarmigan species and redpolls. The ptarmigans manage to survive the rigors of winter by feeding on the buds and catkins of dwarf willows and birches under the snow. It seems a meager diet for such a harsh environment. Redpolls also shelter under the snow, in lemmings' tunnels; their diet consists of seeds.

The comings and goings of the migratory birds are tightly controlled by food or the lack of it. Food is plentiful on the summer tundra but only for a few weeks. To insure that their chicks hatch when food is at its most abundant, many birds, especially the larger species, have to reach their nesting territories and lay their eggs before the snow has all melted and spring growth has begun. Waterfowl (geese in particular) prepare for the stresses of spring by feeding enormously at traditional stopovers on their migratory routes; they lay up sufficient reserves of fat and protein to tide them over the lean period and for the females to lay eggs. Shorebirds are believed to do the same; it is known that ruddy turnstones and red knots do.

The end of summer is also a stressful time for arctic birds; the young of the year must be strong enough for the southward migration before the weather cools to the point that food supplies fail. In some species (e.g., ruddy turnstone, red-necked phalarope) the adults depart first, leaving what food there is for their offspring who travel later.

The conservation of migratory birds' stopover sites is obviously vital to the wellbeing of numerous species of waterfowl and shorebirds. They must be protected from interference at all costs.

6.2 More on Migration: Populations and Destinations

A large proportion of the migratory arctic breeders are birds of wetlands: waterfowl (swans, geese, and ducks), shorebirds (plovers and sandpipers) and sandhill cranes. Wetlands are rich food sources in summer: aquatic plants provide succulent leaves and stems and nutritious seeds; water and wet mud are home for a huge variety of edible invertebrates, the larvae of midges and crane flies being especially abundant. In the Low Arctic, wetlands are everywhere. Even though the precipitation is not great, such water as there is persists: not much evaporates because temperatures are low, and not much soaks into the ground because the active layer is thin and the permafrost below it forms an impermeable barrier. Wet ground stays wet and the birds benefit.

Wetlands become less extensive, however, as you go north. By degrees, wet tundra gives way to stony polar desert; food and cover for birds become scarcer, and the open-water season becomes shorter. Fewer and fewer bird species are able to cope. The number of species of breeding birds falls off markedly, from about 60 species at 60° N to about 20 at 80° N.

Some of the shorebirds come from astonishing distances to breed in arctic wetlands. For example, white-rumped and Baird's sandpipers and some sanderlings come more than 14,000 km, from southern South America. Their journeys are almost as long as those of the fabled arctic tern, which travels from the Antarctic to the Arctic each spring, and back again each fall.

Many of the migrants converge on the Arctic from two or more widely separated overwintering areas. Brant geese breeding in the eastern Arctic overwinter on both east and west shores of the Atlantic; those breeding in the western Arctic (black brant, a recognizably different subspecies) overwinter on the Pacific coast. Ruddy turnstones and red knots that nest in the High Arctic islands (north of Parry Channel, see fig. 5.8) overwinter in Europe and Africa, whereas those that nest south of Parry Channel overwinter in the Americas.

6.3 Protection from the Cold

Lack of food rather than the cold is what makes the Arctic in winter so inhospitable to birds. Nevertheless, the cold is fierce, and some of the permanent residents have evolved special adaptations to cope with it.

For most northern birds, conservation of body heat is not the chief problem; the down between their skin and their outer feathers is a superb insulator, as any user of a down sleeping bag well knows. Down is the best insulator in existence, far better than any other natural covering or any synthetic material. By fluffing up their feathers to increase the air spaces within the down, as they do in extreme cold, birds can increase the insulating effect still more. Some birds increase their insulation for winter by growing extra feathers. Ptarmigans and redpolls also protect themselves by escaping from the cold at times. When they have finished feeding, they dive into the snow and burrow down under it. They then have an insulating blanket of snow above them through which the lethal cold of the wind cannot penetrate.

Protecting the feet from cold is much more difficult than maintaining body warmth; this is especially true of swimming and wading birds (see below) and of land birds that spend much of their time on the surface with their feet touching snow or ice. Snowy owls and ptarmigans have fully feathered legs and feet. In ptarmigans, however, the feathers that grow on the soles of the feet in winter may be more important as snow-shoes than as foot-warmers, since ptarmigans, unlike owls, need to run about on the snow surface as they feed (see sec. 6.6).

The soles of owls' and ravens' feet are knobbly with hardened papillae; these are believed to prevent heat loss by reducing the contact area between the birds' feet and the ground.

Seabirds, waterfowl, and shorebirds are adapted to avoid loss of heat through the feet at all seasons. Cold water chills the feet of all birds that swim or wade; moreover, gulls and alcids (murres and guillemots) spend much time standing on ice. The adaptations work by allowing body heat to be retained even though the feet cool down almost to the freezing point. They function in two ways. First (as in caribou; see sec. 7.4), the blood vessels (arteries and veins) entering the legs are in close contact with each other, allowing the warm blood flowing from body to feet through the arteries to warm the cold blood returning from feet to body through the veins; in this way, blood warmth bypasses the feet, which remain cold. Second, the birds can reduce blood flow to the feet by constricting their blood vessels. The two adaptations combined can cause foot temperature to be as much as 30° C lower than body temperature.

Birds that stay motionless for long periods are not merely enduring the cold: they are protecting themselves against it. When the weather is bitterly cold and food in short supply—which is often the case in early spring—searching for food may be more wasteful of energy than simply staying put. A bird standing still, in a sheltered location if it can find one, will also contrive to conserve heat by tucking its bill under a wing, by standing on one leg while bending the other up against its belly, or by sitting on the ground with both legs covered by the body. Gulls, ducks, and shorebirds are often to be seen in these attitudes; they are doing it to keep warm.

Occasionally such passive forms of protection are not enough: chilled birds must then actively generate new body heat to make up for what is unavoidably lost even with the best heat conservation measures. To warm themselves, birds shiver. A shivering bird looks, and is, cold, but the shivering is not useless; the muscle movements entailed constitute involuntary warm-up exercises. Unlike mammals, birds do not have brown fat to keep them warm (see sec. 7.1).

6.4 Breeding in the Arctic

As winter gives way to spring, the number of bird species in the Arctic rises from 11 to over 100. All are there to breed. To rear their chicks successfully, parent birds must be able to keep them warm, feed them, and protect them from predators; different species have evolved different ways of meeting these requirements. The most striking difference is between the degree of maturity of the chicks at the moment of hatching.

In *altricial* species, the hatchlings are naked, blind, and incapable of walking; they must stay in the nest and rely totally on their parents for both food and warmth. In *precocial* species, the hatchlings are far more mature; their eyes are open, they can walk about and feed themselves with parental guidance, and they are downy enough to stay warm without being brooded much of the time. Between these extremes come some intermediates. In a few species the chicks are *semiprecocial:* they stay at the nest, fed by their parents, in spite of being down-covered and sufficiently well developed to move about. And in a few species the chicks are *semialtricial:* like altricial chicks, they are immobile and must be fed, but they are downy and have their eyes open.

The following table classifies arctic breeders into these four groups:

PRECOCIAL	SEMIPRECOCIAL	SEMIALTRICIAL	ALTRICIAL
Sandhill Crane	Loons	Fulmar	Perching birds
Plovers	Jaegers	Hawks	
Sandpipers	Gulls	Eagles	
Ptarmigans	Murres	Owls	
Ducks	Guillemots		
Geese	Dovekies		
Swans			

An important difference between altricial species and those in the other three categories is in the time it takes for the embryo in a newly laid egg to develop into a young bird capable of flight. This period is divided into two stages: the incubation period, while the embryo develops inside the egg; and the fledging period, from hatching until the chick learns to fly. Fully altricial birds go through both stages much faster than all the others; for them, incubation and fledging each take about 2 weeks. In precocial and semiprecocial species each stage lasts about 3 to 4 weeks. In semialtricial species, surprisingly, the stages are longer still, probably because these are large birds: in golden eagles incubation takes 6 weeks, and 10 more weeks are needed before the young can fly; in gyrfalcons, the respective periods are 5 and 7 weeks. (All these figures are averages.)

Rapid development would seem at first to be a desirable adaptation for birds breeding in the short arctic summer; even so, to judge from their scarcity, altricial chicks are ill adapted to arctic conditions. No doubt this is because of their nakedness and helplessness; to be safe, they must spend their first few days of life in well-built, well-hidden nests, of the kind perching birds (sec. 6.6) typically build, but few perching birds have the ability to build them in the arctic environment.

The semialtricial and semiprecocial species—most of them slow developers that stay in the nest unable to escape predators through a long fledging period—would seem to have the worst of both worlds. In some of them (loons, hawks, some gulls) the period in which the chicks are immobile lasts twice as long as in fully altricial chicks. They are protected from predators—if not perfectly, at least adequately—in a number of different ways. Hawks, eagles, murres, fulmars, and some gulls nest on precipices, safe from earthbound hunters if not from flying ones. Guillemots and (sometimes) dovekies make their nests in concealed crevices. Loons nest at water's edge on marshy lake shores, where the vegetation is tall enough to give good cover. Jaegers are strong and fierce, and well able to drive off marauders. The most helpless chicks in these two groups may well be those of snowy owls, but their mothers guard them carefully, brooding them with scarcely a pause for their first 3 weeks of life.

About 50 percent of arctic birds have precocial chicks; they remain inside the egg, where they don't need to be fed, for about 3 or 4 weeks, i.e., for as long as it takes to mature to the point where they can walk away from the nest as soon as they hatch. But this long incubation means that the eggs, when first laid, must contain nourishment enough (fat and protein) to last until the chicks are ready to emerge; and this, in turn, means that the mother bird must be well nourished herself when she lays the eggs (see sec. 6.1). The slow incubation also means she must lay as early as possible, while the weather is still cold and food for herself scarce or unobtainable. Many sandpipers can shelter their eggs and incubate successfully even when the ground is covered by a late snowfall.

The advantages of being precocial become clear when the young hatch. Because the chicks can walk and run, a family scatters when predators approach, making it unlikely all will be taken. The commonest predators are arctic foxes, ermines, jaegers, and gulls. Of course, when these hunters find a nest with eggs in it, the whole clutch will usually be destroyed, but almost as soon as the chicks are out of the eggs, they can run away from danger. It's dangerous to keep on running, however. A chick's best chance of escape is to "freeze" and become invisible; ptarmigan and sandpiper chicks excel at this. They are better at eluding capture than such other small, cryptically colored beasts as lemmings and voles that, like them, scurry through the grass to escape danger. Lemmings and voles have the disadvantage of not being instinctive "freezers."

Another benefit of scattering is that the chicks of a family don't compete with one another for food. Precocial chicks of different species differ in their need for parental help with feeding. Plover chicks and ducklings, for example, find their own food, but ptarmigan chicks need to have it pointed out to them.

Different families of birds differ in the way they time the incubation of their eggs. Nearly all those with precocial chicks lay their full clutch before starting to incubate; the first-laid eggs lie cold and inert in the nest awaiting clutch completion. Before incubation begins, eggs can survive quite low temperatures undamaged. Only when the last egg is laid does one of the parents start incubating: all the eggs develop together and all hatch synchronously, ensuring that the chicks are all the same age. This is the system used by waterfowl, which sometimes take as much as two weeks to lay a full clutch; and also by the majority of shorebirds (for some exceptions, see the account of the Plover Family in sec. 6.6).

Other bird families time things differently. Incubation starts as soon as the first egg is laid, and continues as more are added to the clutch. As a result, the eggs hatch at intervals, so that the ages and sizes of a family of chicks are spread over a wide range. Birds breeding in this way include the raptors (hawks, falcons, and eagles), the owls, the jaegers and gulls, and the sandhill crane. In uneven-aged families of chicks, the oldest uses its superior size and strength to shove its siblings aside when the parents bring food. Then, if food is in short supply, only the first-born of the family survives while the smaller chicks are neglected; they are usually pushed from the nest, before or after dying of starvation. Sometimes, the largest chick protects its interest by simply killing its sibling or siblings. It is a case of survival of the fittest within a family: if the available food were shared evenly among a family of chicks, in a lean year all might starve.

6.5 More on Breeding: Nests and Eggs

In a good year, birds breed in the Arctic in enormous numbers. Nests are everywhere and, of course, none of them are in trees. A few are on ledges high up on rocky cliffs; a few more are deep in crevices in the rocks. Some, in the shrub tundra of the Low Arctic, are concealed in low shrubs. But the majority are on the ground. This doesn't mean they are necessarily easy to see; those in wet sites may be surrounded by tall sedges and grasses; and most of those on dry, bare tundra contain such wonderfully camouflaged eggs that you can walk right past them without noticing them. Where birds are abundant, it is wise to tread carefully, to avoid stepping on a nest inadvertently.

The nests of the various birds are located in typical sites (note the word *typical:* not all species are consistent in their choice).

On the ground, unconcealed: Plovers, some sandpipers (e.g. Baird's), eider ducks, jaegers, glaucous gull, arctic tern, rock ptarmigan, snowy owl, horned lark, Lapland and Smith's longspurs.

On the ground among concealing plants: Loons, tundra swan, geese, most ducks, sandhill crane, some sandpipers (e.g. white-rumped sandpiper), Ross's and Sabine's gulls, willow ptarmigan, short-eared owl, Harris's and American tree sparrows, water pipit, yellow wagtail, arctic warbler.

In shrubs: Redpoll, gray-cheeked thrush, northern shrike.

In crevices in the rocks: Snow bunting, wheatear.

Close to the seashore: Black guillemot, dovekie.

On ledges on high cliffs: Gyrfalcon, peregrine falcon, golden eagle, rough-legged hawk, raven.

Only on sea cliffs: Thick-billed murre, Thayer's, Iceland, and ivory gulls, kittiwake.

Nests that will be concealed by sedges in high summer may be in plain view at the beginning of the season, before the sedges have begun their season's growth. This is particularly true of goose nests. Two other ground-nesters whose nests can often be seen from afar are the snowy owl and the tundra swan. They build their nests on raised mounds, giving distant views over the tundra. The ability to see approaching predators while they are still a long way off must compensate for the inability of these big, conspicuous birds to hide themselves.

Many ground-nesters' nests, especially those of shorebirds (plovers and sandpipers) are all but invisible. The eggs are cryptically patterned, with black and brown spots and speckles on an olive or tan background. They are laid on bare ground or in a hollow lined with lichen, moss, grass, and dead leaves. Either way, they blend with their surroundings perfectly. The majority of shorebirds lay a clutch of 4 eggs and arrange them with the pointed ends toward the center, making a square pattern just not too large for the parent to incubate (fig. 6.1).

The sitting parent is equally hard to spot, because of her (or his) cryptically patterned plumage. For most species, it is easy to tell whether the eggs are incubated by one parent or both: simply compare the males'

Figure 6.1. Nests of 3 bird species: (a) Baird's sandpiper; four eggs blending with their surroundings. (b) Glaucous gull: Though the eggs are cryptically colored, the nest is fully exposed on a sandy beach. (c) Two common eiders' nests, side by side; note the uncamouflaged eggs, and the lining of down in each nest.

and females' plumage. It can safely be assumed that a brightly patterned parent does not do any of the incubating. Common eiders are a good example of a species in which the cryptically colored female does the incubating while the brightly patterned male leaves her to it. In phalaropes, it is the female who is brightly colored and leaves the drably plumaged male to do the incubating.

Not all ground nests are hard to see. Those of glaucous gulls and common eiders, to take two common species, are often built on bare rock or sand, making them easy to spot (fig. 6.1). These birds often protect their nests from their chief enemies, foxes and ermines, by nesting on small offshore islands. This puts the nest out of reach of the predators once the ice has gone; but it means that nesting has to be delayed until break-up which, in bad years, may come so late that not enough warm weather remains for the raising of a family.

A late season is the commonest cause of breeding failure in the Arctic. Some years are so bad they foil the efforts of nearly all species. Drought is a danger, too: without enough moisture, the sedges and other plants that conceal nests from hunting foxes fail to grow big enough to be useful. The ups and downs of the lemming cycle (see sec. 7.10) determine the breeding success of species that depend on them: snowy owls and pomarine jaegers. In a year with few lemmings, these birds don't breed.

In spite of the difficulties, however, the arctic environment is obviously beneficial for those birds adapted to it. Abundant food is available in good summers; birds of many species evidently find it worthwhile to migrate long distances to reach it.

6.6 A Field Guide to Arctic Birds

PRELIMINARY NOTES

This section is a field guide to all those birds whose only, or principal, nesting areas are in the Arctic.

The format is the same as that of the field guide to arctic plants in section 5.13. As with the plants, the birds are grouped into families.

Under each family heading, the species are first described in sufficient detail to make identification easy. The English name of each species is followed, in square brackets, by a coded list of the regions of the Arctic where it breeds; the code is explained in figure 5.8, which shows the boundaries of the regions and their code letters. Then follows the scientific (Latin) name of the species. See section 5.13 for an explanation of how these names are constructed. Following all the species descriptions are some further notes on the family's characteristics and ecological peculiarities.

As with plants, identification is easier if you have a checklist of birds known to breed in the area you are visiting. Naturalists' lodges can usually supply checklists. Of course, you cannot count on seeing all the species listed: in cool summers, many birds do not even attempt to breed. On the other hand, birds occasionally expand their breeding ranges and turn up in surprising places.

References

Two books with interesting accounts of arctic birds, their life styles, behavior, and ecology are:

Arctic Animals, by Fred Bruemmer (Toronto: McClelland and Stewart, 1986).

Polar Animals, by A. Pedersen; trans. by Gwynne Vevers (New York: Taplinger Publishing Co., 1966).

Two reference books that, though not limited to arctic birds, contain much information about them are:

The Birds of Canada, by W. Earl Godfrey (Ottawa: National Museums of Canada, 1986).

The Birder's Handbook, by Paul R. Ehrlich, David S. Dobkin, and Darryl Wheye (New York: Simon & Schuster, 1988).

The Loon Family (Gaviidae)

Four species of loons breed in the Arctic, three of them only in the Arctic or near-Arctic (fig. 6.2). The fourth, the Common Loon (*Gavia immer*), found all the year round throughout Canada and the northern United States is only marginally an arctic breeder; it breeds no farther north than the southern border of the Barrenlands, the Ungava Peninsula, and southern Baffin Island. All four loon species belong to the same genus, *Gavia.* They are always seen in breeding plumage (the same in both sexes) in the Arctic; they molt into their duller winter plumage on the wintering grounds.

The **Yellow-billed Loon** [AML], *G. adamsii,* is very similar to the common loon; the obvious difference is the pale yellow (almost white), slightly uptilted bill (the common loon's bill is black and straight). The two species are very closely related.

The **Red-throated L** [AHMLB], *G. stellata,* is the commonest and widest-ranging arctic loon. It is smaller than the common loon. The head and neck look paler than those of other loons from a distance; at closer range the red throat patch and somewhat uptilted bill are distinctive. It lacks the black-and-white checkerboard pattern on the back that other loons have.

The **Pacific L** [AMLb], *G. pacifica,* is the same size as the red-throated.

Figure 6.2. (a) Yellow-billed Loon. (b) Red-throated Loon. (c) Pacific Loon.

The pale pearl-gray crown of the head and contrasting dark face and black throat patch are conspicuous from a distance. The species used to be treated as identical with the Arctic L (*G. arctica*) of Eurasia. The two species are now regarded as separate, though they are undoubtedly closely related.

FURTHER NOTES

All loons are fish-eaters and they are one of the bird families most fully adapted to a life on the water. In the air they are powerful fliers, and in the water (fresh or salt) powerful swimmers. To get under water, they usually dive from the surface, but they can also sink from sight with no change of attitude; to do this, they make themselves heavier, by flattening their feathers and expelling air from their lungs.

Swimming power is mostly provided by their large, webbed feet. Their legs are attached far back on their bodies, which no doubt makes for strong swimming but brings the disadvantage that the birds can scarcely walk on land, being able only to inch forward on flattened legs with their breasts close to, or touching, the ground. They rarely let themelves get in a position of having to do this, however; because of their helplessness on land, they nest at water's edge by a pond or lake, or even on a floating raft of vegetation. From such a site, a loon can reach the water quickly without having to walk there. Except for red-throated loons, all loons must reach water in order to launch into flight. They cannot take off from the ground, but have to patter some distance along a water "runway" before becoming airborne. The red-throated loon can, if it must, take off from the ground; it is more nimble than the other species, and when it takes off from water (the preferred method) requires a shorter runway. This means that it can nest by smaller ponds than the other species. In any case, loons do not depend on their nesting ponds for food supplies. Often there are no fish in the shallow tundra pools where they nest, and they fly to larger water bodies—often the sea—to catch fish.

Like ducks and geese, loons are flightless during the molt. But unlike ducks and geese, their flightless stage comes in fall or winter, when they are in temperate latitudes, at sea or on large lakes, where they can easily escape danger by their diving and swimming abilities.

Loons are monogamous; they share incubation of the (usually) 2 eggs, and also share the duty of feeding and caring for the semiprecocial chicks. They are solicitous parents, and sometimes swim with their tiny chicks riding on their backs. Observations in Greenland showed that, in red-throated loons, fathers were much more dutiful as parents than mothers, both before and after the eggs hatched.

The Shearwater and Fulmar Family (Procellariidae)

This family of seabirds has only one arctic representative, the **Northern Fulmar,** *Fulmarius glacialis* (fig. 6.3). Except during the breeding season, fulmars spend their whole lives at sea, flying or swimming without ever touching land. They are to be found in the eastern Arctic, in Baffin Bay and Davis Strait (see fig. 4.1), and for some distance into the channels entering these large bodies of water from the west; a separate population lives far to the west, in Bering Strait and the Chuckchi Sea.

Fulmars have a range of color phases, from light to dark. Light ones are very similar to gulls in coloring—white below and pearl gray above—and at first glance are often mistaken for gulls. Closer inspection makes the differences apparent: compared with gulls, their necks are noticeably short and thick; their flight consists of short spurts of rapid wing-beats (faster than gulls' wing beats) alternating with prolonged soaring, on stiff wings, low over the surface of the sea; and their tubenoses (see below) give their bills a peculiar shape, visible from afar.

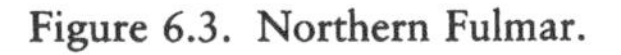

Figure 6.3. Northern Fulmar.

They sit high in the water when swimming, pecking to left and right as they feed. Dark phase fulmars also have these characteristics and are a dark, sooty gray all over.

FURTHER NOTES

Fulmars breed in colonies, usually on towering cliffs overlooking eastern arctic seas, which they share with other seabirds such as murres and kittiwakes. The single white egg, laid on a cliff ledge, takes an unusually long time (almost 2 months) to hatch, into a semialtricial chick (see sec. 6.4). To protect themselves from molestation while incubating, fulmars forcefully vomit stinking stomach oil onto any perceived enemy that comes too close. An intruding bird squarely hit by this liquid missile may die as a result; the oil destroys the water repellancy of the target bird's feathers, whereupon it risks becoming wet to the skin and dying of cold; a large quantity of oil sticking to the target can even cause it to lose buoyancy and drown.

Fulmars get all their food from the sea; they feed at the surface, scooping up plankton of all kinds, squids, jellyfishes, crustaceans, and the like. They have profited immensely from human intrusion into their waters. They follow ships for days on end, feeding on edible garbage thrown overboard; the trails of fish guts left in the wakes of fishing boats are especially attractive to them.

Fulmars, along with albatrosses, shearwaters, and petrels, are the only birds having tubenoses—long, horny tubes enclosing their nostrils. Through these tubes they excrete the salt they absorb on drinking sea water, the only water available to them for most of their lives. The salt goes from bloodstream to nostrils via a pair of big glands at the base of the bill.

The Waterfowl Family (Anatidae): I. Swans and Geese

This enormous family includes swans, geese, and ducks. Swans and geese, the topic of this section, differ markedly from ducks, which are described in the next section.

One species of swan and five of geese breed in the arctic.

The swan is the **Tundra Swan** [ALb], *Cygnus columbianus* (fig. 6.4). The adult is a typical swan, a huge, majestic, long-necked bird of the sea coast. A clean swan is pure white, but because swans spend much of their time afloat in shallow water, probing for water plants rooted in the mud, their heads and necks are often stained. Immature birds are pale gray but have the characteristic swan shape.

Geese differ from swans in having shorter necks. They also have longer legs, enabling them to walk better than swans and to graze more easily on dry land. The five arctic breeding geese are snow goose, Ross's goose, greater white-fronted goose, brant, and the Canada goose.

The **Snow Goose** [AHMlB], *Anser caerulescens* (fig. 6.4), has two color phases, white and blue. The snow geese of the High Arctic are all (with very rare exceptions) white phase; they are also larger than other snow geese and are classified as belonging to a distinct subspecies, the Greater Snow Goose. The smaller snow geese of lower latitudes, known as Lesser Snow Geese, may be white or blue. Regardless of subspecies or phase, a snow goose can be identified with certainty (at close range) if it has a "grinning patch" on its bill, a narrow, dark oval where the mandibles meet (see fig. 6.4). At longer range, a white-phase snow goose could be confused with the much smaller and rarer Ross's goose (see below); both are pure white (unless mud-stained!) except for their chief flight feathers (the primaries) which are entirely black; this means that most of the outer half of each wing is black, not just the tip. A blue-phase snow goose is distinguishable from all other geese, in spite of the indefinite brownish-gray of its body, by having a white head and neck.

Ross's Goose [l], *Anser rossii* (fig. 6.4), is a small version of a white-phase snow goose, without the "grinning patch" on the bill. It is rare, breeding in only a few, small areas on the mainland coast of arctic Canada.

The **Greater White-fronted Goose** [AL], *Anser albifrons* (fig. 6.4), is a grayish-brown goose with less white than its name suggests: the white forms a narrow band around the bill at the front of the face and is not present on immatures which are rather similar to blue-phase snow geese. But no confusion arises if the birds' legs can be seen clearly: white-fronted geese have yellow legs, snow geese have pink.

Brant [AHMLB], *Branta bernicla* (fig. 6.4), is a small, dark goose that, at long range, looks much like a Canada goose. At close range the difference is obvious: though both geese have black heads and necks, the brant lacks the Canada's broad, white chin-strap; instead, it has a less conspicuous, black-and-white "collar." The brants of the western Arctic, which overwinter on the Pacific coast, are darker than those of the eastern Arctic, and belong to the subspecies Black Brant; their breasts and bellies, as well as their necks, are dark. Eastern brants (of the subspecies Atlantic Brant), which overwinter on the Atlantic coast, have pale underparts, separated by a sharp boundary from their dark necks.

The arctic breeding range of the **Canada Goose** [AmLb], *Branta canadensis,* is only the northern edge of its total breeding range, which covers most of northern North America. The species is therefore not described here and, in any case, hardly needs description.

Figure 6.4. (a) A Tundra Swan on her nest. (b) Snow Goose, white phase; *inset:* close-up of head to show "grinning patch" on bill. (c) Snow G, blue phase. (d) Ross's G, showing lack of grinning patch. (e) Greater White-fronted G. (f) Brant, dark, western subspecies.

FURTHER NOTES

Swans and geese, while obviously related to ducks, differ from them in some important respects. First, in swans and geese, the plumage is the same in both sexes and at all seasons. The lack of seasonal change is because these birds have only one annual molt instead of two, as ducks have. But for all of them (and also loons, grebes, and coots), the molt takes place suddenly, with drastic effects. The birds lose all their flight feathers simultaneously and are unable to fly until new ones have grown. Four or five weeks of flightlessness in summer evidently does them no harm. They are large enough and strong enough to fend off enemies without needing to fly for safety, and they don't rely on flying to obtain their food, which consists of lake and marsh plants and aquatic invertebrates.

The second contrast between swans and geese on the one hand and ducks on the other is in their breeding behavior. Swans and geese, unlike ducks, mate for life; both parents care for the young, even though (as with ducks) the mother does all the incubating; and they don't reach sexual maturity until they are 2 or 3 years old.

Because they have been so thoroughly investigated by duck hunters, the population statistics of waterfowl are better known than those of any other group of birds. It turns out that breeding success varies enormously from year to year, and by far the most important controlling factor is weather. In particular, a delayed spring can be disastrous; it may leave insufficient time for swan chicks (cygnets) and goose chicks (goslings) to grow strong enough for the flight south when summer is over. A late spring also means delayed plant growth, with consequent hunger for the parents. In bad years, large numbers of waterfowl don't even attempt to breed; others lay a clutch of eggs and then desert them, probably because of hunger. A snowy winter followed by rapid spring warming can also bring catastrophe. Geese usually build their nests on low-lying ground, which floods if the thaw is sudden. Beside all these weather-related troubles, predation is comparatively unimportant, in the statistical sense if not for individual families. All the usual predators—jaegers, gulls, and foxes especially—take the eggs or young of waterfowl if they can, but are often driven off by the parents, who defend their nests and broods fiercely. The young are precocial (see sec. 6.4).

Swans are usually solitary nesters. Geese tend to breed in colonies if conditions are good and many families are breeding. The worse the season and the fewer the pairs that try to breed, the looser the colonies become, until in a really bad year most nests are solitary. The species differ in their gregariousness. Snow geese form dense breeding colonies in good years and battle with their neighbors to protect their territories. The rare Ross's goose also nests in colonies, sometimes in snow goose colonies. Brants are rather less gregarious. White-fronted geese are the least gregarious of all; though they nest in small colonies occasionally, their nests are more often solitary and well hidden in the tundra vegetation. Solitariness demands secrecy, for there are no neighbors to warn of approaching predators.

Swans and geese nearly always nest close to water, either by the sea itself or by estuaries, broad rivers, marshes, or lakes. Brants always stay close to salt water, nesting only a short way above high tide.

As to the nests themselves, they are hollows in the ground surrounded by "parapets" of sedge, grass, and moss that the bird pulls in toward her as she sits in the middle. A swan usually nests on top of a natural hummock, which helps keep the nest dry. It also provides a lookout 30 or 40 cm tall; with the swan's long neck besides, it ensures quite distant views over flat tundra. Brants line their nests with copious amounts of down; indeed, they use more down than eiders.

For more detail, see *Ducks, Geese, and Swans of North America,* by Frank C. Bellrose (3d ed.; Harrisburg, PA: Stackpole Books, 1980).

The Waterfowl Family (Anatidae): II. Ducks

A great variety of ducks breed in the Arctic, but as soon as early summer is past, conspicuous, colorful ducks are hard to find. The reason is that the males of most species go into seclusion and lose their bright plumage once the drab-colored females have begun to incubate (see below). Therefore, ducks seen late in the season may be hard to identify, even for experienced birders. The following descriptions refer to males while they are still in their bright, so-called *nuptial* plumage, and to females only if they are reasonably easy for a nonspecialist birder to identify.

The great majority of arctic ducks are diving ducks: they submerge completely as they dive down in search of food. Another diagnostic feature is that they patter along the surface for some distance before becoming airborne when they fly up from the water. The other big group of ducks, the dabblers or puddle ducks, have few arctic representatives; only one (described below) has an extensive arctic breeding range. The dabblers are vegetarians, feeding on water plants growing in shallow water, which they reach by tipping up with necks submerged and tails pointing skyward. Therefore they are never found swimming in deep water. They also differ from diving ducks in the way they fly up from the water; they spring directly into the air without pattering.

Ten or eleven diving ducks breed in the Arctic. The most typically arctic are the eiders, with 2 wide-ranging species and 2 species confined to Alaska.

Common Eider [AHMLB, seacoast only], *Somateria mollissima* (fig. 6.5), is one of the widespread species. Both sexes have distinctive sloped profiles, with the forehead and top of the bill aligned. The males's curved white wing feathers, ending over a round white flank spot, make a conspicuous pattern that can be seen from afar.

The **King E** [AHMLB], *S. spectabilis* (fig. 6.5), is as widespread as the common eider and, on the Arctic Islands, nests inland as well as on the coast. Males are unmistakable because of a big orange frontal shield bordered with black on the bill. The females of common and king eiders are streaky brown ducks—hard to tell apart but recognizable as eiders because of the sloped profile.

The 2 other eider species are found only in western Alaska. They are **Spectacled E,** *S. fischeri,* and **Steller's E,** *Polysticta stelleri;* both species are uncommon. The males can be recognized by the unusual patterns on their heads (fig. 6.5).

The three species of scoters, all large, dark ducks (mostly black in the males, charcoal brown in the females), have curiously "lumpy" bills. **Surf Scoter** [Al; chiefly Mackenzie Delta], *Melanitta perspicillata* (fig. 6.5), males have conspicuous rectangular patches of pure white on the forehead and the back of the head; their beating wings emit a loud undulating whistle when they start to fly up from the water. **White-winged S** [Al; chiefly Mackenzie Delta], *M. fusca* (fig. 6.5), males have big white wing patches that are obvious in flight but often concealed when the bird settles on the water; the comma-shaped white marking around the eye is diagnostic, but is hard to see at long range. The least common of the scoters is the **Black S,** *M. nigra* (not illus.), another western Alaska special. The male is all black, with no white markings, and has a bright orange-yellow knob at the base of the bill. Though they breed at high latitudes, scoters are well known to mid-latitude birders who live on the seacoast. All three species winter on salt water, close to shore, on both the Pacific and the Atlantic coast.

The **Oldsquaw** [AHMLB], *Clangula hyemalis* (fig. 6.6), is one of the most abundant arctic ducks, and certainly the most noticeable. It draws attention to itself by its "talkativeness"; flocks seem to converse constantly, using a variety of calls and yodels that carry a long way over calm water. Oldsquaws are also unusual in that both sexes are brightly patterned at all seasons. The male (only) has a long, pointed tail, and in summer a big, circular, white eyepatch which contrasts strongly with his black head and neck; the female's face somewhat resembles a photographic negative of the male's, having a big, dark spot on a white background (fig. 6.6). Flocks of oldsquaws flying fast and low over tundra ponds appear light and dark alternately, as they rapidly tilt their bodies this way and that.

The **Harlequin Duck** [A; lb], *Histrionicus histrionicus* (fig. 6.6), has two widely disjunct breeding ranges: in the west, the arctic coasts of Alaska and Yukon (and thence southward into temperate latitudes); in

Figure 6.5. (a) Common Eider. (b) King E. (c) Spectacled E. (d) Steller's E. (e) Surf Scoter. (f) White-winged S.

Figure 6.6. (a) Oldsquaw (male and female). (b) Harlequin Duck. (c) Red-breasted Merganser. (d) Greater Scaup. (e) Northern Pintail.

the east, southern Baffin Island and the arctic shores of Ungava Bay (thence southward to the Gulf of St. Lawrence). The eastern population is listed as endangered on the 1993 Canadian Endangered Species List. The males are strikingly and uniquely patterned. Within their western range, harlequins are often to be seen flying fast down the valleys of swift rivers and streams and passing low over rapids. (If you skipped the harlequin's Latin name, go back and read it; it describes the bird's behavior exactly.)

The **Red-breasted Merganser** [Alb], *Mergus serrator* (fig. 6.6), is the only arctic-breeding duck with a crested head in both sexes. The crest lies flat when the bird flies and is best seen on swimming birds. This duck is a fish-eater, and its bill is long, narrow, and most unducklike; at very close range, the bill can be seen to have a serrated cutting edge, an adaptation allowing the bird to grip its slippery prey. The pattern of the male is black above (with a green sheen on the head) and white below, with a broad, reddish breast band. The female has a gray body and rusty red, crested head. The related Common Merganser (*M. merganser,* not illus.) is also quite often seen in the Arctic though it seldom breeds there. Like the red-breasted, it has a long, narrow, fish-eating bill; but the male has no crest and is all white below, lacking a red breast band. The females are very similar to red-breasted females. Mergansers (both species) have a habit of swimming along looking downward, with neck arched and face submerged, as they scan the water for fish.

In the two species of scaups, the **Greater Scaup** [Al], *Aythia marila* (fig. 6.6), and the **Lesser S** [Al], *A. affinis* (not illus.), the males are "black at both ends and white in the middle," a rough rule (sometimes the white is gray) that distinguishes them from all other ducks. Their bills are blue. It takes an experienced birder to tell the two species apart. The greater scaup has a round black head, with a green sheen; the lesser scaup has a rather triangular black head, with a purplish sheen. Usually the head shape is indefinite and its color plain black, in which case honesty requires a birder to call the duck a scaup and let it go at that. The same goes for females; they are brown except for a white band surrounding the base of the blue bill. Male scaups retain their bright, nuptial plumage until after other duck species have lost theirs.

The ducks described above are all diving ducks. Several dabbling ducks have breeding ranges that extend from forested latitudes northward for a short distance into the arctic tundra, prominent among them being American wigeon, mallard, green-winged teal, and northern shoveller. They are not described here. The only dabbling duck whose range extends well into the Arctic (even reaching the southernmost Arctic islands) is the **Northern Pintail** [AmL], *Anas acuta* (fig. 6.6). The male (unfortunately not the female) is unmistakable; he has a long, pointed tail, and a long, white neck, from which a narrow white stripe extends upward into the black of the head. The streaky brown female is best distinguished by her unusually long neck.

FURTHER NOTES

Unlike swans and geese, which molt only once a year, ducks molt twice. This allows the males to alternate their plumage, from bright and showy when the need to attract females outweighs the need for concealment, to drab and cryptic when the opposite holds. In this, ducks resemble numerous other birds; among arctic examples are snow buntings, Lapland longspurs, lesser golden and black-bellied plovers, dunlins, and many other sandpipers. But ducks differ from other birds in the timing of their molts. "Ordinary" birds grow their showy breeding plumage in the spring and lose it in the fall; the arctic breeders among them are therefore brightly patterned while in the Arctic, and drab while on their wintering grounds to the south. Ducks differ: the males of most species grow their showy plumage at the end of summer, not the beginning of it, and then fly south to spend their winters displaying to prospective mates—a new one every year, as ducks don't show the lifelong marital fidelity of swans and geese. They wear their drab, nonbreeding plumage for only a short period in summer, while still in the latitude of their breeding grounds (though not necessarily right on the breeding grounds; see below). This explains why colorful ducks are more typical of temperate latitudes in winter than of the Arctic in summer.

Not only is the timing different, so are the results. One of a duck's two annual molts is a complete molt which happens abruptly; the duck loses all its flight feathers simultaneously, leaving it unable to fly until the new ones grow (the same thing happens at the single annual molt of swans and geese). For male ducks, the complete molt is the first molt of the year, but for females it is the second (the timetables are described in more detail below). At the complete molt, a male duck grows drab, cryptic *eclipse* plumage, very similar to a female's fulltime plumage; he is said to "go into eclipse." The eclipse stage lasts only a few weeks, being soon followed by the second molt, a partial molt, which does not impair the duck's flying power as the flight feathers are retained. At this molt he regrows his bright *nuptial* plumage and wears it for the greater part of the year. Females molt twice a year, too, losing their ability to fly after one of the molts. But they have no alternation of plumage, remaining cryptically colored at all times.

Now for the timing: in general terms (details differ from species to species) and for most ducks (except oldsquaw; see below), the molt timetables are as follows—two timetables, as males and females each have their own.

Males desert the females in early summer, soon after the females start incubating their eggs (they incubate unaided) and go through the complete molt. Some species retreat to secluded ponds for this molt and lie low while their eclipse plumage grows. Other species are less reclusive, and flightless ducks can be seen floundering across the surface of the water, propelled by wings and feet combined, in what looks like a human swimmer's butterfly stroke. The second molt, when the male regrows his nuptial plumage, comes at summer's end.

Females do their partial molt in early spring, in preparation for motherhood, and their complete molt (causing temporary flightlessness) in late summer, after their offspring have become independent. The plumage pattern remains much the same, cryptic and drab, all year.

In contrast to other ducks, oldsquaws of each sex have both the breeding and nonbreeding plumages brightly patterned and can be seen in all sorts of intermediate stages, too, as they change from one plumage to another.

In some duck species, males about to molt into eclipse plumage gather together and migrate in big flocks to molting areas a long way from where the females are incubating. They are not searching for a better climate. Rather, they are making for distant tundra ponds where they can go through the flightless period in comparative safety, away from predators lurking around busy breeding areas. Eiders, scoters, scaups, and oldsquaws are often seen on these "molt migrations." Dates and destinations differ from species to species and from year to year, but often one of the busiest migration corridors leads westward along the shores of the Beaufort Sea to molting areas in western Alaska. The sight of huge flocks of migrating sea ducks in mid-summer is surprising until you realize that they are not beginning the southward migration to their wintering territory, but are bound for molting areas which may lie in an altogether different direction.

No doubt because of the food that is available, the majority of arctic ducks are diving ducks, feeding primarily on animal food. They all take crustaceans, mollusks, and (in fresh water) insect larvae. Eiders have bills adapted to clam-eating—the bills are strong enough to crush clam shells; mergansers' bills are adapted to fishing. The Arctic has less to offer dabbling ducks, which rely on plant food for most of their diet. Though tundra pools shallow enough for dabblers to feed are everywhere, except in the Low Arctic they seldom contain much plant life.

A duck whose feeding habits deserve special mention is the harlequin. Harlequins usually nest beside, and feed in, fast-flowing rivers or streams; they are able to forage in these torrents in spite of the strong currents, walking over the rocks of the stream bed just as American dippers do.

The other duck species nest in less dramatic surroundings, on the ground and usually (not always) within easy reach of water. The nests of eiders must be outstandingly comfortable for the chicks, because of the way in which the mother bird lines her nest with eiderdown. The chicks are uncrowded, too, as eiders have smaller clutches than most other ducks. The common eider is the only duck that always nests close to the seashore, often on tiny islands where, once the sea ice has melted, they are out of reach of foxes; tiny islands provide limited space, so several nests are often crowded together (fig. 6.1). The other eider species nest on inland tundra as well as near the coast.

All ducks have precocial young (see sec. 6.4). Clutch size can be large—anywhere from 6 to 11—and at the rate of one egg per day, a mother takes a considerable time to finish laying. She doesn't begin incubating until the last egg is laid, however, with the result that all the eggs hatch simultaneously, and all the ducklings in a family are the same age.

Ducks have brought communal care of their offspring to a fine art. Many species establish "creches": several mothers (the fathers have all deserted by this time!) combine their broods and leave them in the care of one or two adults. This explains the great rafts of ducklings—far too many to belong to one family—sometimes seen together. Among ducks noted for creching are common and king eiders, white-winged scoters, oldsquaws, red-breasted mergansers, and both species of scaups.

Common eiders leave the care of their ducklings to "aunties," nonbreeding females who take over as soon as the eggs hatch. Small flocks of female eiders are to be seen flying hither and thither at high speed at hatching time; these are platoons of aunties attending to their duties, which include the protection of ducklings from predatory gulls and jaegers. A breeding eider female must be glad of their services; she goes without food for the 4 weeks she spends incubating, and is notably weak by the end of her fast.

Another way in which ducks share (or shirk) the burden of reproduction is by brood parasitism: a female lays her eggs (or some of them) in the nest of another bird of the same species. Arctic breeders that are known to do this are white-winged scoter, oldsquaw, and both scaup species. The phenomenon is hard to observe, because the intruder's eggs look the same as those of the nest's owner. In a sense, brood parasitism is a natural form of surrogate motherhood.

For more on ducks, see *Ducks, Geese, and Swans of North America*, by Frank C. Bellrose (3d ed., Harrisburg, PA: Stackpole Books, 1980).

The Plover Family (Charadriidae)

Plovers are short-billed, short-necked birds of tundra and shore that draw attention to themselves wherever the vegetation is low or sparse. When disturbed, they run in swift, short bursts alternating with motionless pauses. Sometimes they are conspicuous, when they are running, "displaying" to distract intruders, or calling. Sometimes they are almost invisible, when they are motionless, on or off the nest.

There are two arctic genera, with two species in each. The two species of the genus *Pluvialis* are the **Lesser Golden Plover** [AhMLB], *P. dominica* (fig. 6.7), and the **Black-bellied P** [hMlB], *P. squatarola* (fig. 6.7). Their English names are misleading, giving the false impression that only one is black-bellied whereas, in fact, both species are. Indeed, black-bellied plovers are less markedly black-bellied than golden plovers. In breeding plumage on their arctic breeding grounds, the differences between them are obvious. Black-bellieds have conspicuously white lower bellies and rumps, and the upper surface of the tail is white with a few black flecks. Their backs and the crowns of their heads are also much paler than those of goldens, and could well be called silver-colored. The two species are easily confused when in nonbreeding plumage except when in flight; then it is easy to see that the black-bellied has a black patch in the "armpit" whereas the golden has not.

The two other arctic plovers belong to the genus *Charadrius*. They are the **Semipalmated Plover** [AMLB], *C. semipalmatus* (fig. 6.7), and the **Common Ringed P** [hb], *C. hiaticula* (not illus.). Where both species occur, they are almost indistinguishable from each other; both are brown-backed birds with a single black breast-band and are much smaller than the two *Pluvialis* plovers. They cannot be told apart unless one has a specimen in hand, when it can be seen that the toes of the semipalmated are webbed for half their length (hence the name), whereas the common ringed has only a vestigial web, between the outer and middle toes. However, there's no risk of misidentifying nesting *Charadrius* plovers except in the very small region where their breeding ranges overlap, along a short stretch of the northeastern side of Baffin Island. Outside Baffin, the common ringed breeds only in the High Arctic, in far northern and eastern Ellesmere Island and eastern Devon Island. Everywhere else, a breeding plover with a black breast-band is sure to be a semipalmated. (The adjective *common* in the name of the uncommon species refers to its commonness in Eurasia.)

FURTHER NOTES

All the plovers are ground-nesters that perform "broken wing" displays to distract you if you get dangerously close to their nests. The black-bellied's display is particularly eye-catching because of the bird's conspicuous white rump. While distracting you, a black-bellied will often settle down, with all the mannerisms of a bird arranging itself on its eggs, where there is no nest. Then, when you come close, it will run off and repeat the maneuver.

Figure 6.7. (a) Lesser Golden Plover. (b) Black-bellied P. (c) Semipalmated P.

Plovers, like sandpipers, have precocial young (see sec. 6.4), but it has been found that the eggs in a plover's nest (there are nearly always 4) do not necessarily all hatch simultaneously as they do in sandpipers' nests. Synchronous hatching would seem to be the best system when the young are capable of scampering away in all directions soon after they emerge from the eggs, and it can be achieved when birds do not begin to incubate a clutch of eggs until the last one has been laid. Black-bellied and Semipalmated Plovers, however, have been seen to start incubating before the laying period is over. The young then hatch at intervals of many hours; just-hatched baby plovers will sometimes be seen exploring their new world close to where their mother (or father) is still sitting, incubating the last-laid eggs.

The Sandpiper Family (Scolopacidae)

Sandpipers (together with plovers) are often described simply as shorebirds. Nineteen members of the family, those that breed only in the Arctic and subarctic, are described below. They can be sorted into the following 4 groups: 4 "big" sandpipers; 10 *Calidris* sandpipers (members of the genus *Calidris*); 2 phalaropes; and 3 "others."

GROUP I

All four of the Large Sandpipers are mainland breeders with restricted breeding ranges; their large size and long, distinctive bills make them easily recognizable.

The **Whimbrel** [Al], (*Numenius phaeopus*) (fig. 6.8), has a long, downcurved bill and striped head.

The **Hudsonian Godwit** [l], (*Limosa haemastica*) (fig. 6.8), and **Bartailed G** [a], (*L. lapponica*) (not illus.), both have up-curved bills. The Hudsonian godwit has a black tail, white rump, and white wing stripes; the bar-tailed godwit has a narrowly barred tail and no wing stripes. Their breeding ranges are believed not to overlap.

The **Long-billed Dowitcher** [A], *Limnodromus scolopaceus* (fig. 6.8), is the smallest of the big four. It has a long, straight bill, and a white rump and back visible only in flight (in the two white-rumped *Calidris* sandpipers, the back is not white, only the rump).

GROUP II

Calidris sandpipers form the largest genus of the family. Small ones, which in their nonbreeding plumage are sometimes hard to recognize, are known collectively as "peeps." Some are easier to identify in the Arctic when they are in breeding finery, which scarcely deserves the name.

Three of the peeps that are hard to tell apart at any season are **Baird's Sandpiper** [AHMlB], *Calidris bairdii* (fig. 6.8); **Least S** [AL], *C. minutilla* (not illus.), recognizable at close range, in a good light, by its greenish-yellow legs; and **Semipalmated S** [AmLb], *C. pusilla* (not illus.), which gets its name from its half-webbed feet, a feature invisible in the field.

Figure 6.8. (a) Whimbrel. (b) Hudsonian Godwit. (c) Long-billed Dowitcher. (d) Baird's Sandpiper. (e) Dunlin. (f) White-rumped Sandpiper. (g) Red Phalarope.

Of these three "featureless" sandpipers, Baird's is the most widespread, being known to breed on all the Arctic Islands. According to some ornithologists, it is the hardest of all to identify. It serves as a standard with which to compare the next seven *Calidris* sandpipers that (in the breeding season) have features by which to identify them superimposed on the basic *Calidris* pattern.

The **Dunlin** [aL], *C. alpina* (fig. 6.8), has an unmistakable black belly.

The **White-rumped S** [hMlb], *C. fuscicollis* (fig. 6.8), has (of course) a white rump, but it is apparent only in flight or when the bird is displaying. Two other white-rumped sandpipers which should not be confused with this one are the stilt sandpiper (described below) and the much larger long-billed dowitcher (all-white back, very long bill; see above). In the lesser yellowlegs (mainly a temperate zone breeder) the tail as well as the rump appears white.

The **Stilt S** [ml], *C. himantopus* (not illus.), has long, yellow-green legs, a chestnut cheek patch, strongly barred underparts, and a white rump.

The **Purple S** [hlB], *C. maritima* (not illus.), seems dark and dumpy (short-legged) compared with other peeps. It is less distinctive in summer when the head, neck, and breast are streaked with dark slate-gray than in winter when the dark color is uniform. Its dark back makes the white wing bars conspicuous; the bill is dark with a yellow-orange base.

The **Pectoral S** [AhML], *C. melanotos* (not illus.), is marked by a sharp boundary between streaked breast and white belly, plus a white eyebrow line. On its breeding grounds, look for males displaying with inflated throats; under the skin are 2 air-filled sacs, which give out a booming sound as the air is exhaled.

Sanderling [aHMl], *C. alba* (not illus.), is less conspicuous on the breeding grounds in summer than in temperate latitudes in winter when its underparts are all silvery white; in breeding plumage, the head and upper breast are orange-rusty.

Red Knot [Hml], *C. canutus* (not illus.), is much larger and sturdier than the others and has a short bill and robin-red breast. It is named after King Canute (he who failed to stop the tides) because he enjoyed them roasted.

GROUP III

Phalaropes are the only swimming sandpipers, enabled to swim by their lobed toes. Females are bigger and brighter than males (see below for more on breeding habits). There are two arctic breeders:

The **Red Phalarope** [aHMlb], *Phalaropus fulicaria* (fig. 6.8), is robin-red all over except for its black cap, big white eye-circle, and mottled back (note the stout bill).

The **Red-necked Phalarope,** also called **Northern Phalarope** [AmLb], *P. lobatus* (not illus.), is recognizable even in silhouette, from the needle-like bill. Its colors are drab, except for a red neck and white chin.

GROUP IV

Each of the three remaining members of the sandpiper family is unique in its own way:

The **Ruddy Turnstone** [HMlb], *Arenaria interpres* (fig. 6.9), is conspicuously and intricately patterned in black, white, and chestnut. The tapering, sharp-pointed bill, very unlike a typical sandpiper's, is used for turning stones but not frequently enough for the action to be a useful identification mark. The bird is unmistakable anyway, from its appearance.

The **Wandering Tattler** [a], *Heteroscelus incanus* (fig. 6.9), breeds by mountain streams. Its back is uniformly dark slate-gray and the breast heavily barred with gray; there is a white streak through the eye. The legs are yellow.

The **Buff-breasted Sandpiper** [ahMl], *Tryngites subruficollis* (fig. 6.9), has underparts uniformly buff all over, and yellow legs. The pure white wing linings are conspicuous during courtship display when the birds gather on "leks" (see notes on sandpiper breeding systems, below).

Besides the sandpipers described above, 4 others that sometimes nest along the southern margin of the tundra are Upland S (*Bartramia longicauda*), Spotted S (*Actitis macularia*), Lesser Yellowlegs (*Tringa flavipes*), and Common Snipe (*Gallinago gallinago*). Because their chief breeding grounds are in temperate latitudes, they should be familiar to southern birders and are not described here.

FURTHER NOTES

Shorebirds (sandpipers and plovers combined) are adapted to exploit the great productivity of the tundra in summer and, being strong fliers, they can migrate long distances to escape its harshness in winter. The summer tundra provides food for them in huge quantities. The majority feast on the countless invertebrates living in the wet soil and in the mud surrounding the thousands upon thousands of tundra ponds dotting the landscape in warm weather. Insect larvae, especially those of midges and crane flies, are particularly important (see sec. 9.3). Shorebirds scurrying hither and thither, pausing to stab and probe the mud in a never-ending hunt for food, must be one of the commonest sights enjoyed by arctic birders.

Phalaropes, alone among shorebirds, use a different feeding technique because they are swimmers. They spin in tight circles in shallow water, creating a vortex which brings food items (insect larvae, water fleas and the like) to the surface. The birds are unusually buoyant, and therefore swim high in the water; this must make it easier for them to see the prey they have stirred up and to stab it accurately.

At egg-laying time, female sandpipers need extra calcium in their diets

Figure 6.9. (a) Ruddy Turnstone. (b) Wandering Tattler. (c) Buff-breasted Sandpiper.

for the formation of strong egg shells. They get it by eating fragments of the bones and teeth of dead lemmings, which they probably obtain by prying open the pellets left by predators like snowy owls and pomarine jaegers, that regurgitate indigestible parts of their prey.

All shorebirds are precocial ground-nesters (see sec. 6.4). Like plovers, sandpipers do "broken wing" displays to distract intruders on their territories; in species with white rump patches, the displays are especially distracting. Some ornithologists believe that a displaying purple sandpiper is (unconsciously) imitating a lemming or vole: it squeals, and its fluffed-up feathers resemble fur.

The majority of arctic sandpipers form monogamous pairs that defend their territories with determination. Ruddy turnstones and red knots will even drive off jaegers that come too close. Both parents incubate the eggs and tend the young, with the father often being more solicitous than the mother.

Exceptions to this common pattern should be noted. In white-rumped and pectoral sandpipers mating is not monogamous and the females do all the incubating. In sanderlings, the female sometimes lays two clutches of eggs; then the male incubates the first clutch and the female the second.

In the buff-breasted sandpiper, a group of several males gather on a *lek,* a tract of ground used as a stage for courtship displays (see fig. 6.9). The males display in a variety of ways, raising and spreading one or both wings and sometimes jumping into the air and fluttering just above the ground; the flash of white wing linings can be seen from a considerable distance against the brown of the tundra. While this goes on, the females watch and judge. Then, having mated with the male of her choice, a female incubates her eggs and rears her young unaided. Unlike the leks of other birds that use them, those of buff-breasteds are not necessarily on the same site year after year.

Phalaropes are unique. The females are larger and more brightly colored than the males and they initiate courtship. Once a couple has mated, the female lays a clutch of eggs and leaves them for the father to incubate and tend, while she flies off in search of another male with whom to start another family. Sometimes she succeeds. But sometimes she will proposition the husband of another female while he is occupied incubating *their* eggs, in which case the sitting male will fly up and attack the intruding temptress and drive her off ferociously; that done, he creeps furtively back to his nest to resume incubating. Polyandry cannot be enjoyed by many females: there are too few males.

When breeding is over, sandpipers waste no time setting off for lower latitudes and warmer climates (unlike ducks, they molt after migrating south). In two species (red knot and ruddy turnstone), the destinations of the migrants depends on where they start from, as mentioned in section 6.2. Those that breed in the northernmost arctic islands winter in Europe; those that breed farther south winter in South America. This is because each of the two species consists of two populations (subspecies) of different origin, that behave independently of each other and cling to their ancestral habits: a northern subspecies, descended from European ancestors, and a southern descended from American ancestors. In both cases the northern subspecies is darker in color than the southern.

The Gull Family (Laridae): I. Jaegers

The Gull Family is represented in the arctic by three related groups of birds, jaegers, true gulls, and terns. Jaegers are described here, gulls and terns in the next section.

Jaegers (genus *Stercorarius*) are conspicuous birds that seem ever-present in the arctic landscape. They are like large, dark-backed gulls with black caps and strongly hooked, cruel-looking bills. Their appearance is menacing, all the more so when they dive-bomb a person intruding in their breeding territory, striking the visitor's head with their claws. They are graceful, effortless fliers.

There are 3 species (fig. 6.10): the **Long-tailed Jaeger** [AHMLB], *Stercorarius longicaudus,* the **Parasitic J** [AhMLB], *S. parasiticus,* and the **Pomarine J** [ahMlb], *S. pomarinus.* The mysterious word "pomarine" should be defined; it is a shortened form of *pomatorhine,* meaning having the nostrils partly covered with a scale, a description that applies to all species of jaegers. All three species occur in two color phases: a light phase, with dark gray-brown back and white underparts, across which (in pomarines and parasitics) there is usually a grayish breast-band; and a dark phase (exceedingly rare in long-taileds) in which the whole body is dark gray-brown.

Adults are easy to identify as to species, regardless of color phase, from the distinctive shapes of their tails. All three species have the two central tail feathers elongated. These feathers are very long, slender, and fluttering in long-taileds; noticeably shorter, stiff, and pointed in parasitics; blunt-tipped and twisted, oarlike, into the vertical plane in pomarines. Long-taileds are the smallest species, pomarines the largest. The immatures lack elongated tail feathers, making it impossible to tell which species they belong to without a specimen in the hand.

FURTHER NOTES

The different species of jaegers resemble one another in many ways: All are seabirds and therefore strong fliers. They come to land only to breed,

Figure 6.10. (a) Long-tailed Jaeger. (b) Parasitic J. (c) Pomarine J.

spending the rest of the year at sea in warmer latitudes of the Pacific and Atlantic Oceans. All are carnivorous pirates, obtaining much of their food by stealing it from other birds such as gulls, terns, and loons; sometimes a jaeger snatches food directly from its victim's bill, and sometimes it bullies the victim into regurgitating and then catches the falling food in midair.

Although more given to piracy than other birds, jaegers capture much of their food for themselves. They feed on lemmings and voles, insects, birds, eggs, carrion, fish, and berries. They are as big a threat as foxes and ermines to eggs and nestlings. A jaeger stealing an egg flies off with it and, away from the nest, drills a neat hole in the shell to get at the contents; this makes a jaeger-eaten egg recognizable; when a fox or an ermine robs a nest, the robber crushes all the eggs at the nest site. Although there is much overlap in the diets of the different jaegers, pomarines and long-taileds feed chiefly on lemmings and other rodents, parasitics chiefly on birds. Parasitics often raid the nesting colonies of colonial nesters such as snow geese, stealing eggs left unattended; parent geese sometimes leave their eggs temporarily, either because they have been scared off by marauding foxes or grizzlies (or visiting humans), or because they are fighting over territory among themselves.

Jaegers have several techniques for hunting active prey: they catch birds in flight by swooping down on them or chasing them; when in pursuit of lemmings, they watch and wait on a hummock that gives a wide view, or hover on beating wings in one spot, kingfisher-style, surveying the ground beneath. Pomarines like to soar low, ready to swoop. Where pomarines or parasitics of both color phases occur, an individual benefits by having the minority color—by being dark where most are light, or vice versa—because prey animals (rodents or birds) are less wary of a predator they are unaccustomed to.

Like other carnivorous birds, jaegers regurgitate pellets of indigestible food material, allowing their diet to be studied in detail.

All three species make shallow, open nests on the tundra, and usually lay 2 eggs. Both parents incubate. The eggs don't hatch simultaneously and the smaller of the 2 semiprecocial nestlings often fails to survive.

The Gull Family (Laridae): II. Gulls and Terns

Seven species of gulls and one of terns nest only in the Arctic. Three other gull species have breeding ranges that, though south of the Arctic for the most part, also reach into arctic tundra. For many species, recognizing immatures is a task for an expert; descriptions below are of adult birds.

The **Glaucous Gull** [AHMLB], *Larus hyperboreus* (fig. 6.11), is recognizable by its very large size plus its lack of any dark feathers. The back is pale pearl-gray, and the wing tips white. Second-year birds are entirely white. Like many members of the genus *Larus,* glaucous gulls have a strong, hooked, yellow bill with a red spot near the tip of the lower mandible, and pink feet.

Figure 6.11. (a) Glaucous Gull. (b) Thayer's G. (c) Mew G.

Thayer's G [HMlB], *Larus thayeri* (fig. 6.11), is like a small, dark glaucous gull, with dark gray (almost black) wing tips, and a dark eye (note this, to distinguish the bird from a herring gull; see below). Some ornithologists treat Thayer's gull as a subspecies of the **Iceland G,** *Larus glaucoides* (not illus.), which resembles a glaucous gull exactly, except for its much smaller size (about one-half the weight). Some of these pale Iceland gulls breed around the coast of southern Baffin Island.

Sabine's G [hMlb], *Xema sabini* (fig. 6.12), is a dark-hooded gull with a forked tail and wing tips "dipped in black ink." The hood looks black from a distance but is actually very dark gray above a black border. Sabine's gull resembles a tern in its tendency to hover in one spot on beating wings, and in its "black" head and forked tail; but a tern has a black cap (not hood, see fig. 6.13), and no black on the wings. The only other dark-hooded gull occasionally found in the Arctic is **Bonaparte's Gull,** *Larus philadelphia,* which sometimes strays north of treeline (where it usually nests in coniferous trees) onto the tundra. Confusion between Sabine's and Bonaparte's is therefore unlikely except near treeline, and not very likely even there because of the contrasting wing-tip patterns of the two species (see fig. 6.12).

Ivory G [h], *Pagophila eburnea* (not illus.), is pure white with black legs. This gull is seen most often in flight over the sea, when its black legs are very noticeable. The only other pure white arctic gull is the glaucous gull in second-year plumage, and its legs are pale pink.

Ross's G [?], *Rhodostethia rosea* (fig. 6.12), is a small gull with a wedge-shaped tail, a narrow black "necklace" when in breeding plumage, and a shell-pink(!) breast and belly. Unfortunately, the color is often difficult to see; it comes from a pigment in the preening oil, which the bird takes with its bill from a gland on the rump and spreads as waterproofing on all accessible feathers. Little is known about this gull's breeding range. (Both the ivory gull and Ross's gull are listed as vulnerable in the Canadian Endangered Species List, 1993.)

The **Black-legged Kittiwake** [hlb], *Rissa tridactyla* (fig. 6.13), is the only gull that is easier to recognize when immature, as the figure shows. Adults are distinguishable from other gray-backed, white-headed, yellow-billed gulls by their black legs, and by their pure black wing-tips unspotted with white. In flight, the black legs are sometimes invisible, tucked away among the white belly feathers.

The 3 gull species whose breeding ranges are partly arctic are herring, mew, and Bonaparte's.

The **Herring G,** *Larus argentatus* (not illus.), which breeds throughout the tundra of the Canadian mainland (as well as farther south), is a southern "replacement" for Thayer's gull, whose breeding range starts where the herring gull's leaves off. (Thayer's extends from parts of the north shore of mainland Canada northward into the Arctic islands.) The two species are closely similar; the herring gull's pale yellow eye is the only consistent identifying feature; its wing tips are black and white.

Mew G, *Larus canus* (fig. 6.11), whose breeding range extends onto the tundra near the Mackenzie Delta and northeast of Great Bear and Great Slave Lakes, is smaller than a herring gull; the bill is proportionately much smaller and lacks a red spot. The legs are yellow-green.

Figure 6.12. (a) Sabine's Gull. (b) Bonaparte's G (wing pattern). (c) Ross's G.

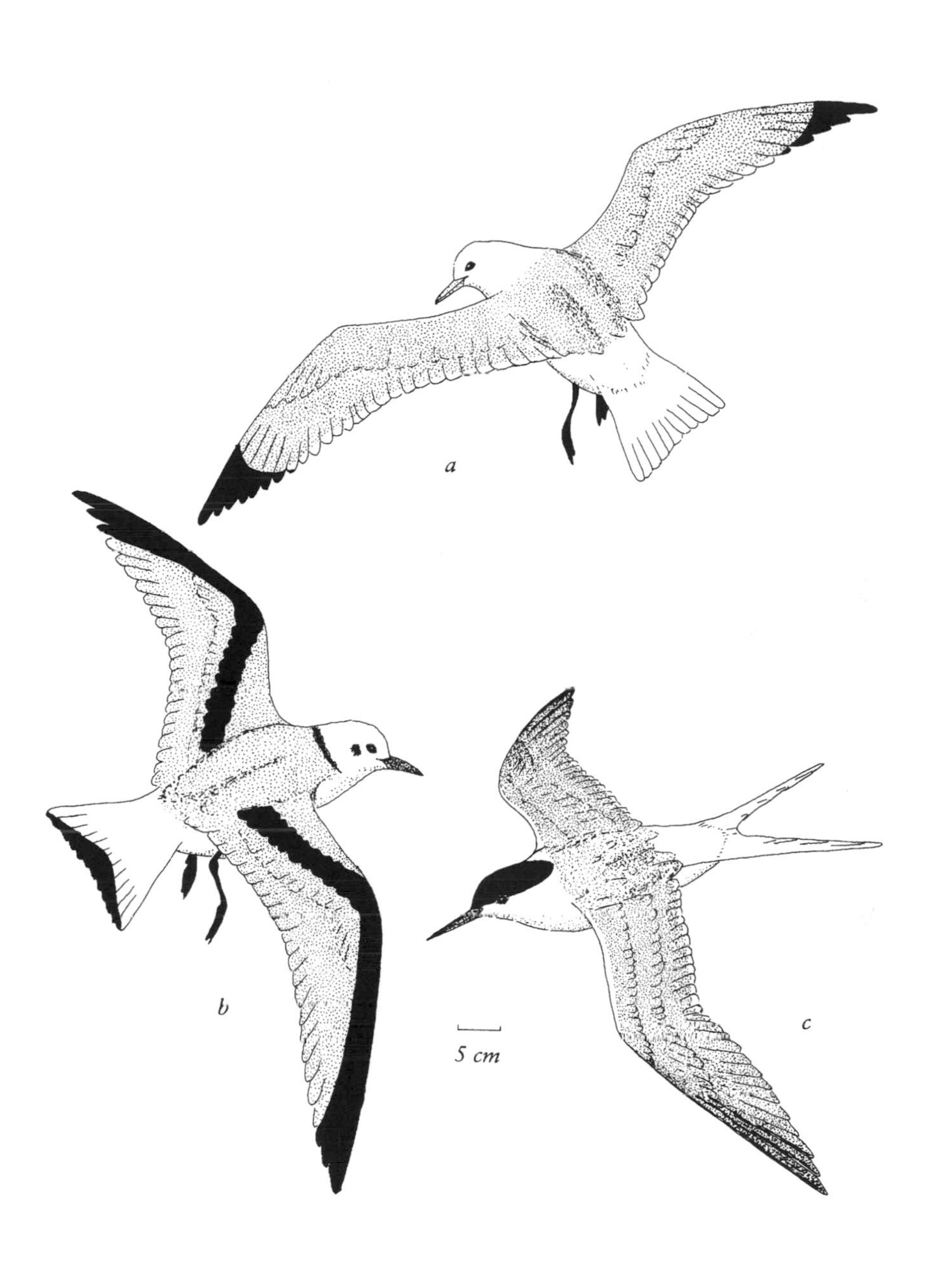

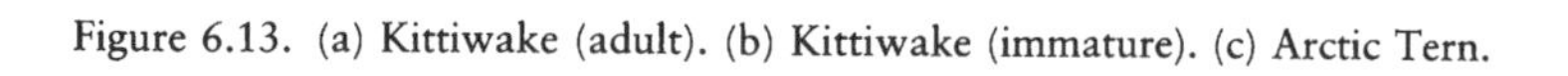
Figure 6.13. (a) Kittiwake (adult). (b) Kittiwake (immature). (c) Arctic Tern.

Bonaparte's gull is described above where it is contrasted with Sabine's gull.

There is only tern in the Arctic:

The **Arctic Tern** [AHMLB], *Sterna paradisaea* (fig. 6.13), is unmistakable because of its black cap (not hood), pale wings, and deeply forked tail. As birders of the temperate zone well know, distinguishing between arctic and common terns, when they occasionally occur together during the arctic tern's migrations, is exceedingly difficult. But no problem arises in the Arctic, where the common tern has never been seen.

FURTHER NOTES

Gulls have a variety of life styles, and few statements are true of all the arctic species. However, all are semiprecocial (see sec. 6.4), all usually nest in colonies (but isolated nests are not rare), and all scavenge for carrion at least occasionally. Beyond that, their habits and habitats are highly variable.

First, nest sites: Thayer's gull, Iceland gull, kittiwake, and (usually) glaucous gull nest on coastal cliffs. Sabine's gull and Ross's gull are ground-nesters that stick to sites near the coast. Herring gull and arctic tern are ground-nesters that nest on both coastal and inland tundra. Ivory gulls are unique: they nest only in the High Arctic, with access to floating sea ice where they get their food; nest sites vary from inland cliffs, to ground sites on offshore islets, to heaps of rock rubble on floating ice islands. The nests of all gulls are at risk from a variety of predators, particularly foxes, gyrfalcons, jaegers, and other gulls. Nesting on offshore islets provides protection from foxes at least, except in years when the sea ice melts so late that "islands" do not become islands until well into the nesting period. Young gulls don't leave the nest until they can fly; this means that (unlike murre chicks) they can fly across shore-fast ice, if necessary, to reach the open sea.

Second, gulls' diets: The four arctic-breeders of the genus *Larus* (glaucous, Thayer's, herring, Iceland) are all omnivores. They eat eggs and chicks of other birds, especially those with nests near their own; also berries, insects, carrion, fish, small mammals, and (near human settlements) garbage. The other five birds on our list (Sabine's, Ross's, and ivory gulls, kittiwake, and arctic tern) are more marine than the *Larus* gulls and rely, most of the time, on a marine diet of fish and plankton organisms and, to a lesser extent, marine carrion and offal. The latter consists of dead seals (mostly the remains of polar bear kills), the placentas of seals and walrus during the whelping season, and their feces at any season; all this can be found on the floating pack ice, a feeding ground not available for nonarctic seabirds. During their nesting period, the ground-nesters among these "marine" gulls (Ross's and Sabine's) search tundra, mud flats and shorelines for insects, burrowing marine invertebrates and dead fish.

The ivory gull and Ross's gull overwinter in the Arctic, which makes them the most truly arctic of all gulls. For much of the year they get their food from the sea close to the ice edge. How these gulls get by, far out on the polar pack in the dark of arctic winter, is hard to discover, but they are believed to be the top members of a food chain whose lower members form an "under ice" community. At the bottom of the chain are vast numbers of microscopic plants attached to the lower surface of the sea ice or floating beneath it (see sec. 4.4); feeding on them are the tiny floating animals of the zooplankton, which include vast numbers of minute crustaceans, especially copepods and shrimplike amphipods. Next up the chain are fish, notably arctic cod and (in the eastern Arctic) the luminescent lantern fish which, shining in the dark of the arctic night, can be easily spotted by gulls flying overhead. The gulls, at the top of this food chain, feed on both the fish and the mini-crustaceans of the zooplankton, which are most abundant right at the edge of the pack ice. The two gull species are believed to overwinter in different regions—ivory gulls at the southern edge of the pack, and Ross's gulls at openings in the pack farther north in the Arctic Ocean. At Point Barrow, the northernmost point on the Alaska coast, Ross's gulls have been seen starting a fall migration in a northeasterly direction(!), heading for the polar pack.

The other astonishing migration—this one well known—is that of the arctic tern, which "overwinters" in the summer of Antarctica, after flying half the way round the world from its breeding grounds in the Arctic.

For more on ivory gulls nesting on sea ice, see "Phantoms of the Polar Pack Ice," by Stewart MacDonald, in *Audubon Magazine* 78, no. 3.

The Murre and Guillemot Family (**Alcidae**)

The members of this family—collectively known as alcids—are black and white seabirds of high northern latitudes, the most abundant seabirds of the northern hemisphere. Outside the breeding season, they spend their lives at sea, either in flight over it or swimming on it. Three of them are plentiful in the eastern Arctic (a map of their breeding sites is given in fig. 4.2).

The **Thick-billed Murre,** *Uria lomvia* (fig. 6.14), is the largest alcid and is present in huge numbers. They breed in a few immense, densely crowded colonies. A visit to a colony at the height of the breeding season is an unforgettable experience; no picture can do it justice since the sounds and smells are as memorable as the sights. The smell can be imagined; the sound—the combined murmurings of thousands of birds—is a continuous roar. The female of each breeding pair lays her single egg on a bare rock ledge on a vertical cliff. The ledge may be narrow, and the egg is saved from rolling off by its almost conical shape, which causes it to roll around in a circle if knocked, rather than off the ledge. Sometimes kittiwakes and northern fulmars share the murres' nesting cliffs.

The **Black Guillemot,** *Cephus grylle* (fig. 6.14), is less abundant. In breeding plumage (but not at other times) these birds are much darker

Figure 6.14. (a) Thick-billed Murre. (b) Black Guillemot. (c) Dovekie.

than murres, being pure black except for a big white wing patch visible on sitting or swimming birds, and white underwing feathers visible when the birds fly over. They have vivid scarlet legs and feet. Rather than being concentrated, like murres, into a few breeding "cities," they nest in comparatively small, scattered colonies, or even as single pairs. They nest in nooks and crannies, at low levels on cliffs, among boulders near the sea, or on rough ground just above the beach.

The **Dovekie,** *Alle alle* (fig. 6.14), is the most abundant alcid and the smallest, being slightly smaller than a robin. Its small size, short neck, and short bill mark it off from all other seabirds. Dovekies are common in and over Baffin Bay and Davis Strait (see fig. 4.1), and farther west in the channels among the eastern islands of the Canadian Arctic. They enter these waters from their breeding colonies in Greenland; so far as is known, dovekies did not breed on Canadian territory before 1984, when, for the first time, they were found breeding in Baffin Island; perhaps their range is expanding. They normally breed in huge colonies, laying their eggs in hollows and crevices on scree slopes or cliffs overlooking the sea.

FURTHER NOTES

Alcids obtain all their food by diving for it into the sea. Thick-billed murres and black guillemots eat small fish—mainly arctic cod, sculpins, and capelin—and various crustaceans in the plankton; dovekies are too small to take fish, and concentrate on plankton animals.

Alcids are anatomically adapted to dive for their prey. Their short, narrow wings enable them "fly" under water as nimbly as they fly through the air above it. Their flight in air is far from effortless, however. To propel their chunky bodies through the air, they must beat their small wings rapidly and uninterruptedly. They cannot soar. The smallness of their wings also makes it hard for them to begin flight, except when the takeoff is from a high cliff ledge, which allows a bird to gather speed in the course of a long drop. They cannot easily fly up from a flat surface, such as an ice floe; to take off from water, they must taxi, "pattering," for a long distance, which calls for a long "runway" free of ice floes.

Because alcids depend on unfrozen seas to obtain their food, their comings and goings are governed by the movements of sea ice. Their breeding colonies have to be within reach of polynyas (big tracts of sea which remain open all winter; see sec. 4.4) or of unfrozen shore leads. Murres have to time their breeding carefully, because their semiprecocial chicks fall from the ledges where they were born before they are able to fly. (Note the contrast with gulls that nest on precipices: gull chicks wait until they can fly before venturing from their birth ledges.) To survive, therefore, murres must have water, not shore-fast ice, at the foot of the breeding cliffs and this means that breeding must not be too early in the year. Late breeding is unlikely to succeed either, because, as fall approaches, food supplies for the chicks quickly dwindle from their summer peak: chicks hatching late in the season go hungry. The optimum time for egg-laying is thus only a small window of time. The females need plentiful food before they can form eggs, so in years when spring comes late and food is scarce, egg-laying may be unavoidably delayed until the window is past.

Because the pack ice drifts with winds and currents, good feeding spots for seabirds—tracts of open water—don't necessarily remain in one place. This means that when parent birds set off from the nest in search of food, they may have no idea which way to go. Thick-billed murres about to leave a breeding cliff on a food-gathering trip are believed to observe incoming birds bringing food home and to follow their route back out to sea, in a direction which presumably leads to a food source. They may fly 100 km each way to bring food to their chicks.

Of the three arctic alcids, only some black guillemots (not all of them) remain in High Arctic polynyas through the winter, along with ivory and Ross's gulls. Thick-billed murres and dovekies spend the winter in Low Arctic polynyas, or over the open sea off Labrador and Newfoundland.

For more on the breeding colonies of thick-billed murres, see "Seabird Citadels of the Arctic," by Tony Gaston, in *Natural History,* April 1987.

The Crane Family (Gruidae)

One species of crane nests in the Arctic. It is the **Sandhill Crane** [AhMLB], *Grus canadensis* (fig. 6.15), a bird familiar to western birders, if not on the ground, then as a migrant flying overhead. Sandhill cranes breed in middle latitudes as well as in the Arctic. They migrate in big flocks, to and from their breeding areas, every spring and fall. Even when they are flying high, they can be heard from afar, calling continuously in the manner of geese and swans.

On the ground, a sandhill crane is a tall, statuesque bird, with long legs and a long neck. Except for a red patch on the front of the head, the bird is uniformly gray (grayish brown in immatures); the red patch is bare skin, not red feathers. The standing bird looks very like a great blue heron and the two are often confused. The distinction is obvious when they fly: a crane flies with its neck outstretched, a heron with its neck folded back, bringing its head just in front of its shoulders. Confusion is impossible in the Arctic; herons never go so far north.

To breed successfully, cranes require plenty of nourishment at stopovers on spring migration, when their fat reserves increase almost fourfold. They nest where the tundra is flat and marshy, and their big nests are often on, or surrounded by, open water.

Eagles, Hawks, and Falcons (Accipitridae and Falconidae)

Two raptors, or diurnal birds of prey as they are called (a description that rather loses its meaning in the land of the midnight sun), breed exclusively in high latitudes; they are the **Gyrfalcon** [AHMLB], *Falco rusticolus* (fig. 6.16), and the **Rough-legged Hawk** [AhMLB], *Buteo lagopus* (fig. 6.17).

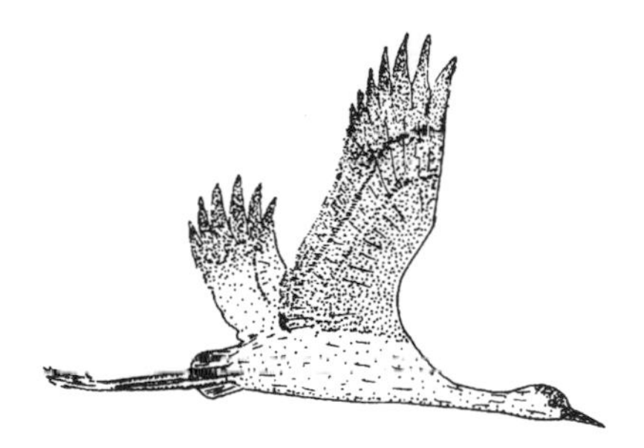

Figure 6.15. Sandhill Crane

Figure 6.16. (a) Gyrfalcon. (b) Peregrine Falcon

Figure 6.17. (a) Rough-legged Hawk, light phase; (*inset:* dark phase). (b) Immature Golden Eagle.

Two other arctic breeders—the **Peregrine Falcon** [ALMB], *Falco peregrinus* (fig. 6.16), and the **Golden Eagle** [Al], *Aquila chrysaetos* (fig. 6.17)—merit special attention here, even though their breeding areas extend far into lower latitudes, because they could be confused, respectively, with the first two.

First note the contrast between the two falcons (of the family Falconidae) and the two others (of the family Accipitridae) (see fig. 6.18). Falcons, as a family, are noted for their long, tapered wings and their narrow tails, a shape that equips them for speed and maneuverability in flight; they are fast, powerful flyers, especially the peregrine. The other two birds (hawk and eagle) are shaped very differently; their wings are broad and their tails fan-shaped, as can be seen when they soar overhead—provided the wind isn't too strong. Wind can modify the shape of a bird's wings; in a gale, the broad wings of a rough-legged hawk can be angled and tapered like those of a falcon.

Figure 6.18. Silhouettes of (a) a falcon, with tapered wings and narrow tail; (b) an eagle or hawk, with broad wings and fan-shaped tail.

The difference between a gyrfalcon and a peregrine falcon is sometimes obvious. A pure white falcon is unquestionably a "gyr" (pronounced jer to rhyme with her), whereas a big falcon with a conspicuous black mustache is unquestionably a peregrine. But gyrs have two color phases, dark and light, and the dark phase (commoner in the western Arctic) is not unlike an immature peregrine (in which the mustache is less noticeable). Confusion is therefore possible except at close range and in a good light, when the fact that dark-phase gyrs are paler than peregrines and considerably larger aids recognition. Also, if a dark-backed falcon seen from above has a narrow white band at the tip of the tail, it is a peregrine; if not, it is a dark-phase gyr.

The other pair are also look-alikes when seen in silhouette at long range. However, the golden eagle is much bigger than the rough-legged hawk (about half again as long); and the rough-legged, even a dark-phase individual, has much more white below than even an immature golden eagle, as shown in fig. 6.17. Adult golden eagles are entirely dark below, indeed, dark all over except when the backs of their heads catch the sun and appear "golden" (strictly, blond). Rough-leggeds have a noticeable habit of hovering at one spot in the sky, with flapping wings and dangling legs, as they search for prey on the tundra. Both rough-leggeds and golden eagles, unlike their more southerly relatives, have fully feathered legs, no doubt as an adaptation to arctic cold; this feature is seldom visible, of course.

Two other raptors occasionally nest close to treeline, but are primarily birds of the temperate zone. They are the Northern Harrier (*Circus cyaneus*) (not illus.) of open country, and the Merlin (*Falco columbarius*) (not illus.) of woods and forests.

FURTHER NOTES

All four arctic raptors are cliff nesters; the location of a nest is sometimes given away by the presence at the foot of a cliff of the bodies of "rejected" chicks, young starvelings pushed from the nest by their older, larger, and better-fed siblings (see sec. 6.5).

Most raptors feed on agile prey. The falcons feed mainly on birds, with gyrs sticking to willow ptarmigans exclusively when a good supply is available. Rough-legs and golden eagles depend more on mammals. But all four can, if they must, use whatever prey is available and not too large. Thus rough-legs eat mostly lemmings and voles, whereas golden eagles take larger prey such as ermines and hares. A golden eagle can kill an animal too heavy for it to carry in one piece and dismember it on the ground (I have seen one kill a Canada goose).

The hunting methods of the different birds match their anatomical adaptations: a peregrine is built for speed and "stoops" on its prey—a bird in flight—striking it from the air with the force and precision of a guided missile. A gyr pursuing ptarmigans flies fast and low, snatching an unsuspecting victim from the ground before it can take cover. An eagle soars as it scans the ground for a victim, and then pounces when it spots one. A rough-leg does much the same except that, besides soaring, it often hovers.

Raptors, unlike insect- or seed-eating birds, depend on cunning and skill to obtain their food; often they get it only at long intervals, but each food item, when it comes, is large. Their semialtricial chicks are adapted to this style of feeding. Whereas small birds' chicks have to be fed little and often if they are to survive, raptors' chicks can wait for hours or even days from one big meal to the next.

The northernmost breeder of all the raptors—the only one to nest in

Ellesmere Island—is the gyr. It is also the only raptor to winter in the Arctic. Only a few, probably those too young to breed, go south for the winter. Although survival through an arctic winter has its problems, gyrs gain some advantages by staying in the north. They can take their pick of available nesting ledges before the other raptors arrive in spring. Also, by nesting very early, they leave themselves time to try again if the first clutch of eggs fails. Young gyrs are voracious and continue to be fed by their parents even after they have grown big and are feathered well enough not to need brooding.

Because they nest before most other birds, gyrs have to feed their nestlings on adult prey, mostly ptarmigans; but by the time the nestlings have fledged and must fend for themselves, ptarmigan chicks (and ducklings, too) have hatched, and provide easy prey for these beginners to hunt. Peregrines nest later than gyrs, when chicks of many bird species are available as food for their young; therefore, peregrine parents do not have to feed their families on adult birds, which are less numerous than chicks and harder to capture.

The Grouse and Ptarmigan Family (Phasianidae)

Two species of ptarmigan (genus *Lagopus*) are the only members of this family in the North American Arctic. Its nonarctic members are grouse, turkeys, quail, pheasants, partridges, and one other ptarmigan, the White-tailed Ptarmigan of the western mountains.

The two arctic ptarmigans are the **Rock Ptarmigan** [AHMLB], *Lagopus mutus,* and the **Willow Ptarmigan** [AhMLb], *L. lagopus* (both in fig. 6.19). Those that live in the Arctic are truly arctic birds, remaining there all year. But the breeding ranges of both species extend southward in the western mountains to about 50° N.

The two species are hard to distinguish where their ranges coincide; that is, nearly everywhere except in the High Arctic north of Melville and Bathurst Islands, where all ptarmigans are almost certainly rock ptarmigans. Both species are white all over in winter, except for the black tail which, in a sitting bird, is often hidden by the folded wings; some (not all) male rock ptarmigans in winter plumage have a black bar through the eye (as in fig. 6.19). The birds are still white at the time of their courtship displays, when the males inflate brilliant red combs above their eyes and "bow" with spread tails before the females.

The summer colors of both species are finely mottled browns and greys, with much white on wings and belly, best seen when the birds fly. The flash of white wings makes a flock of flying ptarmigan noticeable at long range. Their calls—like a noisy clockwork toy running down—can be heard from afar, too.

Both sexes have red combs over the eyes, but the females' are paler. Their legs and toes are fully feathered, in white. The species can be distinguished (with difficulty) by noting the greater up-and-down width of the Willow's bill, and its slightly larger size (hard to judge when only one species is present). In male willows (only) the dark parts of the summer plumage are chestnut-colored, and noticeably brighter than the browns and greys of female willows, and rocks of both sexes. The species

Figure 6.19. (a) Rock Ptarmigan males in winter and early spring. (b) Willow Ptarmigan male in summer.

differ, too, in habitat and behavior: willows occupy damper habitats than do rocks, which prefer drier, stonier uplands. Their winter diets are different: willows eat the buds and catkins of willows, whereas rocks eat those of dwarf birch. Both species have a more varied diet in summer. Finally, willows are much bolder than rocks, which tend to be timid.

FURTHER NOTES

Female ptarmigans, who incubate without help from their mates, grow their cryptic summer plumage as much as 2 months earlier than the males; the latter remain conspicuously white against the dun, snow-free ground while the females, sitting on their eggs, are almost invisible. When the spring molt is in progress, resting birds choose the appropriate surfaces for resting on—a bird in winter white will rest on a snow patch, a bird in summer browns on a dark background. A mixture of plumages is to be seen for much of the summer, with birds in various stages of molt showing different amounts of white. Then, when fall arrives, all ptarmigans, the year's chicks as well as the adults, turn white simultaneously.

The chicks are precocial but are not wholly self-reliant. They scamper along in the company of a parent for their first few weeks, learning what's edible and what's not.

Most arctic ptarmigans migrate a short distance south from their breeding areas for the winter, but they remain in places where winter is bitterly cold. The females go farther than the males; the northernmost winter flocks contain many more males than females, whereas farther south females outnumber males.

The snow is deep and the cold extreme wherever ptarmigans overwinter, but they are adapted to thrive in these conditions. They grow extra feathers on their bodies as protection against the cold. They also grow feathers on the soles of their feet and toes (the upper surfaces are feathered all year), which act as efficient snowshoes and may provide insulation as well. Their claws grow longer in winter, too, adding to the snowshoe effect; when summer comes, the lengthened claws are shed and new ones grown. They find shelter from cold winds by burrowing into soft snow, which they dive into from the air, leaving no tracks to guide their predators. In winter, these are wolves, foxes, gyrfalcons, golden eagles, and ravens.

Ptarmigans have additional enemies in summer, when they are prey for peregrine falcons, rough-legged hawks, northern harriers, goshawks, snowy owls, short-eared owls, parasitic jaegers, and glaucous gulls, as well as for their winter predators. It is no wonder that they need to be well camouflaged and wary at all seasons.

The Owl Family (Strigidae)

The snowy owl is probably the most famous of all arctic birds and is the only owl species found throughout the Arctic; the short-eared owl is an arctic breeder, too, but not in the High Arctic. The northern hawk-owl is a bird to look for near treeline.

The **Snowy Owl** [AHMLB], *Nyctea scandiaca* (fig. 6.20), could never be mistaken for any other bird. As with all owls, snowies have flat faces with both eyes looking forward. The male is slightly smaller than the female. He is almost pure white, while she is spotted or barred with dark brown. These markings camouflage her excellently while the ground is still speckled with snow, as it normally is at the beginning of the nesting season; but later, when all the snow has gone, a sitting snowy owl is visible from a distance against the dark background of the tundra. Snowy owls are usually thought of as being devoid of ear tufts; in fact, they have them, but they are very small and not visible unless the feathers surrounding them are sleek and unruffled.

The **Short-eared Owl** [AL], *Asio flammeus* (fig. 6.20), is also easy to recognize, as it is the only other owl to be found north of treeline. Its color is mostly pale tan, with a noticeable black patch on the underside of the wing, at the wrist. The short ear-tufts are visible only at close range when the owl is settled, but they are more conspicuous than those of the snowy owl. The flight of the short-eared owl, "lilting" up and down at no great height above the ground, is characteristic; once seen, it is not easily forgotten.

The **Northern Hawk-owl,** *Surnia ulula* (not illus.), deserves mention as an "almost-arctic" owl, which nests in holes in standing dead trees as far north as trees grow. It is a dark bird, with an unusually long tail for an owl; it glowering face is notably fierce, even by owl standards. Look for it perched at the very top of a low tree, in daylight or dusk.

FURTHER NOTES

Snowy owls are among the few birds capable of overwintering in the Arctic (see sec. 6.1) though in some winters hunger forces them south. Their principal food is lemmings, whose numbers fluctuate tremendously (see sec. 7.10); so in a bad lemming year, snowy owls avoid starvation by wintering in grassy, open country in temperate latitudes, where mice are plentiful. In such years, they are a familiar sight to thousands of birders. Their nesting grounds are entirely arctic. Their dependence on lemmings means that they are not to be found everywhere in the summer Arctic, however; they can breed only where sufficient lemmings are available. Their nests are often conspicuously placed atop prominent mounds

Figure 6.20. (a) Snowy owl. (b) Short-eared owl. In both cases the female is on nest, and the male flying in.

that become snow-free early in the season and that give the nesting birds a commanding view.

Short-eared owls are smaller than snowies, and much less conspicuous nesters; they probably do not breed on the Arctic Islands and their breeding range extends far south into temperate latitudes. Even in the south, where summer nights are dark, they prefer to hunt in daylight or dusk rather than in complete darkness. This means they are adapted to the continuous daylight of the arctic summer.

As with all owls, the chicks of snowies and short-eareds are semialtricial (see sec. 6.4); the chicks in a nest are of a range of different ages because the mother starts incubating as soon as the first egg is laid. The mother does all the incubating, but the father is not idle: he protects his sitting mate and her clutch from all manner of predators. He also feeds her and, in due course, the older chicks, who leave the nest while the mother is still sitting. The body warmth of young chicks still in the nest keep the last-laid egg warm during the mother's temporary absences. This is not familial affection: if food is scarce, the older chicks eat the younger ones. In snowy owls, the oldest chick may be 3 weeks old when the last egg hatches. When all the chicks have left the nest, both parents feed and protect them.

Snowy owls go through elaborate courtship rituals, in which the male captures a lemming and displays it to his prospective mate, either holding it in his bill or laying it on the ground before her; no doubt this is to demonstrate his ability to provide for her and their progeny. Males also hoot, warning and threatening intruders to their territories.

For an account of snowy owls and their nesting, see *Kingdom of the Ice Bear,* by Hugh Miles and Mike Salisbury (British Broadcasting Corporation, 1985).

The Perching Bird Order (Passeriformes)

The perching birds belong to several different families in a single order (an order is the next higher classificatory rank above family). Perching birds, as a whole, are the "ordinary" birds that make up almost half of all the birds in the world, birds such as robins, sparrows, starlings, warblers, finches, and the like. To an unscientific birder, their chief distinguishing characteristic is their ordinariness. They are land birds; the males of many species are good singers in the breeding season; and their feet are adapted to grasp perches such as twigs, hence the English name of the order. The scientific name is from *passer,* the Latin for sparrow. In some (not all) species, the male in breeding plumage is more brightly colored than the female.

The majority of perching birds are adapted to life in woodlands and forest; comparatively few have taken to the tundra, where perching sites are scarce and nests must be near ground level. As a result, perching birds form a much smaller proportion of the total bird population in the Arctic than elsewhere. About 31 species breed somewhere in the Arctic, and of these, the 13 described below breed only or chiefly in the Arctic. The remaining 17 species (listed at the end of the section) breed farther south for the most part.

The 13 species to be described can conveniently be split into 3 groups: first, 5 species that breed in (but not only in) the High Arctic (north of Parry Channel); second, 6 species with widespread breeding ranges that do not include the High Arctic; third, 2 "western specials."

GROUP I

The High Arctic breeders are all easily recognized. The first three have the most northerly breeding ranges, including northern Ellesmere Island.

The **Snow Bunting** [AHMLB], *Plectrophenax nivalis* (fig. 6.21), is the only perching bird in which the plumage is mostly white in both sexes; the male also has some black, and the female some rusty brown.

In the **Lapland Longspur** [AHMLB], *Calcarius lapponicus* (fig. 6.21), the male has a distinctively shaped black face mask and a bright chestnut nape, but the female is less showy; the long "spur" (hind toe) is only visible at very close range. In the breeding season, the males repeatedly fly up, and then sink slowly to earth on stiffly spread wings singing all the while.

The **Redpoll** [AHmLb], *Carduelis flammea* (not illus.), is a streaky brown finch with a conical bill, made conspicuous (especially in the male) by a bright red cap and pink breast. Some redpolls—the majority, in the Arctic—also have conspicuously white rumps and used to be classed as a distinct species, the hoary redpoll. Redpolls probably do not breed in the western Arctic Islands.

The **Horned Lark** [hMALB], *Eremophila alpestris* (fig. 6.21), is described here because besides breeding nearly everywhere in Canada, its breeding range extends into the southernmost High Arctic—the islands immediately north of Parry Channel. It can be recognized by its black and yellow face pattern, and in flight by its black tail; the "horns" (only in the male) are visible at close range.

The **Northern Wheatear** [A; hB], *Oenanthe oenanthe* (fig. 6.21), is light-colored and distinctively patterned in black, white, and pale gray, with the only other color being buff on the breast. The wheatear has a *disjunct* breeding range, i.e., two different ranges separated by a wide gap. It breeds west of the Mackenzie River and also in the eastern Arctic (chiefly Ellesmere and Baffin Islands), but not in the central Arctic. Most

Figure 6.21. (a) Snow Bunting. (b) Lapland Longspur. (c) Horned Lark. (d) Northern Wheatear. (e) Smith's Longspur. (All males)

wheatears live in the Old World all year, but some come to North America to breed; those breeding in the western Arctic come from eastern Asia, those in the eastern Arctic from Europe and western Asia.

GROUP II

Six species of perching birds have widespread breeding ranges that do not include the High Arctic.

Breeding males of **Smith's Longspur** [AL], *Calcarius pictus* (fig. 6.21), are recognizable by the black and white face pattern and the spotless orange-buff breast. Females have a less contrasty pattern.

The **Gray-cheeked Thrush** [AL], *Catharus minimus* (not illus.), is a thrush like other thrushes (brown head and back, pale breast speckled with brown dots), but it is the only thrush to breed on the tundra. It could be confused with the very similar Swainson's thrush in the treeline zone; good light is needed to see its inconspicuous distinguishing characters—the gray cheek and gray (rather than buff) eye ring. It is a pleasing singer.

The **Northern Shrike** [AL], *Lanius excubitor* (fig. 6.22), is a black, white, and gray, robin-sized bird with a carnivore's hooked beak. It is found in shrub tundra, not far beyond treeline.

The **American Tree Sparrow** [AL], *Spizella arborea* (fig. 6.22), has a chestnut cap and a dark spot in the middle of an otherwise unmarked breast.

The large **Harris's Sparrow** [L], *Zonotrichia querula* (fig. 6.22), is unique in having an all-black face and pink bill. It differs from the slightly smaller Lapland longspur in hopping (rather than running) and in lacking a bright chestnut nape patch.

The **Water Pipit** [ALMB], *Anthus rubescens* (fig. 6.22), is sparrowlike in color pattern with a streaky breast and light eye stripe, but unlike a sparrow in having a thin bill. Like the wagtails (see below), to which it is related, it repeatedly wags its tail up and down. Its white outer tail feathers show in flight.

GROUP III

The third group of perchers are the westerners.

The **Yellow Wagtail** [A], *Motacilla flava* (fig. 6.22), has bright lemon-yellow underparts and a long black and white tail. It constantly flicks its tail up and down and from side to side, making the bird unmistakable.

The **Arctic Warbler** [a], *Phylloscopus borealis* (fig. 6.22), one of the drably colored Old World warblers, is a stray from Siberia found in North America only in western Alaska and (possibly) the westernmost Arctic islands. A nondescript little olive-brown bird with a pale breast and eye stripe, it is recognizable only because it couldn't be anything else.

The following birds breed in warmer latitudes as well as in the Arctic. First the raven, notable because many of them overwinter in the Arctic as well as breeding there; their range extends as far north as Ellesmere Island, in the High Arctic. Ravens can always be identified with certainty,

Figure 6.22. (a) Northern Shrike. (b) American Tree Sparrow. (c) Harris's Sparrow. (d) Water Pipit. (e) Yellow Wagtail. (f) Arctic Warbler.

even at long range, as crows do not occur in the Arctic. Of the remaining species, none is known to breed on the Arctic Islands: they do not go beyond the mainland coast. They are: cliff swallow, bank swallow, American robin, Bohemian waxwing, gray jay, yellow warbler, Wilson's warbler, blackpoll warbler, northern waterthrush, savannah sparrow, fox sparrow, white-crowned sparrow, rusty blackbird and, only in the west, varied thrush, Say's phoebe, and rosy finch.

FURTHER NOTES

The nestlings of all the perching birds are naked and helpless for many days after hatching (i.e., they are altricial: see sec. 6.4). Parent birds therefore conceal their nests as best they may, often a difficult task in the far north, where cover is hard to find; some species use rock crevices, but others have nests fully exposed. In the southern tundra better cover is available and nests can be more easily hidden, for example, in the branches of a shrub, on the ground under one, or even, near treeline, in a stunted tree. Whether above or on the ground, a perching bird's nest is usually of the traditional "bird's nest" form, a carefully built, comfortably lined cup; redpolls, snow buntings, and longspurs like to line their nests with molted ptarmigan feathers.

Although there are (of course!) exceptions, the usual nest sites of the different species are: On the ground, both longspur species, both sparrow species, horned lark, water pipit, yellow wagtail, and arctic warbler (which builds a roofed, oven-shaped nest rather than a cup); in rock crevices or in old burrows of lemming and voles, snow bunting, northern wheatear, and redpolls in the High Arctic; on ledges on high cliffs, raven; in shrubs, redpoll and gray-cheeked thrush; in a stunted tree, northern shrike. Redpolls are unusual in not being territorial; their nests are often closely spaced.

At first thought it seems surprising that so few high arctic breeders conceal their nests in cavities in the rocks or in old rodent burrows, safely inaccessible to all predators except ermines. Ornithologists who compared the breeding behavior of snow buntings and Lapland longspurs discovered the reason: the eggs in buntings' underground nests cool much faster than those under the open sky when the incubating parent (in perching birds, nearly always the mother) is not sitting. This means that a female bunting must stay on her eggs and rely on the male to feed her. His comings and goings draw attention to the nest; and if any accident befalls him his widow finds it hard to cope alone. Lapland longspurs manage differently; the sitting female can leave her eggs long enough to forage for herself without their cooling appreciably in warm weather, and there is no need for the male to bring food to her. Female longspurs and their eggs are adequately protected from predators by their cryptic color patterns.

Nesting birds need large amounts of food for their young; the food must be soft enough for helpless, altricial nestlings to eat, and it must contain the protein they need for growth. The breeding season diet of most perching birds consists chiefly of insects. Wheatears, alone among insect-eating birds, specialize in the big bumblebees common in the Arctic. Yellow wagtails and water pipits often feed near streams and ponds where they find aquatic insects, worms, and snails. With advancing summer, insects become scarce while seeds and berries ripen. Most perching birds switch to seeds as their staple, with wheatears, gray-cheeked thrushes, Harris's sparrows, and wagtails taking berries as well.

Two species of perching birds—northern shrike and raven—are unusual in being carnivorous. Shrikes are unusual carnivores in having comparatively weak feet: they cannot inflict injury with their talons or hold their prey while rending it. The prey is mostly voles, lemmings, big insects and (occasionally) small birds, which a shrike can knock to the ground with a blow from its heavy beak. Once it has captured its prey, a shrike often caches the little carcass by wedging it in the fork of a twig. The other carnivorous perching bird, the raven, is as large and powerful as a gyrfalcon. It preys on lemmings and voles, scavenges carrion (mostly caribou and ptarmigan carcasses), and raids the garbage dumps to be found near any human settlement.

The way in which redpolls (especially the "hoary redpolls" of the far north) sustain life through the intense cold of winter has been carefully studied. They are the smallest birds to overwinter in the Arctic and, being birds, lack the special, rich brown fat of mammals as a source of heat energy (see sec. 7.1). But they obviously get by. Their small size gives them an advantage in allowing them to shelter and feed at least part of the time in the comparative warmth of lemmings' burrows. They have another valuable adaptation: they have a "pocket" leading off the esophagus that functions like a chicken's crop, in which they can store a quantity of seeds. This enables a redpoll to swallow more seeds than it can eat immediately and store them for later; the meal can then be completely swallowed, and digested, at a spot protected from weather and predators.

7 Mammals

7.1 The Arctic as a Home for Mammals

Mammals have such a wide range of different life styles that only a few generalizations are possible. Most of the special adaptations to arctic conditions are peculiar to only one or two families and are described below in the relevant sections. This section notes the contrasting conditions faced by animals living in different habitats, and the varied methods by which different species cope with the harsh environment.[1]

The greatest contrast in life styles is between the marine mammals (seals, walruses, and whales) and the land mammals (all the rest). In winter, when the majority of land mammals are surviving temperatures down to −50° C, the marine mammals are enjoying balmy conditions, submerged in sea water that can never cool below −2° C. The arctic climate is certainly not harsh for them. But although the temperature of their environment hardly differs from that of their relatives in temperate latitudes, they do have one uniquely arctic problem to cope with—access to the air. They must come to the surface at intervals to breathe. Therefore, without gaps in the sea ice, either breathing holes of their own making or polynyas (see sec. 4.4), life in the Arctic would be altogether impossible for marine mammals.

For land mammals winter, too, is a difficult season: cold and hunger are the twin perils they face. First, the cold: the winter coats of all arctic land mammals are renowned for their marvelous insulating qualities. In all species, winter coats are thicker and longer-haired than summer coats.

1. Each species' geographic range is given in coded form, in square brackets after its name; the code is explained in section 5.13, Preliminary Notes.

And some species, for example, caribou and polar bear, have hollow, air-filled hairs; a thick coat of such hairs is an exceptionally good insulator.

Maintaining body warmth is especially difficult for small animals such as lemmings, voles, and shrews. They spend most of the coldest months in tunnels under the snow where the temperature is higher than in the open air and, more important, where they are sheltered from the wind. It is known that voles sometimes keep warm by huddling together.

Many small mammals (unlike birds) have a physiological advantage when it comes to surviving the cold: when temperatures start to drop, they build up internal supplies of *brown fat,* a "richer" fat than ordinary white fat in the sense that it generates more warmth as it is metabolized (equivalently, chemically oxidized, or "burned").

Finding enough to eat in winter is as urgent a problem for arctic mammals as staying warm. When the fresh greenery of summer withers and dies, there seems to be little left to nourish the herbivores. Mammals that hibernate always put on sufficient fat in late summer, while food is still plentiful, to nourish them through the lean season. The only arctic mammals to hibernate, however, are arctic ground squirrels, grizzly bears, and pregnant polar bears (the three species differ greatly in the intensity of their hibernation, as described later).

Active animals like the large ungulates (hoofed mammals) also put on some fat in summer; the amount of their fat reserves, relative to their body weights, though much greater in arctic ungulates than in those of middle latitude, is much less than that of the hibernators. This raises the question: how do muskoxen and those caribou that remain on the tundra all winter with low reserves of body fat manage to survive on sparse, poor quality forage. Scientific research seems to have given the answer for reindeer, and the findings almost certainly apply to caribou and possibly to muskoxen, too. The animals lose considerable weight during winter, but not because of starvation. The weight loss is normal and does them no harm; it is a programmed adaptation that protects them from starvation when food is in short supply. By losing weight, they reduce their need for food; put another way, when the forage becomes insufficient for a heavy animal, the animal "shrinks" to match the food supply. It is believed that the animals put on fat in late summer, not in preparation for winter food shortages, but to supply them with the large amount of energy needed for reproduction—for the rut in males, and for development of the fetus and subsequent care of the calf in females.

Winter forage is low in protein. Far from being a drawback, this is a positive benefit: a low protein diet reduces the amount of water an animal excretes. This reduces the animal's need for water, which in turn reduces the amount of energy it must spend melting snow and ice, the only source of water in winter.

In sum, losing weight and cutting down on excretion seem to be the strategies by which herbivores conserve energy, and conserving energy is the fundamental requirement for surviving the arctic winter.

Winter is kinder to carnivores than to herbivores. They enjoy two advantages not shared by carnivores of warmer latitudes. First, temperatures are low enough most of the time to ensure that meat left lying on the tundra won't putrefy. This means that if, for any reason, a carnivore leaves part of a kill uneaten, the leftovers won't be wasted; they will remain in good condition until a scavenger comes across them. All arctic carnivores will eat carrion when they find it; they don't limit themselves to freshly killed prey.

The biggest benefit of the arctic climate, however, is the existence of sea ice, a unique environment without any parallel in warmer latitudes. It is the hunting ground of polar bears (and occasionally of grizzlies, see sec. 7.2), where they capture seals, their staple food; in effect, a terrestrial carnivore uses the "warm," highly productive environment of the sea as its food source. Some biologists even treat polar bears as marine animals, in spite of their "terrestrial" anatomy, because most of their lives are spent on the frozen sea, eating sea products. Sea ice is equally important for the carrion feeders that clean up after polar bears, especially arctic foxes, and also ravens, gulls and jaegers.

One last point to note on arctic mammals in general: all the land mammals now living in those parts of the Arctic covered by the ice of the last ice age (see fig. 3.5) must have immigrated from outside the area after the ice melted. Judging from their present-day geographic ranges, and from the locations of ice-free land where they could have found refuge while the ice age was at its height, far more must have come from the west than from the east. In spite of the hundreds or thousands of generations in which these animals have had time to spread, many of the smaller species have evidently found Hudson Bay a barrier they could not cross or circumvent, and they are found, now, only on one side of it or the other. Examples of species found on the western shores of the Bay but not the eastern are: arctic ground squirrel, northern red-backed vole, brown lemming, and collared lemming. One species, the Ungava lemming, is found on the eastern but not the western shore of the Bay. Fossils of this species dating from 11,000 years ago, when its present-day range was still covered by the ice sheets of the last ice age, have been found in Pennsylvania. Presumably the species survived the ice age in what is now the eastern United States and then spread north into arctic Quebec, east of Hudson Bay, as the land slowly became free of ice.

References

Three books with much interesting material on the lives of arctic mammals are:

Life in the Cold, by Peter Marchand (Hanover, NH: University Press of New England, 1987).

Arctic Animals, by Fred Bruemmer (Toronto: McClelland and Stewart, 1986).

Polar Animals, by A. Pedersen, trans. by Gwynne Vevers (New York: Taplinger Publishing Co., 1966).

Descriptions of all the arctic mammals, their life cycles, food, behavior, and much else, are to be found in *The Mammals of Canada,* by A. W. F. Banfield (Toronto: University of Toronto Press, 1974).

7.2 The Bear Family (Ursideae)

Bears are the largest living carnivores, and the two largest species, polar bears and grizzly bears, are both inhabitants of the Arctic. These two species, in spite of their markedly different appearance, are closely related. They are descendants of a common ancestor, and they are believed by some scientists to have become distinct from each other as recently as 200,000 years ago. In captivity, the two species can interbreed, and some of the hybrids are reproductively fertile; this shows how similar they are genetically.

It is worth noting that they are the only bears that don't climb trees. Also, that they are dangerous, and as likely to attack as to flee if approached too closely. In open country, with nowhere to hide, attack is the obvious method of defense, especially for animals with the speed and enormous strength of the two arctic bears.

Polar Bear

Polar bears [All coasts], *Ursus maritimus* (fig. 7.1), are the most impressive of all the arctic mammals. Their size and majestic bearing are breathtaking. The males are more than twice the size of the females (average males weigh about 500 kg, females 200 kg).

Polar bears spend most of their lives on sea ice—annual ice rather than multiyear ice (see sec. 4.2) because that's where the seals are (see sec. 7.12). They are adapted to this habitat in many ways. They are protected from the cold by their warm fur and a thick layer of blubber, sometimes protected too well—they become overheated if they exert themselves for long, which explains their usual unhurried demeanor. Their huge paws, with furred soles, serve as snowshoes when they walk

25 cm

Figure 7.1. Polar Bears. A female with two first-year cubs.

and as paddles when they swim. Their cryptic coloration (white or ivory) is not, as with most cryptically colored animals, a protection against predators; polar bears had none before human hunters arrived. Rather, it helps conceal them as they pursue their usual prey—seals, especially ringed seals.

Bears are most often to be found in habitats suitable for seals, near the edge of the sea ice or near leads (channels formed by gaping cracks in the ice). Where seals abound, bears congregate to hunt them. "Still-hunting" is the commonest hunting method: a bear first uses its acute sense of smell to find a seal's breathing hole; then it sits in wait until the seal comes up for air, whereupon the bear pounces; sometimes a breathing hole will be roofed over by a snow crust so strong that the bear must use all its strength to break through it. Less frequently, polar bears stalk their prey. This they do in warm weather, when seals often haul out onto the ice to sun themselves. Bears then stalk them, either by creeping up on them over the ice or, occasionally, by a swimming stalk.

Polar bears eat a variety of other foods. Sometimes they catch swimming seabirds by swimming under them and attacking from below. They have been known to kill belugas (see sec. 7.13) trapped in small pools by moving pack ice; very occasionally they capture young walruses. But polar bears are not exclusively carnivorous: they dive for kelp, and when the sea ice is nearly all melted in late summer, forcing the bears onto the land, they enjoy berries as much as grizzlies and black bears do. They also eat sedges, grasses, and other plants, as well as the lemmings scurrying among the plants. Indeed, polar bears are intelligent and versatile; they adapt their diet to suit changing circumstances and are always willing to experiment. When food is scarce, a bear conserves energy by going to sleep, often in a hollow it has dug in the snow to shelter itself from the wind. If necessary polar bears can even slow their metabolism to tide themselves over a period of food shortage.

Food shortages, when they happen, are not linked to any particular season and since seals don't hibernate, bears (other than pregnant females) don't need to either. In the fall, except on the shores of Hudson Bay, where different conditions lead to different behavior (see below), pregnant females leave the sea ice and come to land. There, within a few kilometers of the coast, they dig themselves dens (caves in snow drifts) in which to spend the last months of their pregnancies and the first months of their cubs' lives. Until they give birth, the pregnant females hibernate in their dens, but the hibernation is extremely shallow: unlike deep hibernators' body temperatures (which drop to near freezing) and pulse rates (which fall drastically), polar bears' body temperatures and pulse rates are only slightly lower when they are hibernating than when they are active.

Sometime in January, the cubs are born. The number of cubs per litter is usually 2, occasionally 1 or 3. The newborn cubs are helpless and tiny, less than 1 kg at birth, but they grow quickly on their mother's rich milk. After a month or two (three in the High Arctic), when the cubs' weights have increased ten- or twenty-fold, mother and cubs emerge from the den, and the mother has a chance to feed—her first chance since entering her den 5 months or so previously. She has vegetarian meals of land plants until the cubs are old enough to be led out onto the sea ice.

Cubs remain with their mothers until their third summer (or second, in Hudson Bay), learning how to hunt for themselves; the mothers must also keep their young at a safe distance from male bears, who tend to cannibalize when opportunity offers. Female bears, followed by their cubs in single file, are to be seen ranging the shore and sea ice in search of food (fig. 7.1). The sizes of the cubs make it easy to judge whether the litter is in its first or second year (in Hudson Bay, nearly all are first year). A mother bear does not mate again until her current litter has left her care; therefore, she will not breed more often than once every 3 years (2 in Hudson Bay).

The polar bears of Hudson Bay are exceptional in several ways because of the comparatively warm climate in which they live. They evidently thrive, being especially numerous around Churchill, where they attract observers from all over the world. Summer warmth causes the sea ice to melt much earlier in summer on Hudson Bay than it does elsewhere in the Arctic, forcing the bears onto the land and depriving them of all opportunities to hunt seals. This is especially hard on the year's breeding females, since they have to manage without seals for as long as 8 months, from the time the ice disappears (July), when they are pregnant, to the time their cubs are big enough to leave the den and go out on the next season's sea ice in the following March.

The warm climate also means there is not enough snow for digging dens when fall comes. Therefore, pregnant bears use dens tunneled into the earth down to the permafrost table. These dens can last a long time and they are also useful in summer. When the sun beats down—temperatures can rise to 30° C—a bear suffering from the heat can find a cool, insect-free resting spot at the end of such a tunnel. As the bear lies cooling itself on the frozen soil, its body heat melts the permafrost, so that as year succeeds year the tunnel gets deeper and deeper; some of the longest ones are many decades old. Most of these tunnels are a long way inland from the coast, some as far as 100 km. This is presumably because ground suitable for tunnelling cannot be found any nearer. Although various theories have been suggested, it is not yet known why Hudson Bay cubs leave their mothers a year younger than cubs do elsewhere.

For a detailed account of the lives and activities of polar bears, see *Polar Bears,* by Ian Stirling and Dan Guravich (Ann Arbor: University of Michigan Press, 1988).

Grizzly Bear

Grizzly Bears [AL], *Ursus arctos* (fig. 7.2), unlike polar bears, are not confined to the Arctic. Their North American range used to extend throughout the western plains and mountains, from the Arctic as far south as Mexico. Now they are extinct over most of this former range, and survive only at those few, remote, mountain sites where enough undisturbed wilderness still remains to sustain a grizzly population. The survivors are listed as vulnerable on the Canadian Endangered Species List, 1993. The chances of seeing a grizzly in the Arctic are still good, however, and on a tract of tundra where the vegetation is no more than ankle high, a grizzly's enormous size is more startlingly impressive than anywhere else. Most arctic grizzlies stay on the mainland, but a few enterprising wanderers cross the sea ice to Banks Island and Victoria Island, and they have been seen out on the sea ice of Parry Channel (see fig. 5.8).

Grizzlies are strikingly different from polar bears, quite apart from the obvious color difference. The contrast is surprising, considering how recently they diverged from their common ancestral species. A grizzly has a much shorter neck and body, a more pronounced shoulder hump, and a different facial profile, concave rather than aquiline. In color, they range from dark brown through tan to sandy buff; a single bear can

Figure 7.2. Grizzly Bears.

have all these colors—for example, a chocolate brown face and spine, with tan flanks, and legs and belly as pale as ivory; color pattern varies so much from one bear to another that individual bears are sometimes recognizable on a second meeting. The grizzlies of the tundra, the so-called barren-ground grizzlies, tend to be lighter than more southerly bears.

Grizzlies also differ from polar bears in being territorial, probably because their feeding areas are on solid ground rather than on shifting, drifting pack ice. Male bears have larger territories than females; a male's territory usually overlaps those of several females, and he will mate with all of them. Fierce fights are apt to break out between males competing for the same territory and its resident females.

Because grizzlies are terrestrial more than marine, their diets are more vegetarian than those of polar bears; they cannot rely on abundant seal meat as polar bears can. Grizzlies eat a variety of plants. They graze on grass and horsetails; they feast on berries, in season; and they are particularly fond of the roots of bear root, a close relative of wild sweet pea (sec. 5.13); bears will tear up the ground to unearth these roots in a way that leaves it looking as though it had been bulldozed. They also tear up the ground to get at "siksiks" (ground squirrels) in their burrows. The meat part of their diet has to be more varied than that of polar bears; it depends on what's available. As opportunity offers, they hunt moose, caribou, Dall's sheep, muskox, hares, lemmings, and voles, as well as siksiks. They eat fish. They can annihilate the eggs of whole colonies of breeding geese. They have no objection to carrion, and have been seen, out on the sea ice, eating seal meat that was presumably polar bears' leftovers. A grizzly discovered out on the sea ice of Parry Channel had killed seals for itself. And adult males have no hesitation in killing and eating cubs of their own species.

Unlike polar bears, in which only the pregnant females hibernate, all grizzlies hibernate in winter. They use dens they dig in the soil (where the unfrozen soil is deep enough) or natural caves among rocks. Their hibernation is more profound than that of polar bears, in the sense that their temperatures, pulse rates, and breathing rates decrease more markedly, and they are not quite so easily awakened; even so, their hibernation is shallower than that of black bears, and black bears are shallow hibernators by most standards.

The number of cubs in both polar and grizzly litters is most often two. Grizzly bear cubs usually stay with their mother for two winters (like the polar bear cubs of Hudson Bay) learning what's good to eat and how to find or catch it, while at the same time being protected from their most dangerous enemy—the cannibalistic male grizzlies.

For more on grizzlies, see "King of the Barrens," by Wayne Lynch, *Canadian Geographic,* May–June 1992.

7.3 The Cattle, Sheep, and Goat Family (Bovidae)

The distinction between the Bovidae and the Cervidae is described in section 7.4 on the Deer Family.

Muskox

Muskoxen [HML], *Ovibos moschatus* (fig. 7.3), are the most truly "arctic" of the northern land mammals (this is not to dismiss polar bears, which many class as marine mammals). Whereas a polar bear resembles other bears, and a caribou resembles other deer, a muskox is unlike any other animal, set apart by its impressively long, curved horns.

To call muskoxen ice-age relics, as is often done, implies, erroneously, that they are the only ice-age relics in the north. In fact, of course, *all* northern animals are ice-age relics; all are descended from ancestors who lived, at least temporarily, in lands bordering the huge ice sheets that covered most of the northern half of this continent 18,000 years ago (see map, fig. 3.5). It is true, however, that the modern muskox is the sole surviving member of quite a large group of ice-age "oxen" that are all now extinct: the fearsome shrub-ox, the woodland muskox, and at least three other species. They belong to the same family (Bovidae) as mountain goats and Dall's and bighorn sheep; domestic cattle, sheep, and goats are also family members.

Muskoxen are the only High Arctic mammals that seek no shelter of any kind during winter blizzards. Their winter coats make them oblivious to the cold. Under the coarse outer hairs, they have a thick, soft, lightweight fleece of the finest wool, which the Inuit call *qiviut;* it can be spun and woven, or knitted, to give a fabric of exceptionally high warmth-to-weight ratio. Over the qiviut is an outer coat of long guard hairs, which reach almost to the ground, forming a heavy skirt that ripples and swings when the animal gallops. This coat is dark brown except for a pale, cream-colored saddle and, in cows and juveniles, a pale patch on the forehead. The fur of the legs, showing below the "skirt," is almost white. In summer, while molting, muskoxen are not nearly as attractive a sight as this description suggests. Sheets and patches of qiviut hang in tattered festoons from their bodies; given enough time, a useful quantity of shed qiviut can be picked up from the ground.

Muskox horns are impressive; from their broad, flat bosses, covering the forehead, they sweep down, out, and up, tapering to sharp tips. Animals with broken horns are often seen; these are usually cows, whose horns are thinner than the bulls'. Probably most breaks occur when a cow catches her horn in a rock crevice.

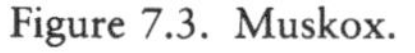
Figure 7.3. Muskox.

The horns are lethal weapons that the animals use as a defense against wolves, almost their only enemy before guns entered their world, though they have been known to fall prey, rarely, to polar bears. When wolves threaten a herd, the adult muskoxen range themselves in a circle with the young in the middle, or in a line with the young behind them; pursuing wolves then face an unbroken row of lowered heads and sharp horns, and if they have not been quick enough to capture a calf before the herd takes up its defensive position the task becomes hopeless and the wolves give up. People, too, should be cautious. Muskoxen, though patient up to a point, will charge if approached too closely, especially if their line of retreat is cut off. Not all charges are bluff; a real charge comes at lightning speed, and muskoxen can (and have been known to) disembowel their tormentors.

Two characteristic muskox traits to observe are "gland-rubbing" and "jousting." When excited, a muskox will rub a gland on its cheek, just below the eye, against its foreleg; this is gland-rubbing, and the reason for it is unclear. Jousting, when two bulls charge and butt each other, is spectacular. From midsummer on, as the rutting season approaches, muskox bulls become more and more belligerent and a bull with a harem of cows will have to defend his supremacy many times against intruding bulls. A battle between two bulls looks as formal as a nineteenth-century duel. To begin, the combatants slowly walk away from each other, without a backward glance. Suddenly, as if responding to a signal, they simultaneously wheel around to face each other and charge. Their heads meet with a crash that resounds across the tundra. Then they walk ponderously back to their starting points and do it again, perhaps many times over, until one becomes groggy and wanders off, admitting defeat.

Muskox diet depends on season. From summer until late spring they graze in sedge meadows; they can dig through snow for their food provided the snow crust doesn't become too hard for them to break. If this happens they shift to ridges blown clear of snow, where they eat arctic willow twigs and whatever else they can find. Willow is their chief food from spring into summer, whereupon they switch back to sedges again.

For a very complete account of muskoxen's lives and behavior, see *The Muskoxen of Polar Bear Pass,* by David R. Gray (Toronto: Fitzhenry and Whiteside, 1987).

Dall's Sheep

Besides muskoxen, only one other horned ungulate is to be found in the Arctic, and that barely. **Dall's Sheep** [a], *Ovis dalli* (fig. 7.4), are inhabitants of the mountain tundra of Alaska and Yukon, and sheep living in the northernmost mountains often stray over onto the northern slopes, where mountain tundra merges, without any dividing line, into the arctic tundra of the coastal plain west of the Mackenzie Delta.

Dall's sheep are obviously related to the bighorn sheep (*Ovis canadensis*) of the mountains to the south. They differ in their smaller size, their pure white coats, and the way in which the rams' horns widen into more open spirals than those of bighorns. Even more closely related to Dall's are the Stone's sheep of southern Yukon and northern British Columbia, which have horns like Dall's and differ only in being grayish brown instead of white. Dall's and Stone's sheep together arc known as thinhorns. The geographic ranges of thinhorns and bighorns do not overlap. Thinhorns are found only in a region which more or less coincides with Beringia, the northwest corner of North America that was not under ice at the height of the ice age (see fig. 3.5); this is strong evidence that they survived the last ice age there. Bighorns survived south of the ice sheets and have since spread only a short distance north. Though descended from a common ancestor, the two groups must have diverged long before the most recent ice age.

Figure 7.4. Dall's Sheep.

Dall's sheep are similar to bighorns in life style. The lambs are born in spring, and ewes and lambs flock together through summer while the rams are at higher pastures. The rut is in late fall, when butting battles between competing rams, which can happen at any time, become much fiercer and more frequent.

The animals are both wary and curious. They watch observers as intently as observers watch them and are easiest to approach when precipitous cliffs, into which they can escape, are close at hand. Their agility enables them to climb where no four-footed predator can follow. Away from such escape terrain, numerous predators threaten them—wolves, grizzlies, and wolverines. And whatever the terrain, lambs are always at risk from golden eagles.

The sheep require minerals from salt licks as an essential part of their diet. Therefore they return again and again, on well-worn paths, to known salt licks, providing places where a chance to observe them is almost assured.

7.4 The Deer Family (Cervidae)

The deer family consists of all cloven-hoofed animals in which the males (and in one species the females as well) have antlers. Two species of deer live in the arctic: caribou and moose.

Antlers differ from horns in being shed and grown anew every year. They consist of bone. Until they are full-sized, antlers are coated with a velvet-textured skin ("velvet") well supplied with blood vessels; and instead of being rigid, they are springy, as shown by the way they sway up and down when an animal trots or gallops. Full-grown antlers become hard, and the deer then scrape off the velvet against anything handy—in forested country, the trunks of shrubs and trees. This ruptures the blood vessels in the velvet, and the antlers are soon red with blood. A caribou or moose with blood-red antlers can be a startling sight.

Horns (such as those of muskoxen and Dalls's sheep, and all the other members the Bovidae) differ from antlers in being permanent: they keep growing for the whole of an animal's lifetime. Both males and females have them. A horn consists of a core of bone sheathed by a layer of fused keratin fibers, the material known as "horn."

Caribou

Caribou [AHMLB], *Rangifer tarandus* (fig. 7.5), are the most numerous large mammals of the Arctic, and the most useful to the Inuit, as suppliers of meat (for people and dogs) and hides (for superlatively warm clothes

Figure 7.5. Barren-ground Caribou

and bedding); in the past, their hides were also used for tents, their sinews for thread, their bones for tools, and their fat as fuel oil.

The majority of North American caribou (the barren-ground caribou) belong to a number of separate herds, populations that stay together, travel together, and stick to their traditional, separate territories. The biggest herds winter in the subarctic forest and in spring migrate northward into the tundra, where the cows bear their young. A couple of months after calving, most of the caribou start a slow return southward, but a small number stay behind, remaining on the tundra all through winter. About the end of October, when they have reached the treeline, the rut takes place; migration stops while mating occupies all the animals' attentions and energies. Then, when the rut is over, migration resumes and the animals continue to their wintering grounds.

For a large part of the year, therefore, the herds are on the move, between wintering grounds and calving grounds that may be more than 1000 km apart. Their travel routes are conspicuous from the air: as you fly over them, you see dozens of roughly parallel trails engraved in the tundra, the result of years of use by thousands of caribou. For much of the time, herd members are spread out, singly or in small groups, over a tremendous area; but three times a year each herd comes together as a whole, forming a dense mob of as many as 100,000 animals. The timetable is roughly as follows: after being scattered in winter, a herd combines for the northward migration in spring; they spread out when they reach the calving grounds and then combine (for the second time) into post-calving aggregations when the calves are 2 or 3 weeks old; They stay together for about 6 weeks, after which the aggregations break up and the caribou begin wandering southward toward treeline; there

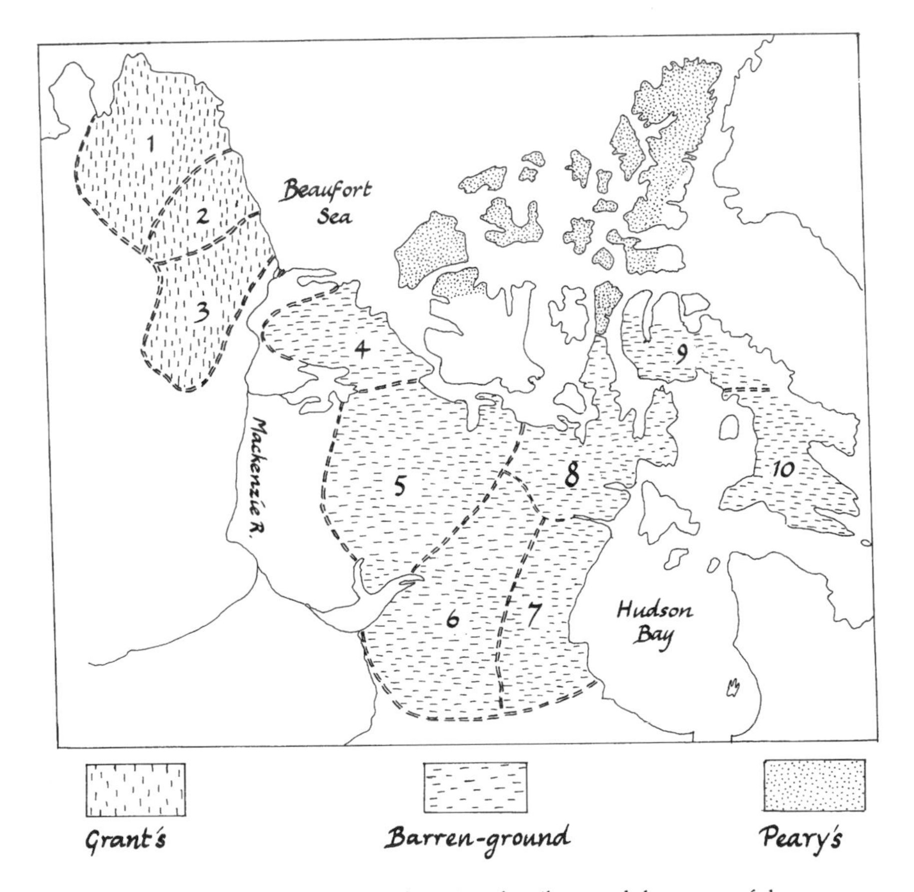

Figure 7.6. The ranges of the three subspecies of caribou; and the names of the separate herds of Grant's and Barren-ground Caribou: 1, Western Arctic. 2, Central Arctic. 3, Porcupine. 4, Bluenose. 5, Bathurst. 6, Beverly. 7, Kaminuriak. 8, Northeastern Keewatin. 9, Northern Baffin. 10, Southern Baffin.

the scattered groups combine (for the third time) for the rut and the journey into the forest which follows it; once there, they spread out, and the cycle starts again.

A big herd in migration is one of the wonders of the Arctic. Thousands—even tens of thousands—of caribou pour across the land, over hills and across valleys as far as the eye can see. But they are not one unit under central command: instead, groups of animals behave independently, following first one leader and then, minutes later, another, with no evident reason for the switch. Then a group will disperse and its members pause to graze, after which new groups will form. A continuous murmur of sound comes from the herds, a mixture of snorts, grunts and bleats. From close range, the sharp clicks made by the bones and tendons of their legs are audible; and now and again comes the thunder of a group galloping. They are easily spooked, so it is important not to disturb them, especially during the post-calving migration; a suckling calf that becomes separated from its mother will not be adopted by another lactating female—on the contrary, it will be rebuffed—and will inevitably starve.

Figure 7.6 shows the territories of the different herds and the names they have been given. Unlike the big, southern herds, those in the arctic islands, Boothia Peninsula, and the northeastern mainland are too far from the forests to winter there and must remain all year on the tundra. Caribou in the arctic islands (excluding Baffin Island) belong to a distinct subspecies, Peary's caribou (*Rangifer tarandus pearyi*), which differ from the barren-ground caribou (*R. t. groenlandicus*) in being smaller and much paler (almost white) in color. The High Arctic population of Peary's caribou is listed as endangered, and the Low Arctic population as threatened, in the Canadian Endangered Species List, 1993. Some zoologists regard the caribou west of the Mackenzie River as forming another subspecies, Grant's caribou (*R. t. grantii*). Two more caribou subspecies exist: the comparatively large, dark, woodland caribou (*R. t. caribou*) of the northern forests south of the Arctic; and the European reindeer (*R. t. tarandus*) which lives in Canada and Alaska only as introduced domestic herds.

Caribou are the only deer in which both sexes bear antlers, but when it comes to growing them and shedding them, the two sexes are out of step. The cows need full-grown antlers for use as defensive weapons during late pregnancy and calving, the only time when a cow threatened by a predator is better able to fight than to flee; cows shed their antlers soon after calving. The bulls need strong, full-grown antlers 5 or 6 months later, for rutting-season battles; they shed them when the rut is over.

Caribou also differ from all other deer in having asymmetric antlers: the forward-pointing brow tine, just above the animal's eye, is expanded into a broad blade on one side while remaining a simple spike on the other. Sometimes the big blade is on the left, sometimes on the right. The "handedness" does not necessarily remain the same throughout an individual animal's lifetime: sometimes (not always) a caribou that was "left-bladed" in one year will be "right-bladed" when the following year's antlers grow. The blade is thought to be an adaptation that evolved for the following reason: during the rut, bulls frequently swing their heads from side to side, threshing the undergrowth pugnaciously; it is believed that the antler blade sweeps aside the twigs and branches of low-growing shrubs that might otherwise injure the bulls' eyes.

A caribou has many adaptations for life in a cold climate. The long, white-tipped guard hairs of the winter coat are hollow, providing good insulation. The crescent-shaped hooves grow longer in the fall, while the pads of the feet, which touch the ground in summer, shrink; in this way the pads are raised above the hard, frozen ground by the time winter comes, and the sharp, protruding edges of the hooves give traction on crusted snow.

The circulation system is adapted to prevent rapid heat loss via the legs in winter, a time the animals spend standing in, or wading through, extremely cold snow. The arteries and veins of the leg are close together. Consequently, the outward-flowing arterial blood gives up its warmth to the inward-flowing venous blood; in this way, the warmth of the arterial blood is conserved in the body instead of being wasted. Body temperature remains at 40° C, while that of the legs is only 10° C.

Yet another adaptation to the cold is that caribou fat (like winter-grade engine oil) remains liquid at low temperatures.

These adaptations mean that the cold of winter is no great hardship for caribou. They are far more severely stressed in summer, when hordes of insects make life miserable and cause serious loss of condition. Mosquitoes in millions suck their blood, and plagues of warble flies and nose bots parasitize them (see sec. 9.5). Caribou find some measure of relief (not much) by crowding onto such snow beds as last into the hottest weeks; they also seek out exposed ridges where wind keeps insects (or some of them) at bay.

Apart from insects, wolves are the most important enemies of the caribou; grizzlies and wolverines are also a threat, and golden eagles can capture newborn calves. Wolves follow the caribou herds on their northward travels in spring and devour many, even though a healthy adult caribou can usually outrun a wolf unless taken by surprise. But in spring the northward-traveling wolves are themselves about to reproduce; the wolves stop and den up for this purpose while the caribou are still traveling. By the time the caribou cows reach their calving grounds, they are far beyond the wolves' denning grounds, and this protects the caribou from attack at the only time they cannot escape by running: the cows when they are giving birth, and the calves when they are less than 3 or 4 days old.

Caribou face two other serious perils. The first is disruption of their calving grounds and migration routes by human intrusion. The second is starvation. This is a risk faced every year by herds that overwinter on the tundra, where they have no food source other than ground vegetation. If a snowy winter is followed by an early thaw and then by a prolonged freeze-up, the ground is soon covered by a thick layer of ice which the caribou cannot break; food of any kind becomes inaccessible. The Peary caribou herds of the high arctic islands suffered a catastrophic die-off from this cause in the 1970s. They have yet to recover.

Caribou eat a wide range of plants—grasses, sedges, and other herbaceous plants, mushrooms, berries, the leaves and twigs of birch and willow and (the best-known item in their diets) lichens. Lichens contain fewer calories than the other plant foods, but make up for this by their abundance. Unlike most animals, caribou produce an enzyme, lichenase, that breaks down lichens making them digestible. Lichens are an excellent winter food because they contain little protein; this means that animals eating them have less need for water than those eating protein-rich food (see sec. 7.1). In winter when lichens form a large fraction of their diet, they recycle the nitrogenous waste material that enters the bloodstream (chiefly urea) back into the rumen where it can be re-used. They are the only ruminants that can do this. Moreover, they add "mineral supplements" to their vegetarian diet by chewing shed antlers, thus recycling calcium that would otherwise be wasted.

For a detailed account of caribou natural history, see *Caribou and the Barren-lands*, by George Calef (Toronto: Firefly Books, 1981).

Moose

Moose (*Alces alces*) are forest animals, but in summer they often stray north of treeline for as far as suitable browse can be found. They are most likely to be found in willow thickets in river floodplains. A favorite willow is the Alaska or Felt-leaf Willow (*Salix alaxensis*), which grows to the size of a small tree and has silvery white leaves and twigs. A moose ambling across the tundra looks even more enormous than a moose at home among trees.

7.5 The Dog, Fox, and Wolf Family (Canidae)

This family is represented in the Arctic by four species: the arctic fox, the red fox, the wolf, and the coyote. The coyote expanded its range into the Arctic within the past century, and is now found in the tundra from the Mackenzie Delta westward. Until more time has passed it can hardly be counted as an arctic animal, however; it is not considered further here.

Arctic Fox

The **Arctic Fox** [AHMLB], *Alopex lagopus* (fig. 7.7), is an obvious fox, although it differs from the familiar red fox of temperate latitudes in having a less "foxy" face. Its muzzle is shorter and its ears shorter and more rounded. Its legs and tail are also relatively short. These characteristics protect the animals against the cold of arctic winter by reducing the area of flesh exposed to the cold. Other adaptations to the cold are their magnificently thick and fluffy winter coats, and the furred soles of their feet. They do not hibernate, being fully equipped to live in the open.

The winter coat of an arctic fox may be white or "blue" (actually bluish gray, ranging from pale to almost black). White foxes are much commoner than blues in Canada and eastern Alaska; blues are as common as whites (or even outnumber them) in Greenland and western Alaska. Regardless of their winter color, all foxes are alike in summer: grayish brown on face, tail, back and the outer sides of the legs, and yellowish white below. But the contrast between their winter and summer appearance is not simply a matter of color: a fox that looks plump in its thick winter fur looks positively skinny in the lightweight coat of summer.

Fox pups are born in late spring and, being helpless at birth—blind, deaf, and toothless—spend their first 2 or 3 weeks in the natal den, usually in a burrow dug in the unfrozen soil of an esker, river bank, or sand dune. The same site is used year after year and becomes honeycombed with branching tunnels. Occasionally, rocky caves are used instead of burrows. Litter sizes vary greatly but ten is an average number. The fox population would explode, of course, if all the cubs prospered. Usually only one or two of the strongest cubs survive; in addition to eating the prey brought by their parents, they also eat their weaker brothers and sisters.

Outside the breeding season, arctic foxes are solitary. They are tremendous travelers: individual journeys of well over 1000 km have been reported. These journeys are made in quest of good feeding spots. In

Figure 7.7. Arctic Fox in winter (above) and summer (below).

winter foxes prey on lemmings, voles, ptarmigans and arctic hares. Summer brings the eggs and nestlings of ground-nesting birds. A rich variety is available at the coast: shorebirds, waterfowl, gulls, and even murres and guillemots when their nests can be reached. Adult ducks and geese, while unable to fly during the molt, sometimes fall prey to hungry foxes.

Before summer's end, some foxes prepare food caches for the lean months, storing their supplies carefully, in neat rows; a cache discovered in Greenland contained the frozen bodies of 42 birds (mostly dovekies, see sec. 6.6) and a pile of dovekie eggs; another contained 50 lemmings and 30 dovekies, all beheaded and arranged tidily side by side.

In winter, many foxes rely on carrion rather than fresh food: they obtain leftover fragments of caribou meat by following hunting wolves on land, or seal meat by following hunting polar bears on sea ice. In early spring, they also feast on fresh young seal: ringed seals bear their young out on the sea ice, in lairs under the snow (see sec. 7.12). A fox will search out a lair, break through the roof, and kill and eat the seal pup inside. A seal pup provides enough food to sustain a fox for weeks; a fox that has made a kill is believed to spend its time, eating and sleeping alternately, in or near the seal's lair for as long as the meal lasts.

In addition to all the foods mentioned above, foxes also eat fish, insects, and berries. Their menu is tremendously varied. But in spite of this, lemmings are unquestionably their chief prey, as shown by the fact that fox populations cycle in time with lemming populations. When lemmings are abundant, so are foxes; when lemming populations crash, so do fox populations. Food supplies are the limiting factor for foxes, even though they are not immune to predators: wolves and polar bears eat them when they can catch them.

To a naturalist, an arctic fox is a delightful animal: active, curious, and a joy to watch. Foxes wreak havoc in nesting birds' colonies, however, and are a threat to rarities such as the ivory gull (see sec. 6.6).

Red Fox

The **Red Fox** [alb], *Vulpes vulpes,* of the temperate zone is seldom thought of as an arctic animal, but it does occur, sparsely, throughout the Low Arctic and Baffin Island. Whereas the arctic fox is bold and inquisitive, the red fox is furtive and reclusive and is therefore less often seen. Its chosen habitat, everywhere in its range, is sparsely populated (or unpopulated) open country, so tundra suits it admirably.

The red fox is larger than the arctic fox and its color doesn't change with the seasons. Not all red foxes are red, however, though red is the commonest color. Some are grayish brown with a black stripe along the spine and across the shoulders in the form of a cross; these are "cross

Figure 7.8. Wolves.

foxes." More rarely they are all black except for some silver "frosting" and a white tail tip; these are "silver foxes." These differences are probably the effect of a single gene; foxes of all three color phases can be found in the same litter of pups.

Wolf

The **Wolf** [AHMLB], *Canis lupus* (fig. 7.8), of the Arctic belongs to the same species as the timber wolf that was once widespread throughout the north temperate zone of the whole world. Now that wolves have been killed off or driven into hiding throughout most of their range, many naturalists go a lifetime without seeing one. But in the Arctic, the chances of a sighting are good: the country is open, and the wolves, though not abundant, are comparatively fearless. Nearly all arctic wolves are light in color—white, off-white, cream, or pale gray.

They are most often seen singly, occasionally in twos, and only rarely in packs. Although, being social animals, most wolves belong to packs, the members of a pack are forced to spread out in search of food: a wolf needs several hundred square kilometers of hunting territory to support itself and, in the breeding season, its pups.

Some wolves remain on a fixed home range all year long; others migrate. How they behave depends on the habits of their prey. The wolves of the Arctic Islands stay put: they feed mainly on Peary caribou and muskoxen. The wolves of the Alaska North Slope and the Canadian Barren-grounds migrate in the wake of migrating caribou herds (see sec. 7.4) and spend winter in the boreal forest. Besides caribou, muskoxen, and moose, wolves can, if the need arises, hunt ringed seals, arctic hare, geese, ducks, lemmings, and voles.

Wolves are highly intelligent. For a pack to surround and kill an animal as large and strong as a caribou or moose takes carefully coordinated strategy; pack members have to behave as team members, not as individuals. They show the same cooperative behavior in rearing the pups. The pups are born in dens, and are blind and helpless for the first week. They emerge from the den a couple of weeks later, but continue until the end of their first summer to be fed, protected, and taught how to hunt, by their parents and other pack members. The parents are reputed to mate for life.

Wolves, especially the pups, are often friendly and playful. They inspire affection and admiration—until one sees them devouring the hind quarters of a still-living caribou. Not surprisingly, they always arouse strong but mixed feelings in human observers because of their close relation to dogs. In any case, to see wolves, and hear their howls, is an unforgettable arctic experience.

For a detailed account of life in a high arctic wolf pack, see *The Arctic Wolf: Living with the Pack,* by L. D. Mech (Stillwater, MN: Voyageur Press, 1988).

7.6 The Weasel Family (Mustelidae)

This is a family of ferocious carnivores with valuable fur (e.g., mink, sable, fisher, ermine). Two members of the family, ermine and wolverine, are found throughout the Arctic, as far as the land extends (northern Ellesmere Island); they are equally at home everywhere in the evergreen forest zone. A third member of the family, the least weasel, is a more southerly animal, with a range reaching beyond the forests only a short distance onto the tundra.

Ermine

Ermines [AHMLB], *Mustela erminea* (fig. 7.9), are typical weasels: small (less than 30 cm), slender, and lithe, with long necks and bodies, and short legs. In summer they are brown on the back, and pale yellow below; in winter they are pure white; the tip of the tail is black at all seasons.

Ermines are often to be seen scampering across the tundra or exploring the crevices in rock piles, and to the human observer are most endearing: they are nimble, playful, and inquisitive. And they are alternately bold and timid, now peeking out from a hiding place to watch you curiously with bright, intelligent eyes, then darting back into safety.

Figure 7.9. (a) Ermine. (b) Wolverine.

While they inspire affection in people, ermines inspire terror in lemmings and voles, their chief prey. When the hunting is good, an ermine will kill more than it needs immediately and store the surplus for future meals. There is no lean season, however, as neither ermines nor their prey hibernate; they are active all winter. When the lemmings are in their winter quarters, in tunnels under the snow, ermines follow them there; their sinuous, snakelike build permits them to follow the lemmings' narrow tunnels. Ermines also take over lemming nests to live in, usurping the rights of the original owner. The protection of a well-insulated lemming's nest may save an ermine from freezing to death in a spell of very cold weather: its slender build gives it a high surface-to-volume ratio, causing it to lose heat faster than would a shorter, plumper animal of the same weight.

Ermines are prey for wolves, foxes, and wolverines, as well as for gyrfalcons and snowy owls. It is said that the black tip of an ermine's tail helps protect it from its predators, especially from predatory birds: the ermine's body is cryptically colored—brown in summer, white in winter—with only the black tail tip showing up conspicuously against the background. This means that a raptor is most likely to pounce on the tail tip, which attracts its attention, giving the ermine an opportunity to escape with no more than a minor injury.

A female ermine bears a litter of 6 (on average) young every year. The young are helpless at first but grow quickly, maturing before the next breeding season.

Least Weasel

The tiny **Least Weasel** [al], *Mustela nivalis* (not illus.), averages a mere 20 cm long. It is seldom seen. It is a miniature version of the ermine, resembling it in most respects except that its winter coat is all white, without a black tip to the tail. It resembles an ermine in behavior, too, being equally fierce in spite of its tiny size. It is the world's smallest carnivore, and is smaller than many of the lemmings and voles it kills and eats.

Wolverine

The **Wolverine** [ahmlb], *Gulo gulo* (fig. 7.9), is legendary for its strength and ferocity. In size, it is almost as long (about 1 m) as a river otter but twice as heavy, weighing about 15 kg. Its stocky body is supported by short, thick legs. Its coat is dark brown except for a pale tan stripe that winds along the flanks and across the forehead. This pale band over the eyes makes it look as if it is wearing a black mask, and the result is a sinister, threatening face. The menace is still more arresting if the animal turns toward the approaching naturalist with a warning snarl. While there are no reports of a wolverine attacking a person, it certainly could if it wanted to. It has no inhibitions about breaking into cabins and camps and trashing them utterly.

Wolverines eat a huge variety of foods. Lemmings, voles, ground squirrels, eggs and nestlings of many bird species, carcasses of all kinds, and various berries and roots make up most of their diet. They can kill much larger prey if they can catch it. They can kill young caribou or moose when the winter snow is soft and deep, for then a wolverine's large, fur-soled feet, acting as snowshoes, enable it to overtake a long-legged animal that sinks into the snow. On snow firm enough to bear its weight, any caribou or moose can easily outrun a wolverine. Caribou do indeed have large enough hooves to "float" on any but the fluffiest snow, but it takes a very strong crust to support a moose.

Wolverines bear their young in early spring, in a den dug in a snowdrift. The cubs (1 to 5 of them) remain in the den, fed by their mother, until summer. Once they emerge, the mother teaches them how to hunt. The family stays together until the following spring.

Wolverines don't turn white in winter. Their fur is renowned for the fact that hoarfrost will not form on it; it is an ideal fur for the edging of parka hoods, because breath won't freeze onto it.

Wolverines are scarce and seldom seen; to meet one is a memorable experience.

7.7 The Hare and Rabbit Family (Leporidae)

Arctic Hare

The **Arctic Hare** [aHMLB], *Lepus arcticus* (fig. 7.10), is the only member of the family to live in the Arctic. The great majority live in the Canadian Arctic. A separate population in far western Alaska, usually considered merely a subspecies of the arctic hare, is sometimes classified as a distinct species and called Tundra Hare (*L. othis*).

The arctic hare shouldn't be confused with the snowshoe hare of the northern forests (it is much larger), and it should *never* be referred to as a rabbit! The newborn young of all species of rabbits start life blind, naked, and helpless, in the warmth of underground burrows. In contrast, newborn hares (leverets) start life fully furred, with open eyes, and capable of running soon after birth. They are born in open nests on the surface of the ground; they don't need, and don't get, much shelter.

Arctic hares are the largest of all hares, sometimes weighing 5 kg or more. They are often remarkably tame. And they have exceptionally thick, soft, lustrous, silky fur, a necessary protection against the rigors

Figure 7.10. Arctic Hares in the High Arctic.

of frigid winter gales when they deliberately seek windswept places where their food plants (mainly arctic willow in winter) are not buried under snow. Like all arctic vegetarians, they have a much more varied diet in summer.

The hares of the High Arctic (north of Parry Channel) are notable for several characteristics. The adults are pure white, except for black ear-tips, all through the year; only the young need the protection of cryptic coloration and are a ground-matching grayish brown for their first few weeks, when most of the snow has disappeared. The hares are gregarious. And, best of all, they often stand up on tiptoes to inspect the view, and then hop away in an upright posture for some distance before dropping down on all fours. It is an astonishing sight.

The hares of the southern islands and the mainland do not share these characteristics. They are white only in winter; in summer they are blue-gray, with only the underparts, legs, scut, and edges of the ears remaining white; the tips of the ears are black. They are less gregarious than the northern hares. And they seldom hop on their hind legs.

Wherever they live, arctic hares give the impression of being contented animals when no predators are around; after resting, or feeding for a while in one position, they stretch unhurriedly and luxuriously before beginning a new activity. When danger threatens, this seeming laziness quickly disappears. The hares are prey to numerous predators: arctic foxes, wolves, snowy owls, and (occasionally) rough-legged hawks and ermines.

7.8 The Squirrel Family (Sciuridae)

Arctic Ground Squirrel

Siksiks, the brief, descriptive, Inuit name for the **Arctic Ground Squirrel** [AL], *Spermophilus parryii* (fig. 7.11), are so like other ground squirrels (e.g., the Franklin, Richardson, Columbian, California, and Washington ground squirrels) that any westerner will recognize them at sight. And as they are the only members of their genus in the Arctic, species identification difficulties don't arise. They are found only in the mainland Arctic, not the islands.

Siksiks are the only arctic mammals that hibernate deeply in winter, in the sense that they become torpid, with metabolism, breathing, and heart rate slowed almost to a stop (bears merely become dormant, not torpid). Siksiks are just big enough to survive in this state of suspended animation without freezing to death. Smaller beasts, like lemmings and voles, simply because they are small, have too high a surface-to-volume

Figure 7.11. Arctic Ground Squirrels ("sik-siks").

ratio to manage it and must remain active to survive. Because they hibernate, siksiks probably experience more daylight during their conscious lives than any other mammal. The world they experience is never dark. All the same, they are diurnal animals, resting in the cool hours; even when the sun never sets, the "nights," while the sun is only just above the northern horizon, are comparatively cool.

Siksiks live in large, busy colonies. Each "family home" is a complicated system of burrows, with numerous side branches and exits to the surface. Some dead-end side branches—at considerable depth but sloping upward so as to trap warm air—lead to hibernation dens which are cosily lined with grasses, lichens, and hair. Siksiks can dig their burrows only where the permafrost table is far below the surface, i.e., where there is a thick active layer of well-drained sand or gravel. Eskers (see sec. 3.3), moraines, sand dunes, and river banks provide suitable sites.

A long-established colony is usually visible from a distance because of the marked effect a crowd of resident siksiks has on the vegetation. Their constant burrowing keeps the soil disturbed, making an environment where plant "pioneers" such as river beauty thrive; and the copious manure keeps the vegetation noticeably lush and green.

Siksiks feed on a multitude of different plants—their leaves, flowers, roots, and seeds—and on meat, too, if given the chance; a hungry male will happily eat a baby siksik. And siksiks are themselves prey to a multitude of predators: grizzly bears, wolves, arctic and red foxes, wolverines and ermines among mammals; snowy owls, rough-legged hawks, gyrfalcons, jaegers, ravens, and golden eagles among birds. Fortunately for the siksiks, few of them live within range of all of these predators.

Unlike many mammal species (particularly lemmings) siksik populations seem to remain relatively stable in size over the years, without alternating "explosions" and die-backs. The probable cause is that population growth, instead of being limited by starvation or predation, is limited by a lack of suitable living space. By summer's end, the number of siksiks in a colony has been augmented by the year's newborns and has many more members than it had in spring. In the fall, all must search for suitable hibernation sites, but owing to a shortage of such sites, not all will succeed. Siksiks that are driven from "home" by their stronger relatives are likely to be losers in the struggle for existence because many will hibernate in unsuitable places and come to grief. Many will drown because they hibernate in burrows flooded by spring meltwater, and some may freeze because they try to hibernate where winter snow is not deep enough to insulate them properly. In this way, apparently, are population explosions prevented.

In preparation for hibernation, siksiks fatten up in the fall; they also lay in a store of food to tide them over the hungry weeks to come in spring, from the time they emerge from their burrows onto the sunlit snow of April or May, until the time when new growth yields an adequate supply of fresh food. Because males emerge from hibernation earlier than females, they have to make bigger caches.

The siksik's name comes from its alarm call, which, in many places, sounds exactly like that. But its "accent" varies from one region to another and in some places the call sounds like "chiz-zik," and is easily mistaken for the chirp of a bird. Siksiks make a variety of other noises too: squeals of terror when a predator pounces; low growls and hisses when they feel bold enough to threaten an intruder; and angry (or anxious?) chatterings like those of red squirrels when they have reason to be defensive or annoyed.

7.9 The Beaver Family (Castoridae)

Beaver

Because of their dependence on wood for lodge building and wood with the bark on for winter food, **Beavers** [al], *Castor canadensis* (not illus.), are forest dwellers for the most part; they are seldom thought of as arctic animals. Some parts of the Arctic are suitable for beavers, however. In parts of the western Arctic where the soil is good enough, isolated patches of balsam poplar trees (a favorite with beavers) often grow in the tundra beyond treeline. Elsewhere, the strip of shrub tundra north of the treeline provides good beaver habitat if tall shrub willows and suitable bodies of water (lakes, marshes, or slow-moving rivers) are available. The willows must be big enough to yield construction material for lodges and dams. And the water must be deep enough not to freeze to the bottom in winter: beavers need space for a food cache that will remain immersed in water however thick the ice becomes.

"Temperate" and "arctic" beavers live notably different life styles. Whereas a beaver living in a mild climate can forage over a wide area in every month of the year, an arctic beaver must spend winter confined to its lodge and a small underwater food cache in front of the lodge. Beavers are not adapted, as truly arctic animals are, to withstand severe cold. Their big, hairless tails and feet would quickly freeze if they ventured into the open air in winter; therefore they stay "indoors," where the temperature rarely drops below −3° or −4° C; they venture out only for short journeys under water.

The contrast between summer and winter conditions in the Arctic means a contrast between summer and winter food requirements. Surprisingly, beavers in the Arctic eat less in winter than in summer. This is because, although they do not hibernate, arctic beavers (unlike temperate ones) are less active in winter than in summer and use up energy more slowly; in winter, they feed at about half the summer rate. Their diet differs with the season, too. Their summer food is fresh, protein-rich "greens"—willow leaves, sedges, waterlilies, and whatever other water plants they can find; they also eat the soft tips of willow twigs. Their winter food, when fresh greens are unobtainable, is bark; they don't eat bark in summer. The best bark comes from sapling willows, and sapling poplars if available, which are cut down and cached before freeze-up.

Where tall willows are common, green alder is likely to be common too. Beavers much prefer willow, however, and eat comparatively small amounts of alder. The result is that when beavers colonize an area supporting a mixed stand of willows and alder, the willows are taken and the alders are left. Although willows can sprout from cut stumps, their regeneration is slow because of the cool climate. The alders, on the other hand, do well: they are not damaged by the beavers, while at the same time, the willows competing with them, for space and soil nutrients, are removed. Eventually the alders succeed in crowding most of the willows out, making the area unattractive to beavers. The beaver colony then has to shift to a new area rich in willows, and begin again. This gives the remaining willows in the abandoned area a chance to recover. It is believed that this process causes the beaver populations of small areas to boom and bust alternately, without much affecting the total population of large regions.

Because of all this, it is worth looking for areas where alders flourish and willows are unexpectedly absent, to see if old, abandoned beaver dams and lodges are near at hand. Alternatively, if signs of departed beavers are discovered first, to see if an abundance of alders and a shortage of willows could account for the beavers' departure.

7.10 The Mouse, Lemming, and Vole Family (Muridae)

Seven members of this family live in the Arctic. They are rodents, as are members of the Squirrel Family (sec. 7.8) and the Beaver Family (sec. 7.9), but *not* the Shrew Family (sec. 7.11). The seven comprise 3 lemming species and 4 vole species. Both these groups are small, mouselike animals, but with much shorter tails and smaller ears (relative to their body size) than true mice. The distinction between them is that lemmings' tails are even shorter than voles'; also, lemmings are larger and tend to be more roly-poly.

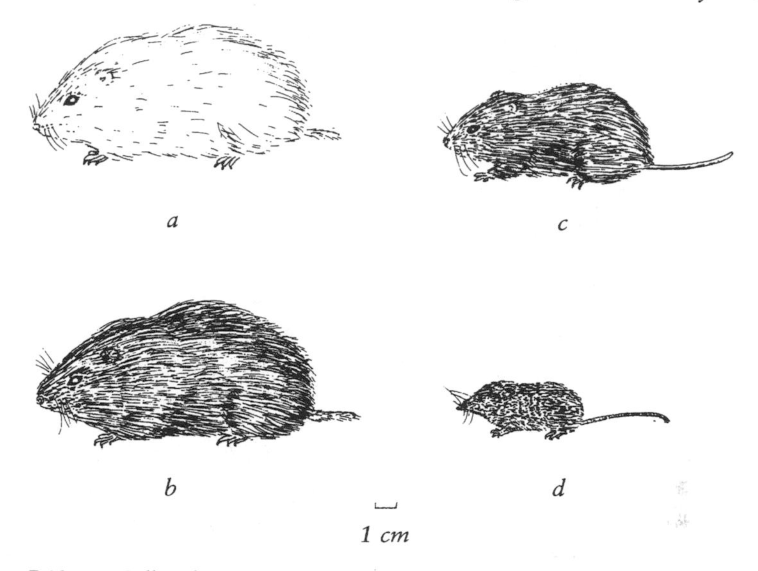

Figure 7.12. (a) Collared Lemming in winter. (b) Brown Lemming. (c) Northern Red-backed Vole. (d) Masked Shrew.

Lemmings

The word "lemmings" applies to several species of animal just as, to take another example, the word "deer" does. The North American Arctic is home to three species of lemmings belonging to two genera, the collared lemmings (*Dicrostonyx*) and the brown lemmings (*Lemmus*). Other lemmings, known as bog lemmings (genus *Synaptomis*), live south of treeline and so are not considered here.

The two *Dicrostonyx* lemmings are the **Collared Lemming,** *Dicrostonyx torquatus* (fig. 7.12), and the **Ungava Lemming,** *D. hudsonius;* between them, they are found everywhere in the Arctic, but their ranges do not overlap: the Ungava lemming is found only in the mainland Arctic east of Hudson Bay (the Ungava Peninsula), while the much commoner collared lemming is found everywhere else. They differ only in the anatomical details of their teeth; the reason for regarding them as separate species is that they are believed to have evolved as separate lineages for a long time (see sec. 7.1). The two species are so similar in appearance and behavior that they will be lumped together in what follows.

The **Brown Lemming** [AMLB], *Lemmus sibiricus* (fig 7.12), is the only member of its genus in North America and is not nearly as famous as its Eurasian congener, the Norwegian lemming, *Lemmus lemmus,* renowned for its mythical mass suicides (more on this below). The range of the brown lemming does not include the Ungava Peninsula.

Collared and brown lemmings are hard to tell apart unless you have a specimen in hand. They are just small, plump "bodies" scurrying around among the low tundra plants. Two characteristics make collared lemmings easily identifiable sometimes: they are the only lemmings in the High Arctic and the Ungava Peninsula, and they turn white in winter, whereas brown lemmings do not. These distinctions are no use over most of the Arctic in summer. Then, habitat is suggestive though not conclusive; collared lemmings tend to avoid damp, marshy ground, whereas brown lemmings prefer it. If you do have a specimen in hand, note the toes on the front feet: a brown lemming's "thumb" bears a vertically flattened "thumbnail," and the first claw (next to the thumb) is much shorter than the three outer ones. The toes of a collared lemming are unremarkable—in summer; in winter two of the front toes of collared lemmings enlarge and harden to form "scraper blades" under the claws, which are useful for tunneling in the snow.

Collared and brown lemmings have different diets, corresponding to their different habitats. Collared lemmings eat dry-land plants, especially willows and arctic dryad; researchers have found that in August, when the lemmings are preparing for winter, these two plants are particularly rich in calories. They also eat much purple saxifrage and various louseworts. Brown lemmings eat the succulent basal parts of sedge and cottongrass leaves growing in wet ground. In winter, when fresh leaves are not available, they eat moss, an indigestible food for most animals but not for brown lemmings.

So much for the differences between collared and brown lemmings. In life styles and behavior, they are similar. They live in underground burrows linked by surface runways in summer, and in chambers linked by runways under the snow in winter. Collareds and browns sometimes share runways. They are active and busy all year, and both species have long breeding seasons. Browns sometimes breed all winter long. Naturally, population sizes build up to "explosion" levels. This is what leads to the so-called "mass suicides" of lemmings. They become so overcrowded that they are forced to depart in droves, in search of food. Those that reach water (fresh or salt) set off confidently to swim across it and then, if a storm gets up or if the farther bank is beyond reach, they drown; the drownings are certainly unintended.

Lemmings' population eruptions come as a blessing to all manner of predators, especially ermines, wolverines, foxes, wolves, gulls, jaegers, falcons, hawks, and owls. Swimming lemmings (they are good swimmers) sometimes fall prey to salmon, pike, and eiders. Indeed all carnivorous mammals and birds, even if they don't specialize in lemmings, eat them at times. Lemmings are at the bottom of numerous food chains; the welfare of arctic animal communities is closely linked to the welfare of the lemmings, whose populations are apt to fluctuate greatly in size. Both species undergo population cycles.

Much research has been done to discover what terminates a lemming population explosion. As lemming numbers increase, their food supplies dwindle and their predators multiply. Also, the lemmings become more quarrelsome among themselves: killings and cannibalism become commonplace. Whatever the controlling factor, each population explosion is followed by a crash, after which the population starts to build again, slowly at first and then with increasing speed. Sometimes populations go through regular cycles averaging four years in length. But the regularity of the cycles is often broken by nonbiological causes: an unusually cold winter will interfere with their breeding; or sudden thawing of a thick snow cover will cause thousands of lemmings to drown.

Besides watching for lemmings themselves, it is interesting to look for traces of them. Signs of their winter living quarters are to be found lying on the tundra. A soft pad of dried grass was the mattress in a collared lemming's sleeping chamber under last winter's snow. Close by, a neat pile of droppings marks the "latrine" chamber, en suite; sometimes it is clear that one latrine served for two sleeping chambers. Brown lemmings' nests are hollow balls of dried grass.

Voles

Voles are not generally thought of as arctic animals, and they don't occur in the Arctic Islands. But they are quite common in parts of the mainland Arctic, and deserve to be noted if only because not every "mouse" you see is necessarily a lemming. The distinction between voles and lemmings is described at the beginning of this section. The four vole species that might be encountered are not easy for an amateur to tell apart because the diagnostic characters depend on skull measurements. Except in the red-backed vole, the coat colors are too variable and indeterminate to be useful for identification.

The **Northern Red-backed Vole** [AL], *Clethrionomys rutilus* (fig. 7.12), lives in shrub tundra throughout most of the mainland Arctic west of Hudson Bay. This vole has a bright chestnut to cinnamon band along its back and a pale tan or cream-colored belly.

The **Meadow V** [l], *Microtus pennsylvanicus* (not illus.), is the common "field mouse" of the temperate zone, an animal of open country whose range extends a short way onto the southernmost tundra.

The **Singing V** [A], *M. miurus* (not illus.), is named for its chirps and trills.

The **Root V** [Al], *M. oeconomus* (not illus.), lives only in the western Arctic (from Bathurst Inlet west) and prefers a damp habitat.

Ecologically, the voles are similar to the lemmings, except that they don't breed in winter. They are fecund, nonetheless, and their numbers fluctuate tremendously in more or less regular population cycles. When they are numerous, they are welcome prey to all the predators that enjoy lemmings.

For details on lemmings' population cycles, see "The Lemming Phenomenon," by Lennart Hansson, *Natural History,* December 1989.

7.11 The Shrew Family (Soricidae)

Shrews are small, furtive animals resembling mice with long snouts. They are rarely seen, and rate mention only for the sake of completeness. Despite their appearance they are not closely related to mice—they are not even rodents. They are classed, with moles, as insectivores. Their principal food is insects and worms, plus baby voles and lemmings when available, eked out with plant material as needed. Like voles, they lead a busy life, active all through the year.

The two species whose range extends into the mainland Arctic are the **Masked Shrew** [AL], *Sorex cinereus* (fig 7.12), and the **Arctic Shrew** [A], *S. arcticus* (not illus.).

7.12 Walruses and Seals (Odobenidae and Phocidae)

Walruses (Odobenidae) and seals (Phocidae) are usually grouped together as *pinnipeds,* marine mammals with flippers. (The pinnipeds also include a third, nonarctic group, the fur seals and sea-lions, in the family Otariidae). Some scientists, however, think that "pinnipeds" are a mixture of unrelated animals that should not be grouped together.

Seals differ markedly from all other pinnipeds in the way they swim. A seal propels itself through the water with alternate strokes of its hind flippers, aided by movements of the hind end of the body; to travel out of the water, on ice or land, it has to "hump" along using only its back muscles and fore flippers, because the hind flippers cannot be swivelled forward into a position where they could be useful.

A walrus, in contrast, swims with its fore flippers, which function rather like oars. This explains why seals have slender shoulders while walruses have massive and powerful ones. On ice or land, a walrus swivels its hind flippers forward under its body, so that they support much of its weight. (Fur seals and sea lions resemble walruses in these characteristics.)

The following subsections treat the two families separately.

The Walrus Family (Odobenidae)

The family has only a single species, the **Walrus,** *Odobenus rosmarus* (fig. 7.13). Two subspecies are recognized, which live, respectively, in the Atlantic and the Pacific. Atlantic walruses live in arctic waters east of Somerset Island, notably in Baffin Bay, Davis Strait, Foxe Basin, and the mouth of Hudson Bay (see map, fig. 4.1); Pacific walruses are found along the Alaska coast and westward along the arctic shores of Siberia. The difference between the subspecies is a matter of size: the westerners are larger. An average Pacific walrus bull weighs about 1.3 tons and has tusks over 60 cm long; an average Atlantic bull weighs about three-quarters of a ton and has tusks of about 35 cm.

Walruses are animals of shallow, icy seas. They mate and give birth on ice floes and, when not feeding, spend much of their time lolling on ice floes, often in big, densely crowded herds. A big herd is noisy: walruses have a loud bark, which carries a long way over calm sea. They seldom haul out on dry land, doing so only if summer winds have temporarily blown all the ice floes away from their feeding waters. Walruses

Figure 7.13. Walrus.

cannot live under thick, solid ice. To make breathing holes, they swim up from below and butt a hole through the icy ceiling with their heads, so the ice must be thin enough or soft enough to allow this.

The vast majority of walruses are bottom feeders, and they do their grazing (as it is called) in shallow waters. Clams and other bivalves are their favorite food. They are believed to stand on their heads to feed, feeling for their quarry with their sensitive mustache whiskers (*vibrissae*) and then sucking out, or nipping off, the meat. They leave the shells intact, eating no more than the clams' siphons and "feet" and wasting the rest; the clams die, since they cannot regenerate the lost parts. Walruses also take whelks, sea cucumbers, and other sea-floor invertebrates. It is reported that old, rogue bulls sometimes develop a taste for larger prey, killing and eating ringed seals. The tusks (present in both sexes) seem to be used mostly for display, and for aggressive threats and defense; it is uncertain whether or not walruses also use them to rake the sand for clams.

Walrus cows are exceedingly conscientious mothers. Until it is weaned, usually at two years old, a calf is constantly protected by its mother and often clings to her, either to her back or cradled in her fore flippers. If a calf is orphaned, it will be adopted by another cow.

The Seal Family (Phocidae)

Five species of seals spend at least some of the year in arctic waters. Two of them (bearded seal and ringed seal) remain in the Arctic all the year round. Two others (harp seal and hooded seal) spend the summer in the Arctic and overwinter farther south. All these four give birth to their pups on the ice. The fifth species, the harbor seal, bears its young on sand bars or rocky promontories; it is not so much an arctic seal as a temperate seal whose range extends into the arctic, and it will not be considered further here.

The **Bearded Seal,** *Erignathus barbatus* (fig. 7.14), is the largest and most walruslike of arctic seals. It is a bottom feeder, eating the same food as walruses, though doing it less neatly; broken clam shells found in their stomachs show that bearded seals crush and swallow whole clams instead of simply sucking off the clams' siphons. Their quill-like whiskers form prominent mustaches (hence the name "bearded"!), which presumably serve, like a walrus's whiskers, as feelers. Beardeds differ from other seals in having square-ended fore flippers, but this is often hard to see. And, unlike walruses, they are never gregarious.

The **Ringed Seal,** *Phoca hispida* (fig. 7.14), is the smallest and commonest of the arctic seals. The species is easy to confuse with the harbor seal in waters where both species live. Its breeding habitat is solid, shore-fast ice. The female ringed seal is unique in constructing a birth lair under the snow in which to bear her pup. Snow deep enough for safe lairs is to be found only on rough ice, downwind of ridges and other irregularities; smooth ice is unsuitable. A breathing hole through the floor of the lair, which the seal keeps open by scratching the ice with the sharp nails on her fore flippers, gives access to food—shrimps and other crustaceans, especially the shrimplike "krill" swimming in the plankton, and a variety of small fish. It also provides a bolt hole, but

Figure 7.14. (a) Bearded Seal. (b) Hooded Seal. (c) Harp Seal. (d) Ringed Seal.

even so, large numbers of pups are dug out of their lairs and eaten by arctic foxes. In winter and spring, mature ringed seals take possession of the shore-fast ice and drive away the adolescents, who move to the broken ice farther from shore, around polynyas (see sec. 4.4). There, the inexperienced young seals are easy prey for polar bears. Ringed seals are, indeed, the chief food of polar bears.

The **Harp Seal,** *Phoca groenlandica,* (fig. 7.14) is slightly larger and only slightly less abundant than the ringed seal. These gregarious seals migrate northward into arctic waters (Baffin Bay, Davis Strait, Foxe Basin, and Hudson Strait; see map, fig. 4.1) for the summer and southward into subarctic seas to overwinter and breed. Their habitat is among floating ice floes, well offshore in deep water for most of the year; the seals dive deep (perhaps as deep as 300 m) in pursuit of food, chiefly small fish. They come in close to arctic shores only in late summer, when the shore-fast ice breaks up. Their pups ("whitecoats") are born in late winter, on ice off the coasts of Labrador and Newfoundland and in the Gulf of St. Lawrence. These are the seal pups whose killing, for their skins, has caused bitter disagreement between those who regard the slaughter as inhumane and those who defend the sealers' right to their traditional way of life.

The **Hooded Seal,** *Cystophora cristata* (fig. 7.14) is notable for the male's big, inflatable proboscis, which gives the animal a distinctive profile. Besides being able to inflate the proboscis, an angry male can also inflate part of the nasal septum so that it protrudes through one nostril as a big, red balloon. Unlike other seals, hoodeds are markedly aggressive. Apart from these unusual characteristics, hoodeds resemble harps in life style. Like harps, they are gregarious in the breeding season; herds of hoodeds migrate north for the summer and south for the winter, where they breed on floating ice floes in the same regions as the harps do, and in parts of Davis Strait as well. Hoodeds are much larger and much less numerous than harps, however, and they feed in deeper water, farther out to sea.

FURTHER NOTES

Seals of all species fast during the breeding period when, except for nursing pups, they all, even lactating females, rely for nourishment on their reserves of blubber. The lactation period tends to be short. The hooded seal carries this tendency to an extreme, nursing her pup for as little as 4 days during which the pup almost doubles its weight; not surprisingly, the mother loses considerable weight. Lactation is also short (about 2 weeks) in harp seals but longer (up to 2 months) in ringed seals.

Thus weaning is earliest for the seal pups (hoodeds and harps) born on floating ice which may break up or overturn at any time; ringed seals, breeding on more stable, shore-fast ice, do things more slowly. The difference may be a result of adaptations to contrasted environments, but the argument doesn't extend to walruses (see above).

For more details on seals see *The Natural History of Seals,* by W. Nigel Bonner (New York: Facts on File, 1990); *Harps and Hoods: Ice-breeding Seals of the Northwest Atlantic,* by David M. Lavigne and Kit M. Kovacs (Ontario: University of Waterloo Press, 1987); and *Seasons of the Seal,* by Fred Bruemmer and Brian Davies (Avenal, NJ: Outlet Book Co., 1991).

7.12 Two Whale Families (Bowhead Whales, Belugas and Narwhals)

Four species of whales may be encountered in arctic waters, three of them truly arctic—they never travel far from icy waters—and the fourth a summer visitor.

The resident arctic whales are the bowhead whale, the beluga, and the narwhal. The first belongs to the family Balaenidae, one of several families of baleen whales. These are whales of gigantic size with *baleen* plates (also called whalebone) in their mouths in place of teeth. The beluga and the narwhal are much smaller whales, and belong to the Monodontidae, one of several families of toothed whales.

The whale that comes only as a summer visitor is not strictly an arctic animal and is not described in detail below. It is the orca (also known as killer whale), a member of the family Delphinidae, another of the families of toothed whales. It stands out from the truly arctic whales because of its tall, triangular dorsal fin (see fig 7.15). The three arctic whales all lack dorsal fins, no doubt as a result of natural selection: a dorsal fin could be injured when a whale surfaced in waters covered with ice floes. Orcas stay away from icy waters, presumably to avoid this risk.

All whales are probably intelligent animals, who communicate with one another, by sound, over long distances—water is a far better sound conductor than air. They also make sounds whose returning echoes, from objects in the water (friends, enemies, floes, obstacles, prey?) enable them to form a sound picture of the space surrounding them. Scientists are justifiably concerned about the possible effects on whales of the din from ships and oil rigs as human activity in arctic seas increases. Whales are sensitive to sounds; their social organization and behavior could be disrupted if their signals were to become drowned out by obtrusive, man-made noises.

Bowhead Whale

The **Bowhead,** *Balaena mysticetus* (fig. 7.15), is an endangered species (Canadian Endangered Species List, 1993). In North American waters, it lives in two widely separated regions, in eastern and western arctic seas respectively, with a gap between. The eastern population (in Baffin Bay, Davis Strait, Foxe Basin, and northern Hudson Bay, see fig. 4.1) was almost exterminated by commercial whaling in the past, and though the whaling ended at the time of World War I, the population has not recovered; its size in the early 1990s was estimated to be about 250. These whales move north and south with the seasons, remaining at all times in cold, arctic or subarctic seas.

The western population is thought to be about ten times as large as the eastern, but even so, it is not large. The western bowheads spend winter in the Bering Sea, south of Bering Strait, and migrate northward through the strait to spend summer in the Chuckchi and Beaufort seas.

The bowhead is one of the truly enormous whales: a big one may be 18 m long, and fully one-third of its huge mass consists of the head. The head needs to be big: within its mouth, with the curiously "bowed" jaws (hence the whale's name) there may be as much as a ton of baleen, horny plates with fibrous fringes that strain food from large volumes of sea water. The food is chiefly krill (technically, *euphausids*), red, shrimplike crustaceans that swarm in millions in the ocean plankton. On this diet a single whale can grow as much as 30 tons of blubber, encasing its body in an insulating, food-storing layer 50 cm thick.

A sighting of a bowhead is most likely to consist of a sighting of its spout or its tail flukes as it "sounds" or dives. Because the blowholes are paired, an exhaling bowhead produces a double, V-shaped spout, visible from a long distance (over 30 km) on a calm, clear day. Bowheads and narwhals are the only whales in the Arctic that regularly show their tail flukes when sounding. Those of bowheads are spectacular (fig. 7.15); from tip to tip, their width exceeds the total length of an average orca. The trailing edges of the flukes are often notched by bites from orcas, which (apart from humans) are the only predators to threaten bowheads. Orcas hunt in packs, and are known to take bowhead calves. Bowheads try to evade pursuit by escaping to waters with abundant floating ice, where orcas dare not follow because of the danger of injuring their vulnerable dorsal fins. When not escaping from orcas, bowheads are rather slow swimmers; they are said to be timid, except when defending their calves.

Figure 7.15. Bowhead whale; below, its tail flukes when sounding. *Inset:* Four whales on the same scale showing sizes. *Top to bottom:* Bowhead, Orca, Narwhal, Beluga.

Another, more insidious, threat to bowheads is competition for food with ringed seals. Ringed seals (see sec. 7.12) have a more varied diet than bowheads—including fish, which bowheads do not eat—but even so, the seals take quantities of krill. It is not known whether seals take enough to deprive bowheads of the food they need for their numbers to increase.

Narwhal

The **Narwhal,** *Monodon monoceros* (fig. 7.16), is the whale whose long, straight tusk was believed to be the horn of the fabulous unicorn. Not until the 17th century was it generally realized that the unicorn horns inlaid with precious stones and stored as treasures in the regalia of European royal families were in fact the tusks of arctic whales.

Though far from common, narwhals are not nearly as rare as bowheads. The early 1990s population is believed to be between 25,000 and 30,000. Their North American range corresponds, roughly, with that of the eastern bowhead population; narwhals have not been found in the central or western Arctic.

The average male narwhal is about 6 m long not counting the tusk; females are only two-thirds as big. Normally, only males have tusks (females with tusks have been seen, but they are abnormal). The tusk is the left member of the single pair of teeth in the upper jaw; it grows forward, through the lip, while the right tooth, failing to grow, remains embedded in the gum. The tusk is from 2 to 3 m long. It is etched with spiral grooves, always twisted counterclockwise as seen from the narwhal's eyes. In the occasional abnormal narwhal with two tusks, both (so far as is known) twist in the same direction, making the pair as asymmetrical as two left feet.

Narwhals vary somewhat in color; generally they are a mottled and spotted gray, lighter below. Their rounded tail flukes show above water when they sound. Like belugas (but unlike bowheads) they have a single blowhole and consequently a single spout.

The purpose of the narwhal's tusk has been, for many years, the subject of intense debate. It cannot be related to food capture, since females don't have tusks. Almost certainly the chief value of the tusk is as a sexual "ornament." It seems to be used, too, for probably not very violent competitive fights. Many males have heads scarred, presumably, by the jabs of a competitor's tusk; pairs of sparring males are quite often seen with their heads above water and their tusks laid against each other, pointing skyward like crossed swords. The fact that narwhals with bro-

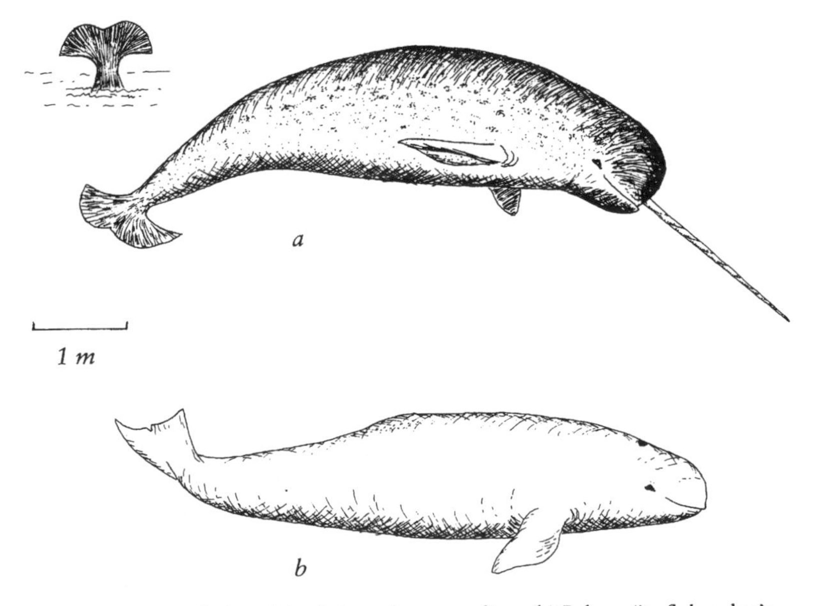

Figure 7.16. (a) Narwhal, and its flukes when sounding. (b) Beluga (its flukes don't appear above water when it sounds).

ken tusks are so often seen is further evidence that they fight. Moreover, they probably defend themselves with their tusks against orcas and Greenland sharks if escape is impossible. Orcas are narwhals' most serious enemy; when they can, they escape by swimming fast, to safety among the ice floes where orcas dare not follow.

The two "small" whales, narwhal and beluga, prefer different habitats: narwhals hunt in deeper water, and are more *pagophilous* (ice-loving) than belugas; they are sometimes found in quite narrow leads in the pack ice. In their seasonal migrations, narwhals keep close to the ice edge. They are fast swimmers and deep divers, scanning the whole water column right down to the sea floor for a variety of foods: arctic and polar cod, halibut, flounders, and other fish, as well as squids and other invertebrates. Even so, groups of narwhals, like belugas, occasionally find themselves trapped in a *savssat,* a pool of water that has become cut off from the open sea by newly formed ice. Until the ice breaks up or melts, the narwhals prevent it from closing in over the pool completely by butting it with their heads.

Narwhals are gregarious, social animals. They communicate with one another by sonar clicks, reminiscent, to one whale specialist, of morse signalling.

Narwhals are sometimes seen motionless at the sea's surface, apparently sleeping, with little more than the blowhole above water.

For more on narwhals, see *Arctic Dreams,* by Barry Lopez (New York: Charles Scribner's Sons, 1986).

Beluga

Belugas, *Delphinapterus leucas* (fig. 7.16), resemble narwhals only in being "small," toothed, arctic whales, but in appearance and behavior they are strikingly different. Their range in the Arctic coincides, approximately, with the bowheads'; a western (Beaufort Sea) population is separated from three eastern populations (in Davis Strait, Baffin Bay, the whole of Hudson Bay, and linking channels) by a gap (a *disjunction*) in the central Arctic. Western belugas are much the same size as narwhals; eastern belugas are somewhat smaller. Belugas in dwindling numbers also inhabit the St. Lawrence River; this population seems doomed beyond hope of recovery, by severe pollution of its waters. The eastern arctic populations are certainly not safe: they are severally listed, on the 1993 Canadian Endangered Species List, as endangered, threatened, and vulnerable.

Belugas always impress aquarium visitors with their pure whiteness (only the adults: the young are gray), and their "smiling" lips. They are the picture of friendly innocence, and it's hard to attribute the smile to a mere anatomical peculiarity. Unfortunately for them it makes them lovable, and much more likely to be imprisoned in zoos than any seemingly disagreeable beast.

Belugas are slower swimmers and shallower divers than narwhals. They spend the summer in shallow, inshore waters including river estuaries (e.g., the Churchill River flowing into Hudson Bay, the Cunningham River flowing into Parry Channel), where they move upstream and down on rising and falling tides. They are sociable animals and travel in *gams* (herds), which can number over a hundred. They are also the most "talkative" of whales, squealing, chirping, trilling, and croaking continually; this has earned them the name "sea canary." They have a full set of sharp teeth and eat much the same food as narwhals except that they don't hunt deep-water prey.

Their chief enemies (apart from humans) are orcas and, very occasionally, polar bears. They risk being caught by bears when they become trapped, sometimes in company with narwhals, in savsatts (see the account of narwhals, above). They also become immobilized, and vulnerable to bears, if they are left stranded by the falling tide, as they sometimes are when foraging close to shore at high tide. In deep water they are in little danger from bears, who are less fully adapted to the sea; belugas are much more agile than bears in the water and much more aware, because of their sensitivity to water-borne sound, of what's happening around them.

Their fondness for estuaries makes belugas the whales most likely to be seen by arctic travelers. One of their reasons for visiting estuaries seems to be that there they find gravel beds where they can conveniently—and no doubt pleasurably—rub themselves; they molt annually, and rubbing themselves on gravel helps scrape off the old skin and uncover the new.

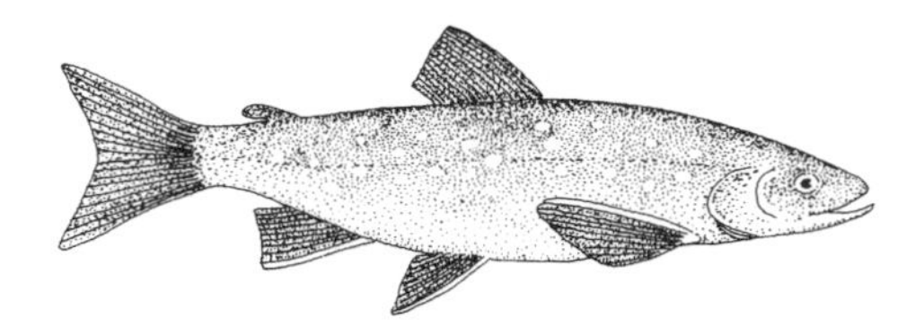

8 Fish

8.1 Fish and Naturalists

Few naturalists take on the study of fish as a specialty. One reason for this is that a fish has to be taken from its natural medium if it is to be examined closely, and this entails injury or death for the specimen. Therefore naturalists who are not anglers seldom pay fish much attention. An arctic traveler is bound to encounter fish now and again, however: fish caught by resident Inuit, by visiting fishermen, or by grizzly bears. This makes it worthwhile to learn enough about arctic fish to be able to identify these catches when you come across them. As members of the arctic ecosystem, fish are as important as any others. The sections that follow will enable you to identify most of the large (30 cm or more) freshwater fish you are likely to come across.

Full descriptions of the life styles and natural history of all arctic freshwater fish will be found in *Freshwater Fishes of Canada,* by W. B. Scott and E. J. Crossman (Bulletin 184 of the Fisheries Research Board of Canada, Ottawa, 1973).

8.2 Fish of the Salmon Family: Pacific Salmon

The most numerous arctic fish, and the most valuable as food, are several members of the Salmon Family. All have an *adipose fin,* a small fleshy fin on the back just in front of the tail fin (see fig. 8.1). Possession of an adipose fin would, in the Arctic, be a certain sign of membership in the family if it were not for the fact that smelts (see sec. 8.4) have them too; luckily, there is little risk of confusing smelts and salmon.

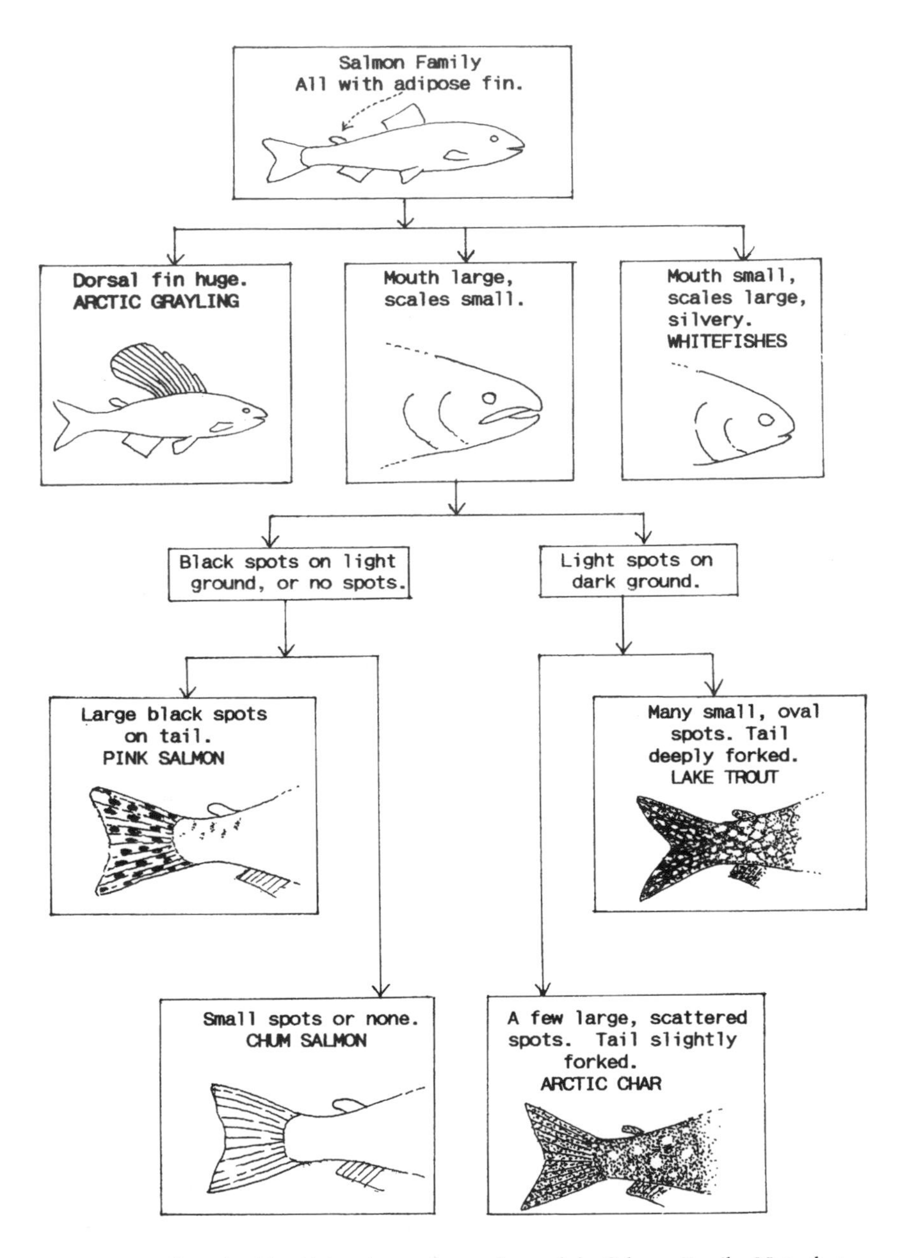

Figure 8.1. Chart for identifying the arctic members of the Salmon Family. Note that besides these fish, rainbow smelt (fig. 8.3) also has an adipose fin.

To identify which member of the Salmon Family a specimen belongs to, consult the chart in figure 8.1. Although many fish in the family are gorgeously colored, colors are not mentioned in the chart because they tend to vary greatly, depending on whether a fish is breeding or nonbreeding, and also on environmental conditions. It is also hard to judge the colors of a fish swimming at liberty under the water.

Now for the varied life styles of these fish. First, consider the two species of Pacific salmon, the **Chum Salmon,** *Oncorhynchus keta* (fig. 8.2) and the **Pink S,** *O. gorbuscha* (not illus.). Like all salmon (Atlantic as well as Pacific), they are *anadromous,* i.e., they spend most of their lives at sea, entering fresh water only to spawn. Not only that, they breed only once in their lives and die soon afterward; in this they differ from Atlantic salmon and their other relatives, trout and char, which live for many years and spawn repeatedly. Chum salmon and pink salmon have many similarities: both are western species spending the saltwater parts of their lives in the Pacific and Arctic oceans and spawning in rivers from the Mackenzie westward. Pinks usually spawn only a short distance inland, whereas chums swim upstream for enormous distances to reach their spawning grounds; they go as far as 2000 km up the Yukon and Mackenzie rivers.

The sexes are alike before they come into breeding condition, but when spawning time approaches, the males undergo extraordinary changes. Their heads lengthen and the jaws curve so that they close only at the tip, like a pair of pincers (fig. 8.2). Their spines become deformed too, especially in pinks, which become grotesquely hump-backed. Breeding chums are recognizable more from their markings than their shape. Both sexes have irregular reddish bars on their sides; seen through water in a good light, the bars look plum-colored.

At the breeding sites, females hollow out *redds* (nests) in the gravel and sand of the river bottom, and there a male and a female discharge their spawn—eggs from the female and *milt* (sperm) from the male. Fertilization is external (i.e., takes place outside the mother's body) but passionate, nonetheless: the male and female settle together in the redd with mouths agape and vibrate strenuously while they spawn. Then the female scoops gravel over the fertilized eggs to protect them from predators such as arctic graylings (see below) or mergansers (sec. 5.7), and remains on guard for the few days of life remaining to her.

Both parents die soon after spawning. Chums and pinks differ in the duration of their life cycles. Chums breed and die when they are three or four years old, pinks when they are two. Because pinks consistently breed and die at age two, they are split into two groups, the even-year breeders and the odd-year breeders, which form separate, unrelated populations. Some rivers are used by spawners of only one of the populations, and therefore contain breeding fish only every other year.

8.3 Other Fish of the Salmon Family

Of all arctic fish, the most highly prized—by anglers and gourmets alike—is the **Arctic Char,** *Salvelinus alpinus* (fig. 8.2). It is also the most truly arctic, with a range extending farther north than that of any other freshwater fish; it has been reported from northernmost Ellesmere Island.

Many arctic char are anadromous like salmon, spending most of the year at sea and entering fresh water only to spawn. But not all char populations behave in this way; some spend their whole lives in fresh water. And, unlike Pacific salmon, char do not die after spawning; they live for several years (occasionally to age 40) and spawn repeatedly. They spawn in the fall, usually in rivers or lakes with a gravelly bottom where the female can make a redd in much the same way that Pacific salmon do. At spawning time, char are spectacularly colorful; the large spots on the sides, the belly, and the lower fins become a vivid orange-red. The front margins of the lower fins remain pure white at all seasons and show clearly when the fish are seen swimming.

Lake Trout, *Salvelinus namaycush* (not illus.), are close relatives of arctic char. In appearance, they differ from them in having the pale spots on their sides numerous, closely spaced, and small, instead of few, scattered, and large; also the tail fin is more deeply forked than in arctic char. Lake trout are less typically "arctic" than arctic char: they are not found north of Parry Channel, and the southern limit of their range extends much farther south. Again unlike char, they are not anadromous; they never enter salt water and are only rarely found in mildly brackish water. They spawn in the fall, always at night, over a rock- and boulder-strewn lake bottom where the fertilized eggs can fall into protected crannies among the rocks. They have been known to live for more than 20 years.

Perhaps the most famous arctic fish is the **Arctic Grayling,** *Thymallus arcticus* (fig. 8.2). It is much smaller than the fish so far described, seldom exceeding 35 cm, but it is remarkable for its colors (note the plural)—purple to blue-black on the body, with brilliant red (or sometimes emerald green!) spots on the dorsal fin—and for the large size of the dorsal fin itself, which opens out like a fan. Sad to say, this magnificence is not obvious when the fish is seen under water, alive. Then, unless the male is threatening a rival at spawning time, the dorsal fin is folded flat, and the color looks mouse-brown. Graylings spawn in spring, at break-up time, and prefer to use small streams. They spawn in daylight and the

Figure 8.2. (from bottom to top) (a) Chum Salmon (spawning male). (b) Arctic Char. (c) Arctic Grayling. (d) Lake Whitefish.

male covers the female with his extended dorsal fin while love-making is in progress. No redd is constructed; the eggs and sperm mingle and unite in the water, and the fertilized eggs then simply settle to the bottom. Graylings live all their lives in fresh water, and like all other arctic freshwater fish, are in deep water under ice all winter.

The most confusing members of the Salmon Family are the whitefish or ciscos. They are easy to recognize as "whitefish" in a general sense but hard to identify as to species. Indeed, experts disagree on how they should be classified. The commonest species is **Lake Whitefish,** *Coregonus clupeaformis* (fig. 8.2), recognizable from its silvery color, large scales, and small mouth (see fig. 8.1). This species is found in the northern United States and throughout mainland Canada, but at only one site (Cambridge Bay) in the Arctic Islands. The vast majority live in fresh water, in large lakes and rivers, but in the Arctic they are also found in the sea close to shore.

At least 7 other species of *Coregonus* live in arctic waters, as well as other whitefish belonging to different genera. The best known of these is **Inconnu,** *Stenodus leucichthys* (not illus.). It is a western fish, whose range extends only a short distance east of the Mackenzie Delta. It is anadromous and is considerably bigger than lake whitefish, sometimes reaching a length of 75 cm (the average length for lake whitefish is about 35 cm). Indeed, its large size is what distinguishes inconnu from all other whitefish species.

8.4 Some Other Arctic Fishes

Some other arctic fishes (not in the Salmon Family) may come to your notice:

Rainbow Smelt, *Osmerus mordax* (fig. 8.3), is a small silvery fish, with an adipose fin, but too small and slender to be mistaken for any member of the Salmon Family. The reason for mentioning rainbow smelts here is that, though small, they are an important food fish; because they swim in schools, they can be netted in quantity. Most smelts are anadromous though a few live permanently in fresh water. In the Arctic, they are found only along the mainland coast from Bathurst Inlet (at 107° W longitude) westward. They spawn in streams and rivers when the ice goes out in spring, spending the rest of the year at sea close to shore, or in lakes not far inland.

Burbot, *Lota lota* (fig. 8.3), is an inland fish of temperate latitudes whose range extends north of treeline from Hudson Bay westward. Its distinctive fin pattern makes it easy to recognize. Burbot is the only completely freshwater member of the cod family. Its relationship to cod is shown by its barbels, a long one on the chin and a little one over each nostril. Burbot spawning is a dramatic affair. It takes place under the ice in midwinter, only in the dark of night, when a dozen or so fish are reported to unite in an intertwined, writhing mass to discharge their spawn. Burbots are regarded as pests because of their predatory habits; they feed on the young of "desirable" fish (char, trout, and whitefish),

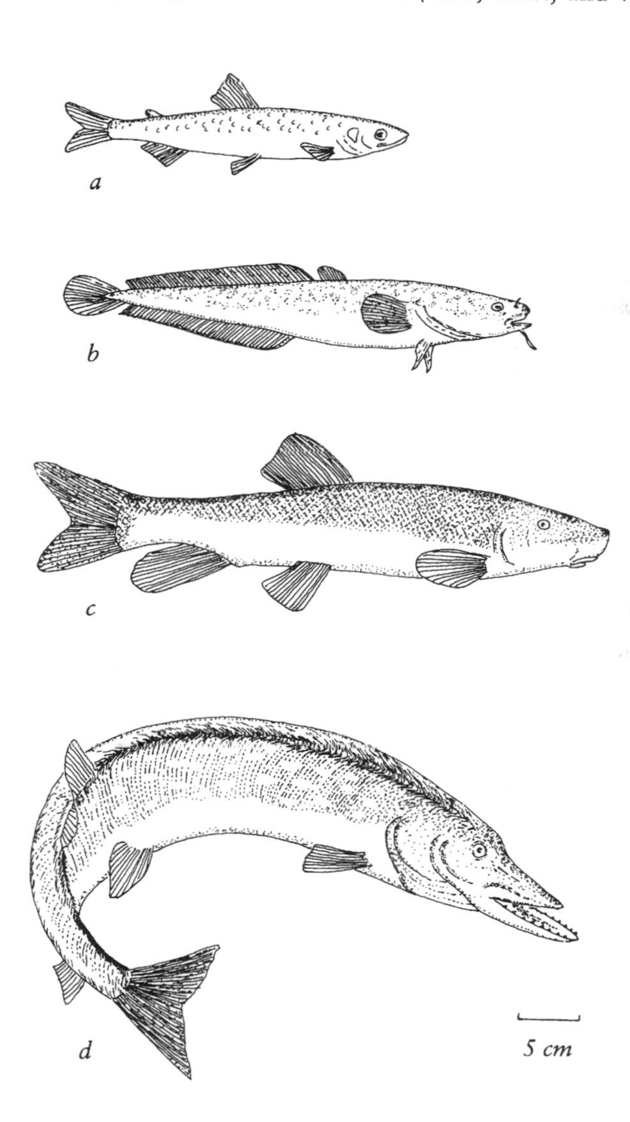

Figure 8.3. (a) Rainbow Smelt. (b) Burbot. (c) Longnose Sucker. (d) Northern Pike.

besides taking quantities of food that would otherwise be available to nourish the desirable fish.

Longnose Sucker, *Catostomus catostomus* (fig. 8.3), is recognizable by its "sucker" mouth, with protruding lips pointing downward. This sucker is found throughout most of the mainland Arctic, usually in lakes but occasionally in the brackish water of estuaries. It spawns in spring, in shallow streams, and is a colorful fish at spawning time with bright pink-red markings on both sexes. It is a bottom feeder, using its mouth like a vacuum cleaner to suck up the multitude of edibles—chiefly insect larvae and tiny crustaceans—that live on lake bottoms.

Northern Pike, *Esox lucius* (fig. 8.3), is recognizable from its long head and protruding lower jaw and from the fact that its single dorsal fin is placed so far back. Pike live in fresh water throughout the mainland Arctic, and spawn in shallow water in spring, immediately after break-up. They are as fierce as they look. A pike can catch and eat a fish as much as one-half its own size, and it doesn't limit itself to cold-blooded prey. Besides taking fish, pike also feed on ducklings and lemmings when they're available.

9 Insects

9.1 Insect Life in the Arctic

The summer Arctic is alive with a variety of insects. Only a few of the many species are pests; most species are harmless and play important roles in the workings of arctic ecosystems, especially as food for birds and as pollinators for flowering plants. The naturalist who ignores the insects (except to swat them) is missing a big segment of arctic natural history, a segment full of interest. Arctic conditions are as big a challenge to insects as to other animals, and the various ways in which insects have become adapted to survive and thrive in spite of the difficulties are worth noting. Because they are cold-blooded, their survival strategies are entirely different from those of mammals and birds.

The conditions that make life difficult are, first, the severe cold of winter and the occasional, unpredictable cold spells of summer; second, the shortness of the season when active life is possible; and third, the almost endless wind.

Insects as a group have several ways of coping with cold. During the prolonged, intense cold of winter, some freeze solid without ill effect. Others cool and become torpid without actually freezing. This means that for a large fraction of each year, any insect is in a state of dormancy or suspended animation analogous to hibernation in mammals. These bouts of dormancy come as interruptions in an insect's life cycle. Freezing weather can cause temporary halts at any point in the sequence of stages that most insects go through—eggs, larvae, pupae, and adults—slowing down the whole development process.

A well-known example is the **Arctic Woolly Bear** caterpillar (*Gynaephora groenlandica,* fig. 9.1), the immature form of a moth. It is common in the High Arctic and in no danger of being overlooked, for it is a big, active "woolly bear," densely clothed with long, rusty-orange hairs. Its life as a caterpillar goes on and on, for the astonishing total of 14 years. Through all the winters it is frozen solid and inert; actual development occurs only in summer, when it resumes life at the point where it left off the previous summer. When the caterpillar has finished growing, it pupates in a silken cocoon and finally emerges as an adult moth in early summer. The moths, once they have emerged, must complete their lives quickly, as they are no longer frost-proof—freezing would kill them. All they have to do is reproduce, which doesn't take long because they waste no time eating; the reserves of energy they accumulated as caterpillars are sufficient for the whole of their short lives as breeding adults.

Many insects can cool down to far below 0° C without their liquid contents turning to ice. They do this by forming natural chemicals within their cells that lower the freezing point of the liquids in the cell in the same way that antifreeze lowers the freezing point of water in an automobile radiator. Indeed, in some insects the chemical is ethylene glycol, identical with automobile antifreeze.

Another way to battle the cold is to avoid it, by parasitizing a warm-blooded mammal and sharing its warmth, as the warble flies and nose bots of caribou do (see sec. 9.5).

Although insects are cold-blooded, this does not mean that their temperature never rises above that of their surroundings. Far from it. Most insects warm up by basking in the sun. Butterflies do it (sec. 9.6), woolly

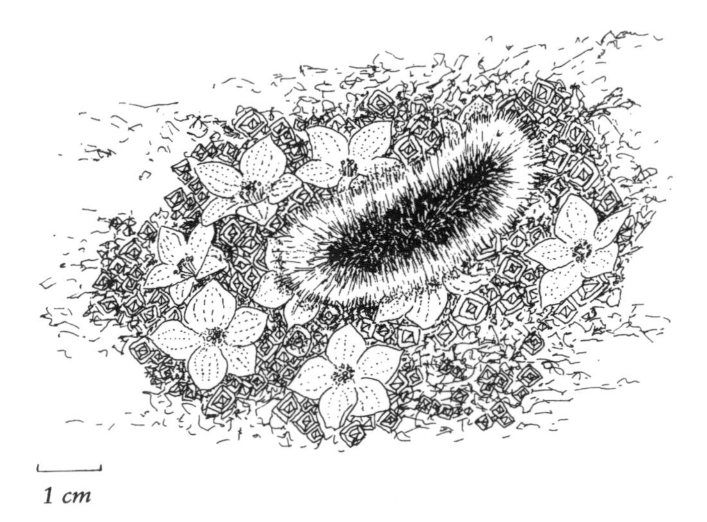

Figure 9.1. The Arctic Woolly Bear Caterpillar.

bears do it, and so do the numerous insects that like to rest motionless at the center of a flower where the sun's warm rays are focussed on them (see sec. 5.8). And bumblebees do more than simply soak up external heat; they generate their own, in the form of energy produced by shivering. Some insects (woolly bears and bumblebees) conserve the heat they absorb or produce, with their insulating layers of "fur."

Another facet of the arctic climate that greatly affects insects is the shortness of the summer growing season. Woolly bear caterpillars, taking 14 years to get through the larval stage of their life cycle, are an extreme example of the slowing down of life that is one of the possible responses to short summers. Some insects respond quite differently; they get through life faster than their temperate-zone relatives. In other words, it takes a shorter time for adults to develop from eggs. Among insects using this strategy are bumblebees and mosquitoes; details are given in sections 9.2 and 9.4, respectively.

The strong winds of the Arctic are yet another hazard insects must face. To avoid being blown away, flying insects fly as close to the ground as possible. Others crawl from place to place rather than fly. Some, notably certain species of mosquitoes and midges, have become evolutionarily adapted to life in the wind. They have lost the power of flight (their wings are congenitally reduced or atrophied), but to make up for it, they have become fast, efficient crawlers.

For more on woolly bear caterpillars and their indifference to cold, see "Caterpillars on Ice," by Olga Krukal, in *Natural History,* January 1988.

9.2 Insect Pollinators: Bumblebees and Flies

Of all the insects that visit arctic flowers—and there are many—bumblebees are the most conspicuous. They are impressively big, astonishingly so to a first-time visitor to the Arctic. Their size shows how bountiful arctic flowers are as a source of the nectar and pollen needed by insects adapted to depend on them. Nectar, which is mostly dissolved sugars, provides the energy these insects need; pollen provides the protein.

Only two bumblebee species inhabit the High Arctic, and one is a "cuckoo bumblebee" (see below). The two species are hard to distinguish unless you see both together—one is larger than the other—and neither of them have English names. Their Latin names are *Bombus polaris* and *B. hyperboreus* (the cuckoo).

Bombus polaris is the smaller of the two and the first to emerge in spring after hibernating through the winter in frozen soil. Queens, already impregnated, are the only bumblebees to survive the winter and, come spring, each one founds a new colony. Because arctic summers are so short, they have little time in which to do this; eggs take time to hatch out their larvae, the larvae (grubs) take time to grow and form pupae; the pupae take time to develop into adults. In temperate bumblebees, a queen lays several successive batches of eggs that all develop into sterile females (workers) before, in her final batch, she produces fertile females (next year's queens-to-be), and drones (fertile males) to mate with them. A *Bombus polaris* queen hasn't time to lay numerous batches of eggs. She has time for two batches at most, with drones and future queens in the second batch or even, occasionally, in the first. Consequently, the queen has only one batch of workers, or occasionally none at all, to help her feed and rear the parents of the next generation; she has to do some, or all, of the work herself.

The foregoing describes life in a *Bombus polaris* colony, provided a bumblebee queen of the other, larger species, *B. hyperboreus,* doesn't intervene. *Bombus hyperboreus* is a cuckoo bumblebee (formally, "a social parasite"). When an impregnated queen starts her spring activities, beginning a bit later than the *B. polaris* queens, she hunts out a growing *B. polaris* colony and invades it. If all goes well, she kills the *B. polaris* queen and "enslaves" the victim's worker daughters to serve as her workers. She lays her eggs so that all of them yield grubs that develop into queens and drones, and these grubs are fed and reared by the enslaved *B. polaris* workers. Thus a *B. hyperboreus* queen produces no workers of her own; she leaves the care of her offspring entirely to workers of the other species.

Another adaptation enabling bumblebees to flourish in spite of the shortness and coolness of summer is their capacity to generate heat for themselves, in spite of being "cold-blooded." This allows them to become active while the weather is still cool and, later, to warm the eggs and the grubs that hatch from them, as they develop in the nest. The heat is generated by rapid contractions (shivering) in the bumblebees' flight muscles; the insect can raise its temperature as high as 35° C, almost as high as human body temperature. These muscles account for the large size of arctic bumblebees. Also, a large body retains heat longer than a small one, and the dense "fur" covering bumblebees' bodies provides effective heat insulation.

The energy used up in shivering all comes, of course, from the nectar in arctic flowers. Bumblebees visit a variety of flowers, but not all. They are noticeably choosy. Some studies have suggested that they prefer fragrant flowers to scentless ones and are less influenced by color and showiness. Travelers in the Arctic who know the plants can collect useful natural history data by careful observation of the activities of bumblebees, particularly by noting what flowers they visit. For more on bee pollination, see section 5.8.

Another group of insects that pollinate arctic flowers are flies, in the collective sense. The name covers all two-winged insects; besides the obvious ones (houseflies, bluebottles and their like), it includes mosquitoes (see sec. 9.4) and a variety of other insects that are tremendously important in tundra ecosystems because they serve both as pollinators of plants and as food for birds. They are described in that context in the next section (sec. 9.3), but their function as pollinators must be stressed here. Their fondness for flowers as warm places in which to bask is most obvious on cool, sunny days, when the center of a bowl-shaped flower facing the sun is the warmest place to be; as they move from flower to flower, they carry pollen with them. Indeed, according to many entomologists, flies are far more important than bumblebees as pollinators in the Arctic. This seems an unfair comparison, however; bumblebees are represented by only two species, but scores of different species rank as "flies."

For more on arctic bumblebees, see "The Antifreeze of Bees" by Bernd Heinrich, *Natural History* (July 1990).

9.3 Insects as Food for Birds: Flies

When it comes to being useful to unrelated organisms, flies excel. The preceding section mentioned their usefulness as pollinators. They are absolutely indispensable, as food, to vast numbers of arctic birds, especially the waders (sandpipers and plovers). From the waders' point of view, the most important flies (in the broad sense) are crane flies (of the family Tipulidae) and nonbiting midges (family Chironomidae).

Crane flies, as adults, often draw attention when they sun themselves in flowers. There are many species, and large ones are sometimes called daddy longlegs, a name also used for harvestmen (wingless, spidery creatures that are neither insects nor spiders but a unique group on their own). A crane fly has exceedingly long, fragile legs, and looks like a giant mosquito. But it is a gentle, slow-moving beast that neither bites nor pesters people by flying in their faces. Before reaching maturity, most species of crane fly in the Arctic are fleshy larvae that live in mud, providing a feast for all the waders pattering over wet tundra and along the shores of tundra ponds, ceaselessly probing the mud for a meal. Larvae of other crane fly species live in fresh water rather than mud, where they are accessible to diving ducks (chiefly scaups, see sec. 6.7), phalaropes, and waders that forage in the shallows of tundra ponds.

Even more numerous than crane flies, and consequently more important as bird food, are nonbiting midges. They are the most varied and abundant of all arctic insects. Usually, they first draw attention to themselves by hovering in dense mating swarms; a swarm drifts through the air slowly, like a little gray cloud, but the insects forming the cloud bob up and down within it at tremendous speed. Close inspection of an individual midge shows it to be rather like a tiny mosquito with extravagantly bushy antennae. Like crane fly larvae, midge larvae live in mud or freshwater and are eagerly sought by wading birds. The ones in water are also food for fish. Arctic char would starve without them.

Midges are not single species but whole families of species. Many other fly families, for example, the mosquito family (Culicidae) have similar life cycles and are a tremendous benefit to birds. But mosquitoes (plus blackflies), in their capacity as pests, deserve a section to themselves.

9.4 Mosquitoes and Blackflies

The Arctic is famous for its fearsome mosquitoes; for many people, mosquitoes are the only arctic insects worthy of note. This attitude would have been understandable among early explorers; nowadays, with a bug jacket impregnated with modern insect repellents and (if necessary) a head net, nobody need be seriously inconvenienced by the pests. Instead, mosquitoes can be appreciated, as an important food supply for shorebirds and as interesting animals in their own right.

The ability of such fragile, delicate creatures to flourish under arctic conditions is remarkable. Consider the hardships they face: strong winds often make flight difficult for them; the short summer season gives the females little time to develop eggs and lay them before winter sets in; and sometimes the females have trouble finding vertebrate animals as a source of the blood most mosquitoes need for the development of eggs. Arctic mosquitoes are adapted to cope with all of these problems.

First, the wind: in windy weather, mosquitoes (as well as other lightweight insects such as midges) crawl to their destinations instead of flying. The destination is often a flower, as nectar is one of their chief food sources, for males often the only source.

Second, the short warm season: a female mosquito lays her eggs in summer and needs to place them where they will hatch as early as possible the following year. Since immature mosquitoes (larvae to begin with, then pupae) live in water, the eggs are laid on the margins of a pond, just above the summer water level; there they remain, under the snow, all winter. Then, when the snow melts at the beginning of the following summer, the meltwater raises the water level in the pond, submerging the eggs so that they can hatch. Small, shallow ponds are used, where the ice melts early and the water warms up quickly. If a pond has high banks, a mosquito with eggs to lay prefers the south-facing bank on the north shore of the pond, because that is where the snow melts earliest. The eggs soon hatch, and the young take a little less than a month to mature; they can grow and develop at temperatures as low as 1° C.

Third, blood: female mosquitoes (but not males) feed on the blood of a variety of animals, birds, reptiles, and amphibians as well as mammals. In most species of mosquito, the females must have a blood meal before they can lay eggs. Arctic species, however, can produce eggs *autogenously,* i.e., they are able to use the food reserves accumulated while they were living in water, as larvae, to develop at least a few eggs. A meal of fresh blood certainly allows a female to lay a larger number of eggs, but by laying some autogenous eggs, she will leave descendants even if she cannot get blood.

When newly formed adults emerge from their pupae to begin their lives out of water, they fly first to flowers that provide nectar. Nectar contains sugar, and mosquitoes of both sexes use it as an energy source. A favorite flower in most places is the ubiquitous and beautiful arctic dryad (described in sec. 5.13). Mosquitoes also use flowers as sites for resting and basking in the sun's warmth.

Mosquitoes are valuable plant pollinators, as can be seen in many places in the Low Arctic where northern bog orchids (see sec. 5.13) grow. In most flowers, the pollen is simply a loose powder in which the pollen grains don't stick together. But orchid pollen is different: the grains do stick together, forming little masses of pollen, each the size of a pinhead or smaller; it is these pollen-masses that are carried from flower to flower by pollinating insects, including mosquitoes. Where bog orchids grow, you will often see a mosquito, flying or settled, with an orchid's pollen-mass attached to its head. It has picked up the pollen while it was sucking the nectar from one orchid flower and will deposit it in the next flower it visits.

Adult mosquitoes, as well as larvae, are food for other animals. Large numbers are eaten by yellowjackets, which are common in the Low Arctic. It is a pleasure to watch a yellowjacket capture a mosquito in flight, or pounce on a mosquito settled on your hand. The mosquito is defenseless, and the yellowjacket quickly crushes and folds it into a compact package that is easy to carry to her underground nest, to feed to her grubs.

In parts of the Low Arctic (roughly, south of 70° N latitude), blackflies rival mosquitoes as bloodsucking pests, just as they do in the boreal forest. But farther north, their adaptations to arctic conditions are even more remarkable than those of mosquitoes. In temperate latitudes, blackflies mate on the wing, and the females must have a blood meal before they can develop eggs, which they deposit and glue to rocks in running water. In the far north, blackflies mate on the ground, when a male and a female happen to meet each other as they crawl around. And the adults don't feed at all; they have stored all the nourishment they need for adult life, including egg development, during their larval life in flowing water, and the adult mouth parts are useless.

This stripped-down life has been carried to an extreme in some High Arctic blackflies. In these species, there are no males. The females need neither blood for nourishment, nor sperm from a male: they simply reproduce parthenogenetically. Some of them lay eggs soon after emerging into the air from the submerged pupa (the passive stage, between the larval and adult stages). In other species, the most extreme of all, no adult insect ever emerges from the pupa; instead, the eggs develop inside the pupa and are liberated into the water when the pupa dies and disintegrates.

9.5 The Warble Flies and Nose Bots of Caribou

Anyone who grumbles about mosquitoes should consider the fate of the caribou. Besides being attacked by hordes of mosquitoes, caribou are victims of two other, truly revolting insects: warble flies and nose bots.

Both are "flies" in the entomologist's sense (two-winged insects). The adult insects could easily be mistaken for bumblebees at first sight, but a closer look reveals one of the many differences: adult warbles and bots have no mouths (see fig. 9.2). They do not need them as they do not eat. In their immature (larval) form, as maggots that parasitize caribou, they feed so voraciously that they store up all the nourishment they will need for the rest of their lives. Indeed, the maggots lay up food reserves not only for their adult selves, but also for the offspring (eggs or live-born larvae) that the female adults will produce.

Both parasites belong to the same insect family, the Oestridae.

The **Caribou Warble Fly,** *Hypoderma tarandi* (fig. 9.2), is sometimes seen on the wing, in summer, when caribou are in the neighborhood. The flies are much like bluebottles in shape and size, but their dense "fur," in black, orange, and yellow stripes, makes them look like bumblebees at first glance. They lay their eggs on caribou hairs, usually on the legs or belly; the eggs stick to the hairs and are not easily brushed off. When the larvae (maggots) hatch, they burrow into the caribou and tunnel their way up inside it until they are just under the skin of the

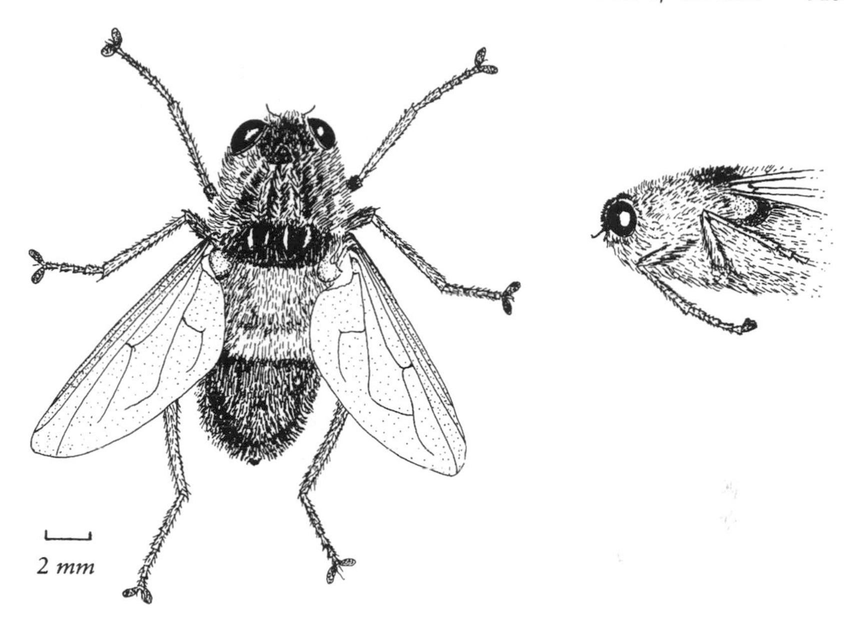

Figure 9.2. The Caribou Warble Fly. The profile of its head (on the right) shows that it lacks any visible mouth parts.

back, fairly near the spine; once they arrive, in early fall, the maggots cut breathing holes through the skin of the caribou's back and remain just under the holes while they develop. A single caribou may be infested with more than 2000 maggots, all deriving their sustenance from, and seriously weakening, the unfortunate host. Moreover, the breathing holes easily become infected and abscessed. The maggots continue to grow all through winter, reaching a length of 1 cm or more. Late in the following spring, they emerge through their breathing holes, drop to the ground, and pupate (i.e., become inactive pupae, in which the final, adult form of the insect develops). About a month later, a new generation of adult flies emerges and the cycle begins again.

The **Caribou Nose Bot** (*Cephenomyia trompe*) is in some ways an even more repulsive parasite. It is a large, hairy fly that deposits its young in a caribou's nostrils. The young are live, active maggots, already hatched inside the mother fly's body. The maggots migrate through the caribou's nasal passages and collect at the opening to its throat, where they live and grow all through winter; by spring, the close-packed mass of maggots forms a big lump in the caribou's throat, sometimes becoming large enough to obstruct its breathing.

It is hard to imagine the torment that nose bots cause. Infested caribou can be seen (and heard) repeatedly coughing and sneezing in their efforts to dislodge them. Eventually the caribou succeed, but this is merely because the bots are adapted to be expelled from their hosts when the time comes for them to complete their development. This they do as pupae, on the ground. That done, the new generation of adult bot flies emerges, and the cycle starts again.

9.6 Butterflies

Butterflies are not at all uncommon in the Arctic, and though they lie low when the weather is cold or windy, they are sometimes plentiful when the weather is warm, sunny, and relatively calm. Your attention may first be caught by the shadows of butterflies, zigzagging across the tundra. They fly low, since the vegetation is low; and the stronger the wind the lower they fly, so as to remain in the layer of calm air close to the surface.

Most butterflies are hard to identify unless they are caught and killed. Many species can only be identified conclusively by examining their genitals with a dissecting microscope. For nonspecialists, it is far more satisfying to recognize the family (or genus) that a living butterfly belongs to and to observe the way its behavior adapts it to its habitat, than it is to try to match a specimen with one of a batch of almost identical pictures.

Delicate though they seem, arctic butterflies are tough. Their caterpillars can freeze solid in winter and then thaw out and resume life when the weather warms again. The growth that a caterpillar can complete in one summer, especially a short, cold summer, is small. It may require several years for a newly hatched caterpillar to develop to an adult butterfly, taking time out intermittently, as succeeding winters bring life to a halt temporarily (see sec. 9.1).

Most arctic butterflies belong to one of the following four families: the Whites and Sulphurs (Pieridae); the Brush-footed Butterflies (Nymphalidae); the Satyrs (Satyridae); and the Blues and Coppers (Lycaenidae).

The first of these families is represented in the Arctic by several **Sulphur Butterflies** of the genus *Colias*. The undersides of the wings (the sides seen when a butterly settles with wings folded) are sulphur yellow, often suffused with lime green near the base of the wings close to the body. On the middle of the lower wing there is often a small silvery spot outlined with pink, and the wings are fringed with pink scales. This is not as spectacular as it sounds; the pink trim is barely visible without a hand lens. The upper sides of the wings are yellow (orange in the male Greenland Sulphur) and often have a dark band along the outer edges.

The caterpillars of some sulphur butterflies feed on blueberry leaves, those of others on milk-vetch. The adult butterflies, however, don't necessarily remain near the plants their offspring require; they can be seen almost anywhere, and are to be found in the harshest High Arctic environments. They display an interesting mannerism: when they settle with folded wings on a sunny day, they shuffle around until the sun's rays come from one side and then lean so that the ray's strike the wings perpendicularly (fig. 9.3). They are believed to do this in order to absorb as much of the sun's warmth as possible.

The distinguishing characteristic of the whole family of Whites and Sulphurs is that, in both sexes, all six legs are fully developed and functional. In this they differ from all other butterfly families.

In the Brush-footed Butterflies, the forelegs, in both sexes, are merely stubs covered with hairlike scales: hence the family name. Several members are to be found in the Arctic, all belonging to a group of several species collectively known as **Lesser Fritillaries.** These butterflies usually

Figure 9.3. Some arctic butterflies. (a) Greenland Sulphur *(Colias hecla);* (b) Polaris Fritillary *(Clossinia polaris);* (c) Banded Alpine *(Erebia fasciata);* (d) High Mountain Blue *(Agriades franklinii).*

settle with their wings spread, showing the upper sides. In every lesser fritillary but one, the upper sides of the wings are orange, intricately patterned in black (fig. 9.3). They are all very similar to one another, making species identification difficult; adding to the difficulty is the fact that a single species can have separate populations, in different areas, which vary among themselves in pattern as markedly as different species vary. The exceptional fritillary is the aptly named Dingy Arctic Fritillary, which has a dark gray-brown background color. The caterpillars of the several species feed on a variety of plants: arctic willow, arctic dryad, viviparous knotweed, blueberry, bearberry, and crowberry.

A lesser fritillary, like a sulphur, will settle in a way that maximizes the warmth it can absorb from the sun, but its technique is different. It settles with wings spread and then, if the sun is shining, it will turn itself until its head points away from the sun; this tilts the spread wings at the best angle to be warmed.

The Satyr Family of butterflies is represented in the Arctic by two genera, the **Alpines** (*Erebia*) and the **Arctics** (*Oeneis*), each with several species. They are not colorful butterflies, being patterned in black, brown, gray, tan, and various shades of charcoal, but they are distinctive (as a family), nonetheless, because of their mode of flight which is notably jerky; they bob up and down erratically, making it hard to judge where they are going and where they will land. When they finally settle, they promptly disappear, because of the cryptic patterning on the undersides of their folded wings. The Banded Alpine (fig. 9.3) is an example; the lighter bands on its blackish-brown wings are noticeable in flight but have the effect of breaking up the butterfly's outline when it is settled and motionless.

An arctic or alpine usually settles with its folded wings tilted almost horizontal in a way that minimizes the size of the shadow it casts on the ground. This adds to the camouflage effect of the wing pattern and makes the settled insect almost impossible to detect. Recall that sulphur butterflies are believed to settle with tilted wings in order to get the most warmth from sunlight. If these theories are true, it means that similar maneuvers have evolved in two groups of butterflies for quite different purposes, a fascinating conclusion but not, of course, proven.

The two genera of satyrs can be recognized by noting that arctics are paler in color than alpines. But the different species within each genus are difficult to distinguish. In all members of the family, the forelegs are reduced and nonfunctional in both sexes, but they are not quite so stubby as the forelegs of the brush-footed butterflies. The caterpillars feed on grasses and sedges.

The **Blues** and **Coppers** are the fourth butterfly family with several arctic representatives. In this family, the forelegs are reduced and useless in the males but normal and functional in the females. The names are misleading: some coppers are blue, and some blues are copper-colored; or a species may have blue males and copper-colored females. Small blue butterflies are unquestionably members of this family, and nonblue members are recognizable by their smallness compared with other butterflies (see the High Mountain Blue in fig. 9.3).

Blues and coppers flutter around weakly, until they are alarmed; then they escape at speed. Their caterpillars are found on various plants: those of the Flame Copper (the most brightly colored copper) feed on mountain sorrel, and those of the High Mountain Blue on diapensia. These butterflies can be found in the most barren and inhospitable arctic landscapes, where they look small and frail, but the frailty is obviously apparent rather than real.

Index

(Main references are in boldface)